APPLIED ELECTROTECHNOLOGY FOR ENGINEERS

C H LAYCOCK
Senior Lecturer
School of Engineering Information Technology
Sheffield Hallam University

In return for the patience
and understanding of my
wife and my former students

Contents

Preface

In writing this book I have attempted to fill a gap in the literature for a text on electrical engineering for other engineering students. Professor Sutcliffe attempted this for electronics in his standard text written in 1964. To the best of my belief no one else has taken a cold realistic look at the wide field of electrical engineering today and set down what he thinks is the minimum essential knowledge for a practising mechanical, production or civil engineer.

In composing their degree and H.N.D. courses many institutions have produced syllabuses for this subject but hitherto students have had to rely upon the standard electrical-students' texts for their study. These often present the material in an indigestible way for the non-specialist and accordingly I have only taken the basic concepts of the subject to sufficient depth to provide a sure foundation for later study. My aim has been to simplify to the utmost without hindering any deeper study that the student may later undertake. Previous electrical study to the level of O1 Electrical Engineering Science of the Ordinary National Certificate or to G.C.E. Advanced Level physics is assumed.

Many electrical colleagues may feel that I have given scant attention to many favourite examination topics, for instance, the transformer phasor-diagram and the subject of a.c. bridges. My defence is that these academically elegant subjects are not essential to a first undergraduate study for these students. Indeed, even for electrical students, the classical a.c. bridges have lost importance compared to the transformer ratio bridge in industrial practice.

Similarly in electronics no attempt has been made to consider transistor parameters rigorously. After an introduction to transistor action via the load line the reader proceeds straight to a black-box consideration of amplifiers.

The two main areas where electrical engineering impinges on other engineering studies are in the fields of motive power and instrumentation. The two halves of this text cover most of these topics. In compiling both the worked and unworked examples I have drawn freely from my teaching notes compiled over some years from many sources. I may thus unwittingly be infringing copyright, in which case due acknowledgement will be made at the earliest opportunity, as will any other errors of omission or commission.

I wish to thank all my colleagues in the Department of Electrical Engineering at Sheffield Polytechnic for their valuable comments and criticisms. I also wish to thank the governors of that institution for granting me a period of leave to complete this work. Especial tribute is due to Mrs Jennings of that department and to Mrs Hellett and Mr Fellows of Skye for their patience when deciphering my manuscript.

A final tribute is due to the wettest Hebridean December in fifty years which prevented all activity other than writing!

Glendale, Isle of Skye C. H. LAYCOCK

1 Behaviour of Electrical Components

'A day's work – getting started' – Gaelic proverb

The purpose of this chapter is twofold. First we must revise and consolidate our knowledge of the electrical behaviour of resistors, inductors and capacitors[1]. Second we must investigate the behaviour of these components in first-order combinations before we confine our studies to the commonly encountered steady-state conditions of chapter 2. This knowledge, complemented by an understanding of electrical devices from chapter 3, will allow us to proceed to a study of the main areas of electrical application for engineers. These are electrical machines and utilisation (chapter 4), power supplies (chapter 5) and instrumentation (chapters 6 to 9).

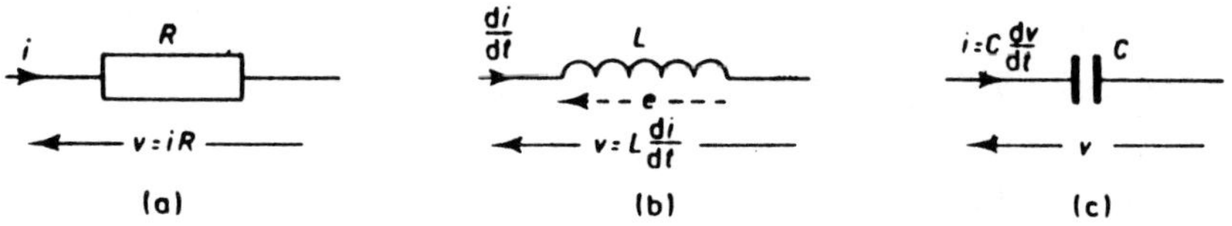

FIGURE 1.1

Voltage-current relationships for pure electrical components

1.1 Behaviour of Pure Circuit Components

Figure 1.1 shows a pure resistor (zero inductance), a pure inductor (zero resistance) and a pure capacitor (zero leakage resistance) each excited by an instantaneous voltage v.

For the resistor (by Ohm's law)

$$v = iR \qquad \text{or} \qquad i = \frac{v}{R} \tag{1.1}$$

where i is the instantaneous current and R the resistance measured in ohms.

In the inductor there will be a back e.m.f. e induced by any current change di/dt. Lenz's law states that these are related by the expression

$$e = -L\frac{di}{dt}$$

where L is the self-inductance measured in henrys. The negative sign shows that the back e.m.f. opposes the proposed current change. In order to ensure this change therefore the instantaneous applied voltage v must exactly cancel the back e.m.f.

$$v = -e = L\frac{di}{dt} \qquad \text{or } i = \frac{1}{L}\int v\,dt \tag{1.2}$$

The capacitor's fundamental equation is $v = q/C$ where q is the instantaneous charge in coulombs and C is the capacitance in farads. Because the charge on the capacitor is produced by the current i flowing into it we may write

$$q = \int i\,dt$$

$$v = \frac{1}{C}\int i\,dt \qquad \text{or } i = C\frac{dv}{dt} \tag{1.3}$$

1.2 Graphical Solutions of Simple Excitations on Purely Parallel or Series Circuits

Step and Ramp Excitation

The nature of purely series or parallel circuits allows graphical solutions if simple current or voltage excitations are applied since the same excitation is common to all components. The graphical solution for (i) step and (ii) ramp voltage functions applied to an *RLC* parallel circuit and step and ramp current functions applied to an *RLC* series circuit is shown in figure 1.2. These curves are drawn by simply applying equations 1.1 to 1.3. For step excitation $v = V$ or $i = I$ after $t = t'$ and for a ramp v or $i = Kt$ after $t = t'$.

Sinusoidal Excitation

For a capacitor

$$i = C\frac{dv}{dt}$$

hence if

$$v = V_m \sin \omega t$$

$$i = CV_m \, d(\sin \omega t)/dt$$

$$= V_m \omega C \cos \omega t$$

$$= \omega C V_m \sin\left(\omega t + \frac{\pi}{2}\right)$$

which is a sinusoidal current waveform of maximum value

$$I_m = \omega C V_m$$

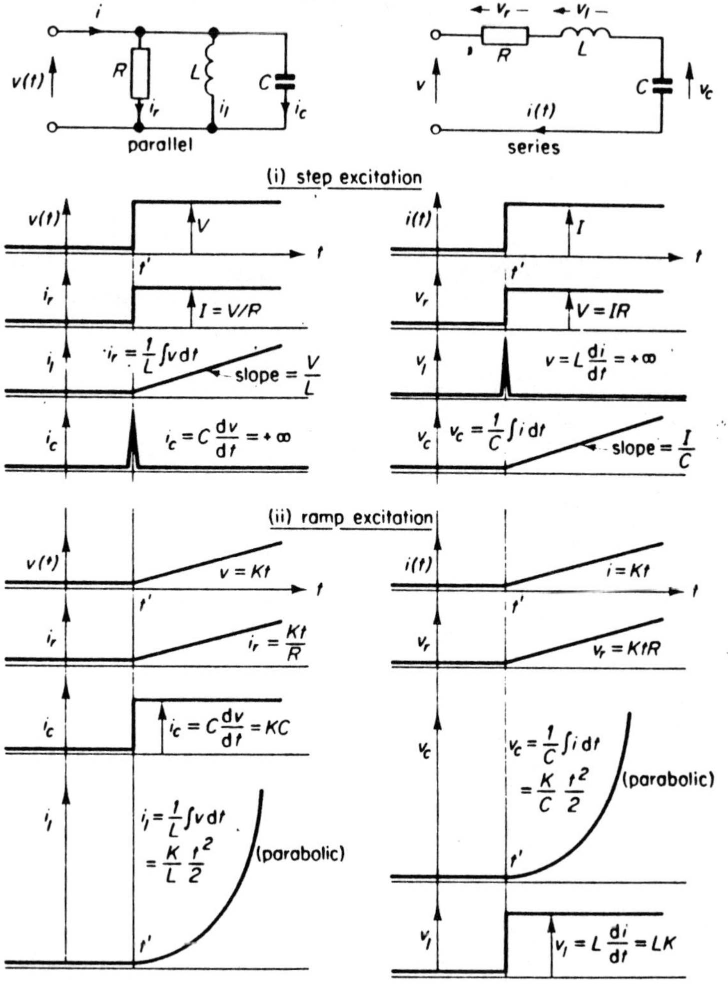

FIGURE 1.2

Graphical solution of step and ramp excitations. Total current i may be obtained by graphical addition of i_r, i_l and i_c; total voltage v may be obtained by graphical addition of v_r, v_l and v_c

This current waveform *leads* the voltage by $\pi/2$ radians or 90°.

For an inductor

$$v = L\frac{di}{dt}$$

hence if

$$i = I_m \sin \omega t$$

$$v = LI_m \, d(\sin \omega t)/dt$$

$$= I_m \omega L \cos \omega t$$

$$= \omega L I_m \sin \left(\omega t + \frac{\pi}{2}\right)$$

which is a sinusoidal waveform of maximum value

$$V_m = \omega L I_m$$

In this case the current waveform *lags* the voltage by $\pi/2$ radians or 90°.

For a resistor a sinusoidal current waveform merely produces a voltage waveform of maximum value $V_m = I_m \times R$ rising and falling *in phase* with the current.

These equations cause the waveform diagrams for *RLC* parallel and series circuits to be as shown in figure 1.3.

Note that the following mnemonic is useful to remember the phase relationship

Capacitor – *I* leads *V*

C I V I L

Inductor – *I* lags *V*

Reactance

The ratios of the maximum values of current I_m and voltage V_m are useful in calculations. For a capacitor

$$I_m = \omega C V_m$$

hence

$$\frac{V_m}{I_m} = \frac{1}{\omega C}$$

this ratio is called the *capacitive reactance* X_c and being a voltage : current ratio has the *ohm* as a unit, thus

$$X_c = \frac{1}{\omega C} \text{ ohms}$$

For an inductor

$$V_m = \omega L I_m$$

hence

$$\frac{V_m}{I_m} = \omega L$$

which is called the *inductive reactance* X_l

$$X_l = \omega L \text{ ohms}$$

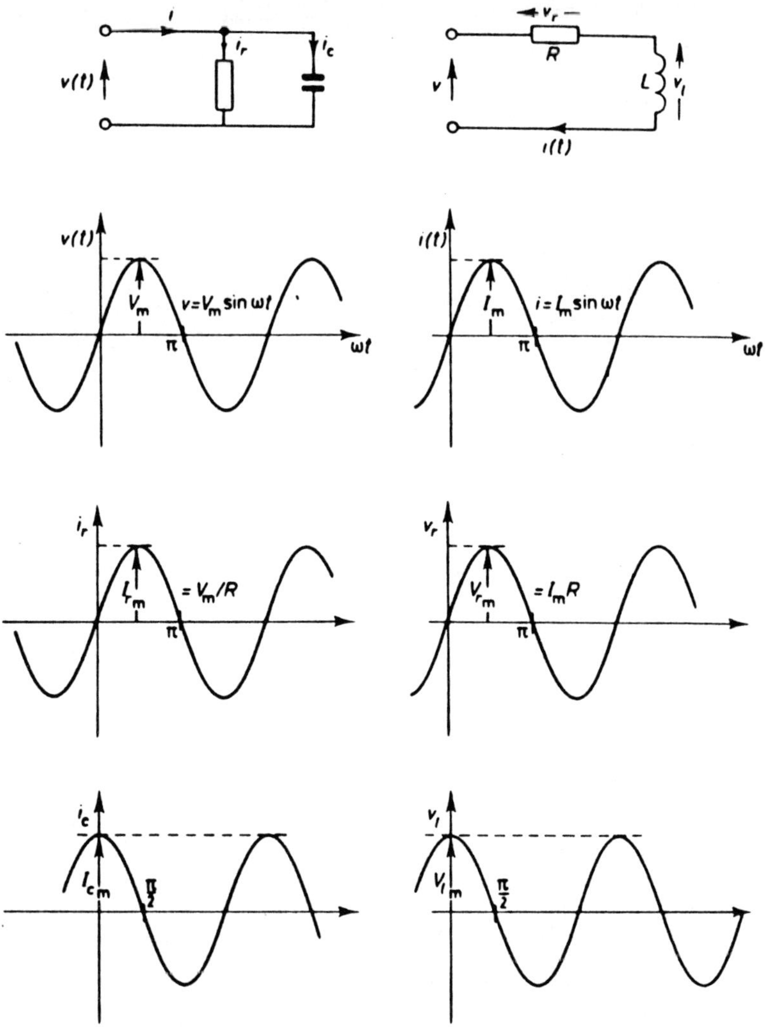

FIGURE 1.3

Graphical solution of sinusoidal excitations applied to purely series and purely parallel circuits

The behaviour of *R*, *L* and *C* combinations under sinusoidal excitation is investigated fully in chapter 2.

Example 1.1

Determine graphically the total current *i* in the circuit shown in figure 1.4a during the first half-second after application of (i) a 10 V step voltage and (ii) a ramp voltage given by the equation $v = 10t$ after time $t = 0$.

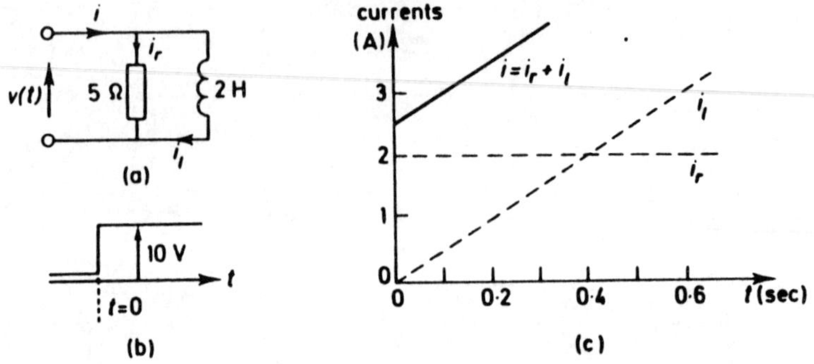

FIGURE 1.4

Example 1.1(i)

(i) After $t = 0$

$$i_r \text{ is constant} = \frac{v}{R} = \frac{10}{5} = 2 \text{ A}$$

$$i_l = \frac{1}{L}\int v \, dt = \frac{1}{L}\int 10 \, dt$$

$$= 0.5 \times 10t = 5t$$

that is, i_l will rise linearly from the origin to a value of $5 \times 0.5 = 2.5$ A after 0.5 s. Thus the total current i will be given by the graphical summation of i_l and i_r as shown in figure 1.4c.

(ii) If $v = 10t$ then

$$i_r = \frac{v}{R} = \frac{10t}{5} = 2t$$

which is a straight line through the origin.

$$i_l = \frac{1}{L}\int v \, dt = \frac{1}{2}\left[\frac{10t^2}{2}\right]_0^t$$

$$i_l = 2.5t^2$$

that is, a parabola through the origin whose value after 0.5 s will be 2.5/4 or 0.625 A. Again i is obtained by graphical summation as shown in figure 1.5.

Example 1.2

Determine the period, maximum value and phase of the voltage waveform across a series combination of a 10 Ω resistor and a 10^3 μF capacitor when a current

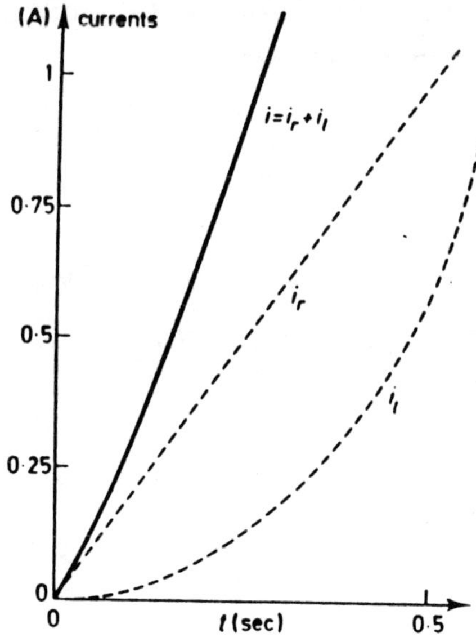

FIGURE 1.5

Example 1.1(ii)

given by the equation $i = 5 \sin 100t$ has flowed through them both for many cycles.

Period T

$$i = I_m \sin \omega t = 5 \sin 100t$$

therefore

$$\omega = 2\pi f = 100 \text{ rad/s}$$

and

$$T = \frac{1}{f} = \frac{\pi}{50}$$

$$T = 0.0628 \text{ s}$$

The total voltage will be the graphical sum of the resistor voltage v_r and the capacitor voltage v_C. From the above, the maximum current, I_m, is 5 A.

Resistor Voltage v_r will be *in phase* with i and since

$$V_m = I_m \times R$$

$$= 5 \times 10 = 50 \text{ V (figure 1.6b)}$$

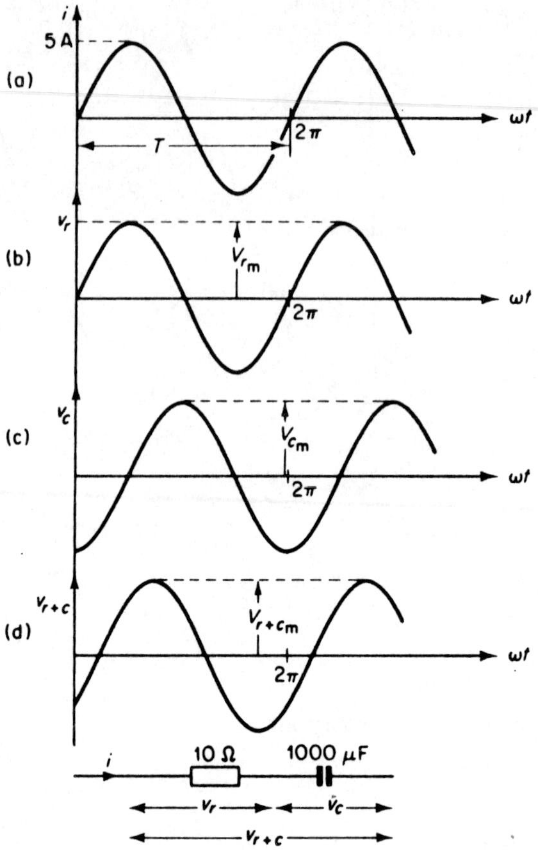

FIGURE 1.6
Example 1.2

Capacitor Voltage v_c – The current will *lead* the capacitor voltage by 90° or $\pi/2$ radians

$$V_{c_m} = I_m X_c = I_m \times \frac{1}{\omega C}$$

$$= \frac{5}{100 \times 10^3 \times 10^{-6}} = 50 \text{ V (figure 1.6c)}$$

The total voltage v_{r+c} is obtained by graphical addition of v_r and v_c. Scaling from figure 1.6d, the maximum value is 70.7 V lagging the current by 45° or $\pi/4$ radians.

1.3 Algebraic Solution of Step Functions on RL and RC Circuits

A. Variable Increasing

Assume S in figure 1.7 closes at $t = 0$, and that $v_c = 0$ and $i_l = 0$ when $t = 0$. Applying Kirchhoff's laws

RL circuit

$$V - L\frac{di}{dt} = iR$$

$$-L\frac{di}{dt} = iR - V$$

$$-\frac{L}{R}\frac{di}{dt} = i - \frac{V}{R}$$

$$-\frac{R\,dt}{L} = \frac{di}{i - I_F}$$

RC circuit

$$V - v_c = iR$$

$$\text{since } i = C\frac{dv_c}{dt}$$

$$V - v_C = RC\frac{dv_c}{dt}$$

$$\frac{dt}{RC} = \frac{dv_c}{V - v_c}$$

where $I_F = V/R$ is the final current as $t \to \infty$

$$-\frac{Rt}{L} = \ln(i - I_F) + C \qquad\qquad \frac{t}{RC} = -\ln(V - v_c) + B$$

where C and B are constants of integration.

Inserting the initial conditions of $i = 0$ and $v_c = 0$ when $t = 0$

$$C = -\ln(-I_F)$$

$$-\frac{Rt}{L} = \ln\left(\frac{i - I_F}{-I_F}\right)$$

$$\exp\left(-\frac{Rt}{L}\right) = -\frac{i}{I_F} + 1$$

$$i = I_F\left[1 - \exp\left(\frac{-Rt}{L}\right)\right]$$

$$B = \ln V$$

$$\frac{t}{RC} = \ln\left(\frac{V}{V - v_c}\right)$$

$$\exp\left(\frac{t}{RC}\right) = \frac{V}{V - v_c}$$

$$v_c = V\left[1 - \exp\left(\frac{-t}{RC}\right)\right]$$

since $v_c \to V$ as $t \to \infty$

$$v_c = V_F\left[1 - \exp\left(\frac{-t}{RC}\right)\right]$$

where V_F is the final value of v_c.

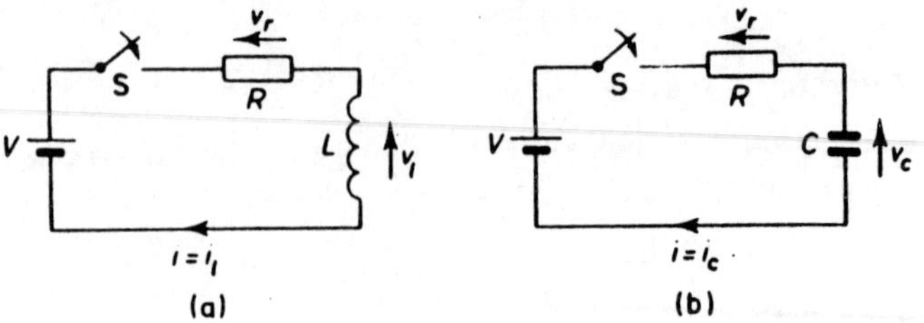

FIGURE 1.7

First-order circuits (variable increasing)

Graphs of these increasing currents and voltages are given in figure 1.8.

Because in each case the index of the exponential must be a pure number (dimensionless) L/R and CR must both have the units of t, that is seconds. Hence L/R and CR are called *time-constants* τ.

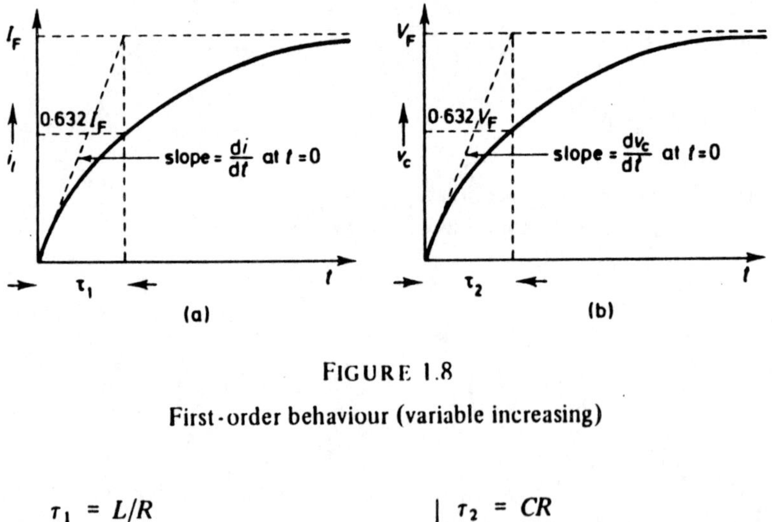

FIGURE 1.8

First-order behaviour (variable increasing)

$$\tau_1 = L/R$$

$$i = I_F \left[1 - \exp(-t/\tau_1)\right]$$

After $t = \tau_1$, $i = I_F (1 - e^{-1})$

$$i = I_F (1 - 0.368) = 0.632\, I_F$$

$$\tau_2 = CR$$

$$v_c = V_F' \left[1 - \exp(-t/\tau_2)\right]$$

After τ_2, $v_c = V_F'(1 - e^{-1})$

$$v_c = 0.632\, V_F$$

Hence we may define the time-constant as the time taken for the variable to reach 0.632 of its final value; alternatively

$$\frac{di}{dt} = \frac{I_F}{\tau_1} \exp(-t/\tau_1) \qquad \frac{dv_c}{dt} = \frac{V_F}{\tau_2} \exp(-t/\tau_2)$$

hence at $t = 0$, the slope of these curves is

$$\frac{di}{dt} = \frac{I_F}{\tau_1} \qquad \frac{dv_c}{dt} = \frac{V_F}{\tau_2}$$

$$\tau_1 = \frac{I_F}{\text{initial slope of graph}} \qquad \tau_2 = \frac{V_F}{\text{initial slope of graph}}$$

This gives an alternative definition of time-constant as the time needed for the variable to reach its final value if it increased linearly with its *initial* rate of rise.

The variable requires an infinite time to reach its final value, but in practice we may assume that it does so after a *settling time* of five time-constants. If

$$x = X_F \; [\, 1 - \exp\,(-t/\tau)\,]$$

at $t = 5$

$$x = X_F \; (1 - e^{-5})$$

$$x = X_F \; (1 - 0.007)$$

$$x = 0.993 \; X_F$$

the assumption is true to within 1 per cent. The concepts of time-constant and settling time occur and are important in mechanical and fluid as well as in electrical systems.

B. Variable Decreasing

Assume S is in position A long enough for transients to have died away, hence at $t = 0$, $i = I_{IN}$ and $v_c = V_{IN}$. At $t = 0$, S moves instantaneously to position B (figure 1.9).

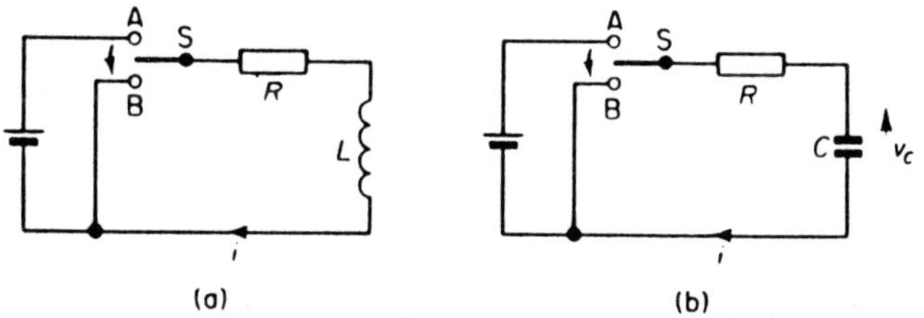

FIGURE 1.9

First-order circuits (variable decreasing)

$-L\dfrac{di}{dt} = iR$	$-v_c = iR$
$-\dfrac{R}{L}dt = \dfrac{di}{i}$	$-v_c = CR\dfrac{dv}{dt}$
$-\dfrac{Rt}{L} = \ln i + Q$	$-\dfrac{t}{CR} = \ln v_c + P$

where Q and P are constants of integration. Inserting the initial conditions

$t = 0,\ i = I_{IN}$	$t = 0,\ v_c = V_{IN}$
$Q = -\ln I_{IN}$	$P = -\ln V_{IN}$
$\dfrac{-Rt}{L} = \ln\dfrac{i}{I_{IN}}$	$\dfrac{-t}{CR} = \ln\dfrac{v_c}{V_{IN}}$
$\exp(-Rt/L) = \dfrac{i}{I_{IN}}$	$\exp(-t/CR) = \dfrac{v_c}{V_{IN}}$
$i = I_{IN}\exp(-Rt/L)$	$v_c = V_{IN}\exp(-t/CR)$

These curves are shown in figure 1.10.

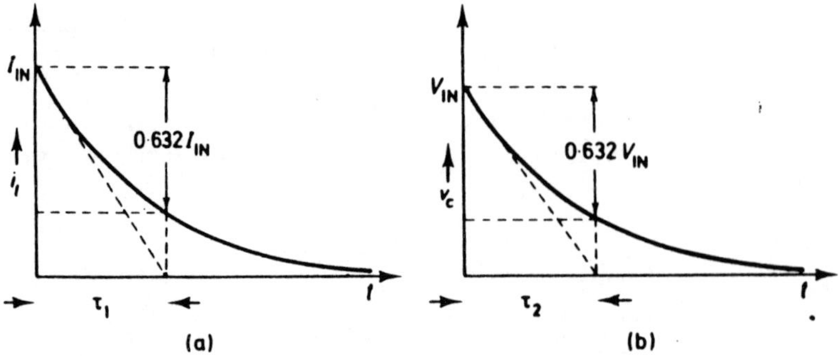

FIGURE 1.10

First-order behaviour (variable decreasing)

1.4 Other Circuit Combinations

Circuit arrangements containing both inductance and capacitance are known as second-order circuits. Their behaviour during the period immediately after the application of voltage or current may be complex, containing exponential terms describing their transient behaviour. The full analysis of these circuits from their differential equations or by Laplace transform methods is outside the scope of this text but is fully treated in standard works[2,3]. After the settling time of these exponential transient terms the long-term steady-state behaviour may often be predicted more simply. Perhaps one of the most commonly encountered excitations is the sinusoidal one. This is partly because of its ease of generation and partly because any periodic non-sinusoidal waveform may be expressed by Fourier analysis in terms of sinusoidal quantities (section 6.8). Steady-state sinusoidal (a.c.) analysis will thus comprise our next area of investigation.

1.5 Problems

1.1 Commencing with the elementary electrical behaviour equations for pure components, show that the energy stored in a charged capacitor is $CV^2/2$ J and that stored in a fluxed inductor is $LI^2/2$ J.

1.2 A voltage waveform rises linearly from zero to 50 V between $t = 0$ and $t = 2$ ms and then returns linearly to zero at $t = 4$ ms. It is applied to a pure capacitor of 50 μF. Sketch the current waveform and calculate the maximum current and the maximum charge.

1.3 A current rises linearly from zero to 250 mA in 2 ms, remains constant for 2 ms and then falls linearly to − 250 mA in 4 ms, remains constant for 2 ms and returns linearly to zero in 2 ms. It flows through a pure inductor of 500 mH. Sketch graphs of the induced e.m.f. and applied voltage. What is the value of the latter 1 ms after the start of the waveform?

1.4 The current waveform of example 1.3 passes through a series circuit consisting of a 4 Ω resistor, a 10 mH inductor and a 750 μF capacitor, all in series. Sketch the voltage waveforms across each component, calculate the maximum voltage across each component and the maximum capacitor charge.

1.5 (i) A series circuit consists of a 2 Ω resistor and a 10 H inductor. What is the time-constant? A constant voltage of 50 V is applied to the circuit. What is the current at 5 s after the voltage application?

(ii) A 100 μF capacitor discharges through its own leakage resistance from 250 V to 200 V in 70 s. Calculate its leakage resistance.

2 Steady-state A.C. Circuit Analysis

The behaviour of circuits in the first few cycles after the application of an alternating voltage or current is complex. The general solution consists of two parts, a transient and a steady-state response. In many applications the settling time during which these transient effects subside is short compared with the effects that we are investigating and hence only the steady-state response is required. With this proviso clearly in mind we may use this simplification to give more convenient solutions than those obtained from differential equations.

2.1 Power in Resistive A.C. Circuits

The power dissipated by resistor R in figure 2.1a is easily expressed for this d.c. case as I^2R or VI or V^2/R. In the a.c. situation of figure 2.1b, however, the solution is less clear. Perhaps the power could be obtained by substituting the peak values of voltage V_m or current I_m directly into the d.c. equations above. This is unlikely since the alternating waveform only attains these numerical values very briefly twice per cycle. During the remainder of the period the voltage or current is less than these peak values and therefore one would expect the *mean power* over a full cycle to be less than either $V_m I_m$ or $I_m{}^2R$. Another guess would be that the mean power is related to the mean voltage or current so let us obtain expressions for these mean values.

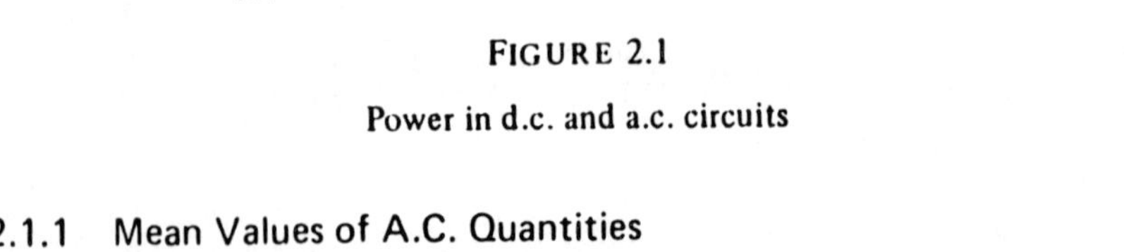

FIGURE 2.1

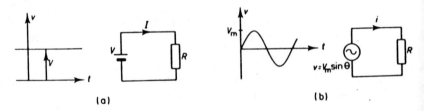

Power in d.c. and a.c. circuits

2.1.1 Mean Values of A.C. Quantities

Consider the current waveform $i = I_m \sin\theta$ shown in figure 2.2a. The normal methods of calculus show that the mean value of this function over a full cycle (0 to 2π radians) is given by

$$I_{mean} = \frac{1}{2\pi}\int_0^{2\pi} (I_m \sin\theta)\, d\theta$$

$$= \frac{1}{2\pi} \times I_m \left[-\cos\theta \right]_0^{2\pi}$$

$$= \frac{I_m}{2\pi} \left[+1 - (+1) \right] = 0$$

The mean value of a sinusoidal current or voltage over an integral number of cycles is thus zero which might have been inferred from figure 2.2a because the areas above and below the x-axis are equal.

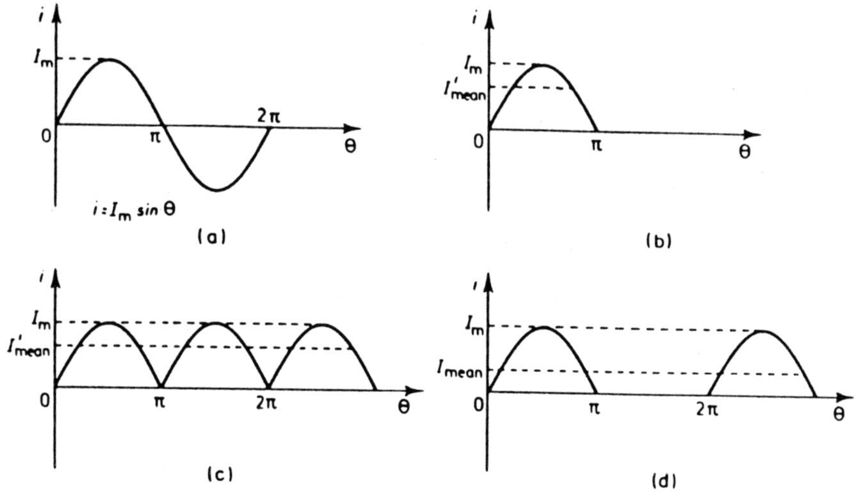

FIGURE 2.2

Mean values of a.c. waveforms

To prepare for power-supply studies in chapter 5 we shall derive expressions for the mean values of an alternating waveform over the period from 0 to π and hence for the full-wave rectified waveform of figures 2.2b and c.

$$I'_{mean} = \frac{1}{\pi} \int_0^{\pi} (I_m \sin\theta)\, d\theta$$

$$= \frac{I_m}{\pi} \left[-\cos\theta \right]_0^{\pi}$$

$$= \frac{I_m}{\pi} \left[-(-1) - (-1) \right]$$

$$= \frac{2I_m}{\pi} \text{ or } 0.637\, I_m \qquad (2.1)$$

The mean value of the full-wave rectified waveform of figure 2.2c is clearly also $2I_m/\pi$.

The half-wave rectified waveform must have a mean value of half the above since the function is equal to zero for half the time. Therefore its mean value is I_m/π (figure 2.2d).

2.1.2 Effective Values of A.C. Quantities

The previous section has clearly shown that the power developed in an a.c. circuit is not directly related to the mean current or voltage (zero) because this would imply that resistor R in figure 2.1b would emit heat during the positive current half-cycle (0 to π) and absorb it from π to 2π. We know, however, that the resistor always *emits* heat irrespective of the current direction – this is confirmed by the equation $p = i^2R$ which shows that the instantaneous power p is always positive irrespective of the direction of i.

Let us therefore derive an expression for the *effective value* of an a.c. waveform defined as that d.c. voltage or current which would produce the same power in the load R hence

d.c. case

Mean power $= (I_{eff})^2R$

a.c. case

Instantaneous power $= p = i^2R$

$$\text{Mean power} = P = \frac{1}{2\pi}\int_0^{2\pi} i^2R \, d\theta$$

$$P = \frac{R}{2\pi}\int_0^{2\pi} I_m^2 \sin^2\theta \, d\theta$$

$$= \frac{I_m^2R}{2\pi}\int_0^{2\pi} \frac{1 - \cos 2\theta}{2} \, d\theta$$

$$= \frac{I_m^2R}{4\pi}\left[\theta - \frac{\sin 2\theta}{2}\right]_0^{2\pi}$$

$$= \frac{I_m^2R}{4\pi}[2\pi - 0]$$

$$P = \frac{I_m^2R}{2}$$

Equating the d.c. and a.c. powers

$$I_{eff}^2R = \frac{I_m^2R}{2}$$

or

$$I_{eff} = \frac{I_m}{\sqrt{2}} = 0.707\, I_m \qquad (2.2)$$

Similarly the effective value of a voltage waveform can be shown to be

$$V_{\text{eff}} = \frac{V_m}{\sqrt{2}} = 0.707\ V_m$$

Because this value was obtained by taking the square root of the (function squared), the effective value of an a.c. waveform is commonly known as the *root-mean-square* value (r.m.s.).

The power in a resistive a.c. circuit can now be written directly as $I_{\text{eff}}{}^2 R$ or $I_{\text{r.m.s.}}{}^2 R$ or $V_{\text{r.m.s.}}\ I_{\text{r.m.s.}}$ or $V_{\text{r.m.s.}}{}^2/R$. These equations correspond directly with the d.c. case. Since all electrical energy is used for its power-transmission ability, the r.m.s. or effective values are the ones having the greatest physical significance. Consequently, unless otherwise stated, it is always assumed that voltages and currents are r.m.s. values and the simple symbols V and I will mean r.m.s. or effective values.

This has no effect on our definition of reactance given in section 1.2 since if $X = V_m/I_m$ it is also equal to $0.707\ V_m/0.707\ I_m$.

The r.m.s. values of non-sinusoidal alternating waveforms cannot easily be obtained from the maximum values by the above formulae. They must be obtained from first principles using calculus where the equation of the function is known or by determining the mean square value of the waveform graphically if the function is discontinuous or very complex.

Example 2.1

Calculate the r.m.s. value of the waveform shown in figure 2.3.

From the figure the mid-ordinate values at 1 ms intervals and their squares are

t (ms)	0.5	1.5	2.5	3.5	4.5	5.5
v (volts)	1	3	2	−1	−3	−2
v^2 (volts^2)	1	9	4	1	9	4

By the mid-ordinate rule the r.m.s. or effective voltage is

$$V = \sqrt{\left(\frac{v_1{}^2 + v_2{}^2 + v_3{}^2 + \ldots + v_n{}^2}{n}\right)}$$

$$V = \sqrt{\left(\frac{1 + 9 + 4 + 1 + 9 + 4}{6}\right)} = \sqrt{\frac{28}{6}}$$

$$= 2.16\ \text{V}$$

2.1.3 Form Factor, Peak Factor and Instrument Errors

The terms 'form factor' and 'peak' or 'crest factor' are frequently met in electrical literature; they are defined as follows.

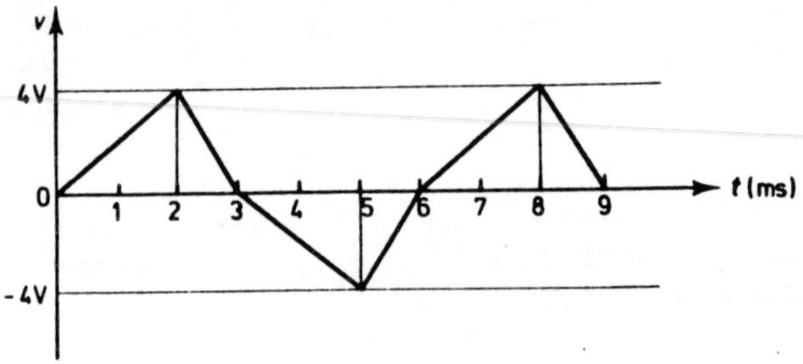

FIGURE 2.3

Example 2.1

$$\text{Form factor} = \frac{\text{r.m.s. value}}{\text{mean value}}$$

$$\text{Peak factor} = \frac{\text{peak value}}{\text{r.m.s. value}}$$

For a sinusoid these are constant

$$\left.\begin{aligned} \text{form factor} &= \frac{0.707\ V_m}{0.637\ V_m} = 1.11 \\ \text{peak factor} &= \frac{V_m}{0.707\ V_m} = 1.414 \end{aligned}\right\} \quad (2.3)$$

For other waveshapes these factors must be calculated individually.

Many electrical instruments respond either to (current)^2 or (voltage)^2. They will thus read correctly the r.m.s. value of any waveshape within their frequency range; among such instruments are moving-iron, electrodynamic, hot-wire and thermocouple types. Most a.c. electrical measurements are, however, made with rectifier-moving-coil instruments (for example the Avometer). These instruments do not indicate the r.m.s. value of a waveform, merely the mean rectified value V_{mean}. But their scales are calibrated in r.m.s. units *assuming a sinusoidal waveform* and thus a form factor of 1.11. Wherever there is the slightest room for doubt the waveshape of the voltage monitored by them must be verified with an oscilloscope. If it is not sinusoidal the meter readings are subject to large errors.

2.2 *RLC* Series and Parallel Circuits

2.2.1 Use of Phasor Diagrams

For those readers unacquainted with phasor diagrams a short résumé is given. Instead of laboriously drawing the waveform diagram of figure 2.4a to show the relationships between waveforms of the same frequency, the more convenient phasor diagram shown at figure 2.4b may be employed.

The two phasors V_{1m} and V_{2m} revolve anticlockwise (by convention) about

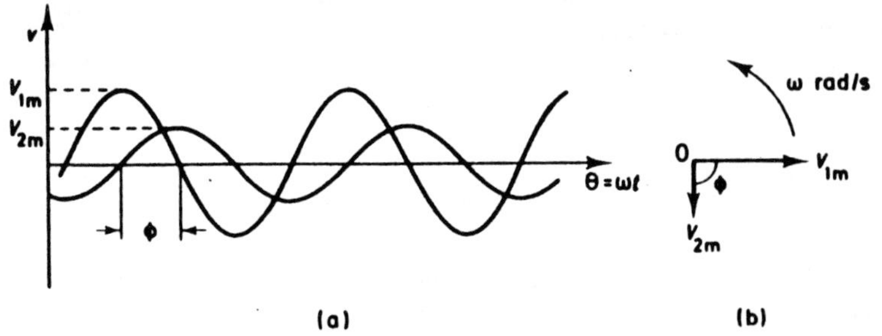

FIGURE 2.4

Phasor representation

centre O at an angular velocity equal to the angular frequency ω of the waveform. The projection of the tips of these phasors on to the vertical through O is thus capable of giving the instantaneous value and hence, if the frequency is known, contains all the information in the waveform diagram. Voltage and current phasors may be added or subtracted graphically by completion of the parallelogram as in figures 2.5a and b respectively, or by calculation. For example, the sum of V_1 and V_2 in figure 2.5a will have components

$$V_{T_{vert}} = V_1 \sin\phi_1 + V_2 \sin\phi_2$$

$$V_{T_{horiz}} = V_1 \cos\phi_1 + V_2 \cos\phi_2$$

They may also be multiplied or divided by normal polar methods

$$V = V_1 \times V_2 = V_1 V_2 \underline{/\phi_1 + \phi_2}$$

or

$$V = \frac{V_1}{V_2} = \frac{V_1}{V_2} \underline{/\phi_1 - \phi_2}$$

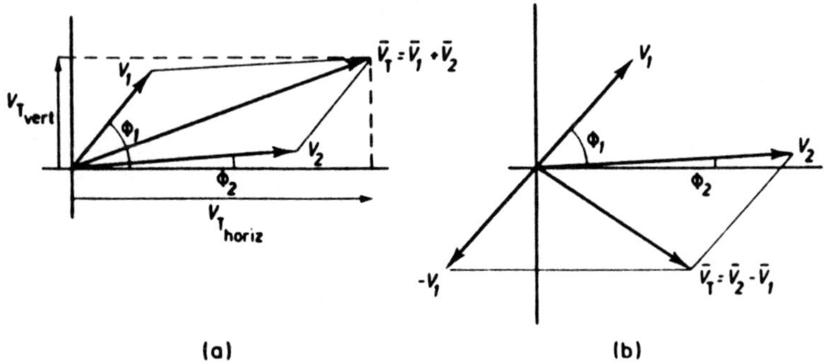

FIGURE 2.5

Phasor addition and subtraction

Notice that since resistance and reactance are both the ratios of two phasor quantities – voltage and current – resistance and reactance will also be phasor quantities from the above equations.

The angle between V_1 and V_2 is known as the *phase angle* between them – it must always be stated which voltage lags or leads the other.

Needless to say, phasors can only be used for sinusoidal quantities and, because of the overriding importance of the r.m.s. as opposed to the maximum values of these quantities, the phasor is nearly always drawn equal in length to the r.m.s. value. This only has the effect of scaling down the phasor diagram to $1/\sqrt{2}$ of its former size, the angles being unaffected.

2.2.2 *RLC* Series Circuit

Consider the circuit of figure 2.6; since *R*, *L* and *C* are in series it is the r.m.s. current *I* that will be common to all three. The current phasor *I* is thus made the reference phasor and is shown horizontal.

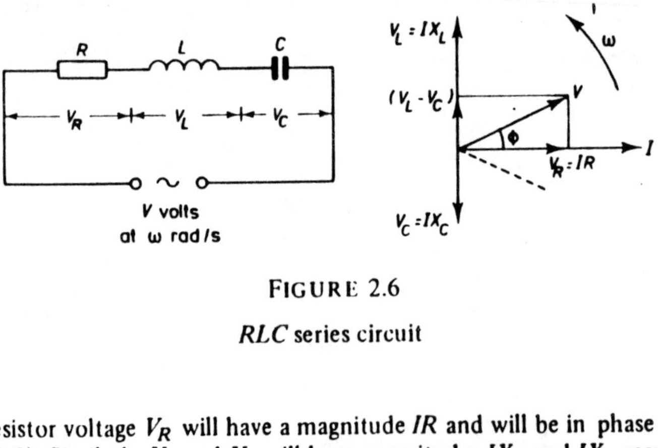

FIGURE 2.6

RLC series circuit

The resistor voltage V_R will have a magnitude IR and will be in phase with I (section 1.2). Similarly V_L and V_C will have magnitudes IX_L and IX_C respectively in directions indicated by the mnemonic in section 1.2. To obtain the total voltage V, phasor addition of V_R, V_L and V_C gives V as shown where

$$|V| = \sqrt{\left[V_R^2 + (V_L - V_C)^2\right]}$$

or

$$|V| = |I| \times \sqrt{\left[R^2 + (X_L - X_C)^2\right]}$$

Thus the ratio $|V| \,/\, |I|$ is called the *impedance Z*, where

$$Z = \sqrt{\left[R^2 + (X_L - X_C)^2\right]} \qquad (2.4)$$

which being the ratio of voltage to current can be expressed in phasor (or polar) form and has units of ohms. Also the phase angle between I and V is ϕ

$$\phi = \arctan \frac{V_L - V_C}{V_R} = \arctan \frac{X_L - X_C}{R} \tag{2.5}$$

and in this case, since the current lags the voltage, the phase angle ϕ is also lagging.

Notice that in this case X_L was assumed to be greater than X_C, making V_L exceed V_C. If V_C had proved greater than V_L, their sum $(V_C - V_L)$ would have been downwards, giving a leading phase angle as shown by the dotted line.

Example 2.2

Calculate the current, the phase angle and hence the voltages across R, L and C in figure 2.6 if their respective values are 10 Ω, 1 H and 10 μF, the supply being 240 V at 50 Hz.

$$X_L = \omega L = 2\pi f L$$

$$= 100\pi \times 1 = 314\ \Omega$$

$$X_C = \frac{1}{\omega C} = \frac{1}{2\pi f C}$$

$$= \frac{10^6}{100\pi \times 10} = 318\ \Omega$$

Since $X_C > X_L$

$$Z = \sqrt{[R^2 + (X_C - X_L)^2]}$$

$$= \sqrt{[100 + (318 - 314)^2]}$$

$$= \sqrt{116} = 10.76\ \Omega$$

Hence current $I = V/Z = 240/10.76$

$$= 22.3\ \text{A}$$

$$\phi = \arctan \frac{X_C - X_L}{R} = \arctan \frac{4}{10}$$

$$\phi = 21^\circ\ 48' \textit{ leading} \text{ since } I \text{ leads } V$$

$$V_R = IR = 22.3 \times 10 = 223\ \text{V}$$

$$V_L = IX_L = 22.3 \times 314 = 7000\ \text{V}$$

$$V_C = IX_C = 22.3 \times 318 = 7090\ \text{V}$$

Notice that the capacitor and inductor voltages are many times larger than the supply voltage. This may represent a source of danger in such circuits where X_L is almost equal to X_C. This effect, known as *voltage multiplication*, will be investigated quantitatively in section 2.4.1.

Note also that the equations given for the *RLC* series circuit are equally valid for two-component circuits. For example, in the *RL* case merely put X_C to zero and for the *RC* case put X_L to zero.

2.2.3 *RLC* Parallel Circuit

Consider the circuit of figure 2.7; since *R*, *L* and *C* are in parallel it is the voltage *V* that is common to all three. The voltage phasor *V* is thus made the reference phasor which is conventionally drawn horizontally.

The resistor current I_R will have a magnitude V/R and will be in phase with *V*. Similarly I_L and I_C will have magnitudes V/X_L and V/X_C respectively in the directions shown. Phasor addition of I_L and I_C gives $(I_L - I_C)$ which when combined with I_R gives the total current *I* as shown (assuming $X_C > X_L$).

By Pythagoras

$$|I| = \sqrt{\left[I_R^2 + (I_L - I_C)^2\right]}$$

or

$$|I| = |V| \times \sqrt{\left[\frac{1}{R^2} + \left(\frac{1}{X_L} - \frac{1}{X_C}\right)^2\right]}$$

Hence

$$|Z| = \frac{|V|}{|I|} = \left\{\sqrt{\left[\frac{1}{R^2} + \left(\frac{1}{X_L} - \frac{1}{X_C}\right)^2\right]}\right\}^{-1} \ \Omega$$

and

$$\phi = \arctan \frac{I_L - I_C}{I_R} = \arctan \frac{(1/X_L) - (1/X_C)}{1/R}$$

the phase angle lagging in this case since $I_L > I_C$ ($X_C > X_L$).

Again if the circuit is only a two-component one, say *RL*, we can use the above equations putting X_C equal to infinity. Similarly for *RC* parallel circuits we can put X_L to infinity.

Example 2.3

Calculate the individual currents, the total current, the phase angle and the magnitude of the impedance for the circuit shown in figure 2.7 if $R = 10\ \Omega$, $L = 0.01$ H and $C = 800\ \mu$F. The supply is 240 V at 50 Hz.

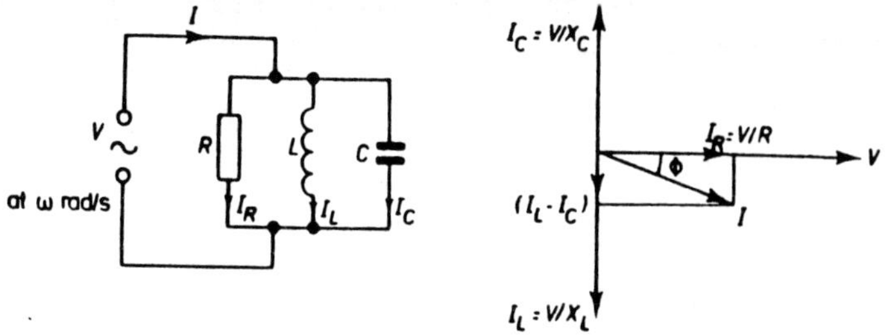

FIGURE 2.7

RLC parallel circuit

$$I_R = \frac{V}{R} = \frac{240}{10} = 24 \text{ A}$$

$$X_L = 2\pi fL = 100\pi \times 10^{-2} = 3.14\ \Omega$$

$$I_L = \frac{V}{X_L} = \frac{240}{3.14} = 76.4 \text{ A}$$

$$X_C = \frac{1}{2\pi fC} = \frac{10^6}{100\pi \times 800} = 3.98\ \Omega$$

$$I_C = \frac{V}{X_C} = \frac{240}{3.98} = 60.3 \text{ A}$$

$$I_L - I_C = 76.4 - 60.3 = 16.1 \text{ A}$$

Therefore

$$I = \sqrt{\left[I_R^2 + (I_L - I_C)^2 \right]}$$

$$I = \sqrt{[576 + 260]} = 28.9 \text{ A}$$

$$\phi = \arctan \frac{I_L - I_C}{I_R} = \arctan \frac{16.1}{24}$$

$$\phi = \arctan 0.670 = 33^\circ\ 49' \text{ lagging}$$

The magnitude of the impedance $|Z| = |V|/|I|$

$$|Z| = \frac{240}{28.9} = 8.31\ \Omega$$

Notice that both the capacitor and inductor currents are considerably larger than the total current. This *current-multiplication* effect will be considered in section 2.4.2.

The impedance Z in example 2.3 may be expressed fully in polar form as

$$Z = \frac{V}{I} = \frac{240 \angle 0}{28.9 \angle -33°49'}$$

$$= 8.31\ \Omega \text{ at } 33°\ 49'$$

2.2.4 Use of the j Operator

While the use of the above formulae with phasor diagrams is satisfactory for simpler series and parallel circuit problems, more complex examples require a less laborious method. The reader will be aware that any phasor can be expressed in cartesian form as well as by the polar co-ordinates used above. The rectangular cartesian co-ordinates may be quoted in complex form allowing the phasor to be drawn on an Argand diagram. Figure 2.8a shows the phasor whose polar co-ordinates are $R \angle \theta$ drawn on an Argand diagram. The rectangular co-ordinates can be seen to be

Imaginary axis	$(+R \sin\theta)$
Real axis	$(+R \cos\theta)$

which may be expressed as the complex number

$$R \angle \theta \equiv R \cos\theta + jR \sin\theta \tag{2.6}$$

Conversely any phasor specified by the complex number $a + jb$ as in figure 2.8b may be returned to polar co-ordinates by the use of Pythagoras' theorem and trigonometry

$$a + jb \equiv \sqrt{(a^2 + b^2)} \angle \arctan \frac{b}{a} \tag{2.7}$$

Any voltage, current or impedance phasor may thus be completely specified by a complex number, written as $\dot{V}$, $\dot{I}$ and $\dot{Z}$ respectively.

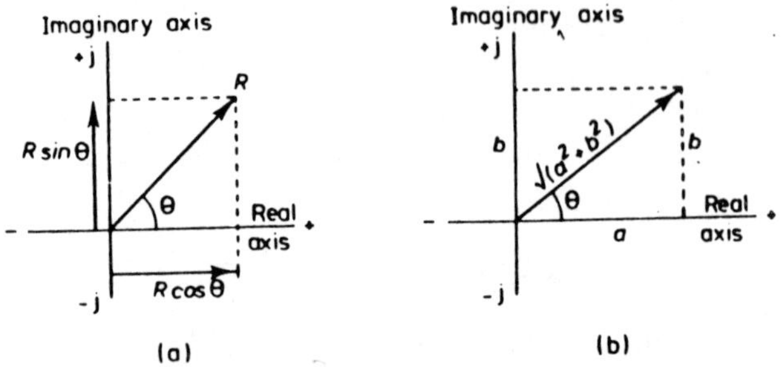

FIGURE 2.8

Conversion between polar and complex notation

Example 2.4

A resistor of 3 Ω and an inductor of reactance 4 Ω are placed in series across an a.c. voltage source of 250 V $\angle 0^\circ$. Express both the voltage and current in complex form.

$$\text{Impedance } Z = \sqrt{(R^2 + X_L{}^2)}$$

$$= \sqrt{(9 + 16)} = 5\ \Omega$$

$$\text{Current} \quad I = \frac{V}{Z} = \frac{250}{5}$$

$$= 50\text{ A}$$

$$\text{Phase angle } \phi = \arctan \frac{X_L}{R}$$

$$= \arctan \frac{4}{3}$$

$$= 53^\circ\ 7' \text{ lagging}$$

The current lags the voltage since the circuit is an inductive one. Converting V and I to complex form

$$\dot{V} = 250 \angle 0^\circ \equiv 250 \cos 0 + j250 \sin 0$$

$$\equiv (250 + j0)\text{ V}$$

$$\dot{I} = 50 \angle -53^\circ\ 7' \equiv 50 \cos(-53^\circ\ 7') + j50 \sin(-53^\circ\ 7')$$

$$\equiv (50 \times 0.6) + j(50 \times -0.8)$$

$$\equiv (30 - j40)\text{ A}$$

Let us for a moment consider the impedance phasor $\dot{Z}$ in the above problem

$$\dot{Z} = \frac{V}{I} = \frac{250 \angle 0}{50 \angle -53^\circ 7'} = 5 \angle 53^\circ\ 7'\ \Omega$$

If we convert this to complex form

$$\dot{Z} = 5 \cos 53^\circ\ 7' + j \sin 53^\circ 7' = (3 + j4)\ \Omega$$

Examination of this complex impedance reveals that the real term is equal to the resistance of the series circuit while the imaginary term is equal to the inductive reactance

$$\dot{Z} = R + j X_L \tag{2.8}$$

It is left as an exercise to the student to show that the complex impedance for an *RC* series circuit is

$$\dot{Z} = R - \mathrm{j}X_C \tag{2.9}$$

so that the complete impedance expression for an *RLC* series circuit is

$$\dot{Z} = R + \mathrm{j}(X_L - X_C) \tag{2.10}$$

Impedances in Series

Consider the two impedances $\dot{Z}_1 = a + \mathrm{j}b$ and $\dot{Z}_2 = c + \mathrm{j}d$ shown in figure 2.9a. Because resistors or inductors in series may be added arithmetically the circuit

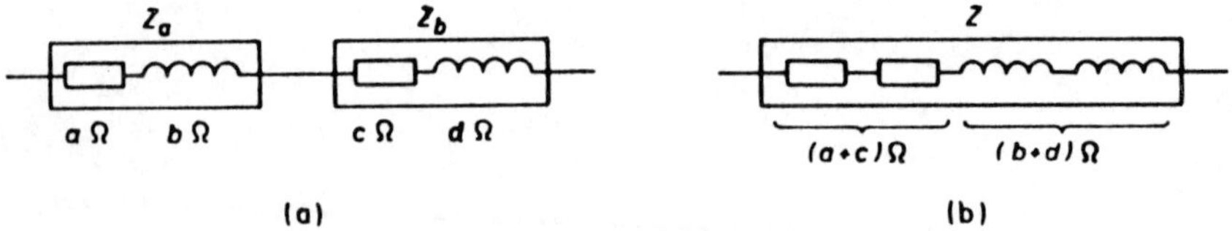

FIGURE 2.9

Complex impedances in series

may be redrawn as in figure 2.9b. The complex impedance of Z is thus $(a + c) + \mathrm{j}(b + d)$. The total impedance of two series impedances is thus the complex sum of the individual impedances.

$$\dot{Z} = \dot{Z}_a + \dot{Z}_b$$

Impedances in Parallel

The total resistance of two parallel resistors is given by $1/R = 1/R_1 + 1/R_2$. Similarly as long as complex notation is used, for two impedances Z_1 and Z_2 in parallel (figure 2.10a)

$$\frac{1}{\dot{Z}} = \frac{1}{\dot{Z}_1} + \frac{1}{\dot{Z}_2} \quad \text{or} \quad \dot{Z} = \frac{\dot{Z}_1\dot{Z}_2}{\dot{Z}_1 + \dot{Z}_2}$$

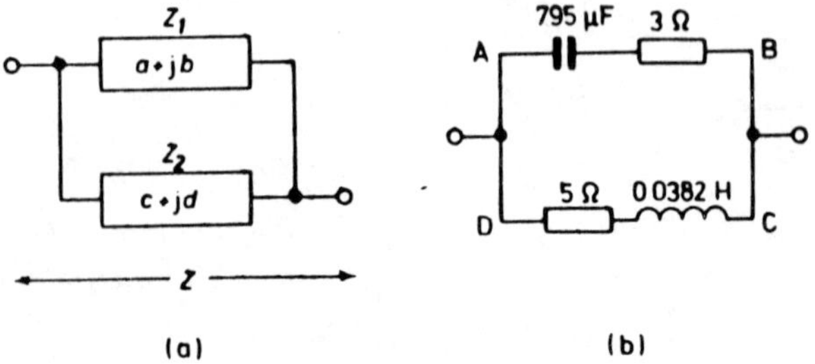

FIGURE 2.10

Complex impedances in parallel

Example 2.5

Calculate the complex impedance of the series - parallel circuit shown in figure 2.10b at a frequency of 50 Hz.

$$X_C = \frac{1}{2\pi fC} = \frac{1}{2\pi \times 50 \times 795 \times 10^{-6}}$$

$$= \frac{10^6}{100\pi \times 795} = 4\ \Omega$$

$$X_L = 2\pi fL = 2\pi \times 50 \times 0.0382$$

$$= 12\ \Omega$$

$$\dot{Z}_{AB} = (3 - j4)\ \Omega$$

$$\dot{Z}_{CD} = (5 + j12)\ \Omega$$

$$\dot{Z} = \frac{\dot{Z}_{AB}\,\dot{Z}_{CD}}{\dot{Z}_{AB} + \dot{Z}_{CD}} = \frac{(3 - j4)(5 + j12)}{(3 - j4) + (5 + j12)}$$

$$= \frac{(15 + 48) + j(36 - 20)}{8 + j8} = \frac{63 + j16}{8 + 8j}$$

$$= \frac{1}{8} \times \frac{(63 + j16)}{1 + j}$$

Rationalising

$$\dot{Z} = \frac{1}{8} \times \frac{63 + j16}{1 + j} \times \frac{1 - j}{1 - j}$$

$$= \frac{1}{8} \times \frac{63 + j16 - j63 + 16}{1^2 + 1^2}$$

$$= \frac{1}{16} \times (79 - j47) = (4.94 - j2.94)\ \Omega$$

2.3 Power in the General A.C. Circuit

2.3.1 Power Factor

Consider the general case of a sinusoidal voltage and current separated by a phase angle ϕ. If we assume for the moment a leading phase-angle (capacitive circuit) we may write the voltage and current expressions as follows.

$$\text{Instantaneous voltage } v = V_m \sin\theta$$

$$\text{Instantaneous current } i = I_m \sin(\theta + \phi)$$

So at any point in the cycle the instantaneous power is

$$p = vi = V_m I_m \sin\theta \sin(\theta + \phi)$$

Since

$$\sin A \sin B = \tfrac{1}{2}[\cos(A - B) - \cos(A + B)]$$

$$p = \tfrac{1}{2} V_m I_m [\cos\phi - \cos(2\theta - \phi)]$$

Therefore the mean value of this expression over an integral number of cycles is

$$\text{Average power} \quad P = \frac{1}{2\pi}\int_0^{2\pi} \tfrac{1}{2} V_m I_m [\cos\phi - \cos(2\theta - \phi)]\, d\theta$$

$$= \frac{1}{2\pi}\int_0^{2\pi} \tfrac{1}{2}(V_m I_m \cos\phi) - [\tfrac{1}{2} V_m I_m \cos(2\theta - \phi)]\, d\theta$$

Examining the integral terms, $V_m I_m \cos\phi$ is a constant since the phase angle ϕ does not vary throughout the cycle. The second term $[\frac{1}{2} V_m I_m \cos(2\theta - \phi)]$ is a cosine term which goes through two complete cycles as θ goes from 0 to 2π. Its average value must therefore be zero (see section 2.1.1). Thus

$$P = \frac{1}{2\pi}\int_0^{2\pi} \tfrac{1}{2}(V_m I_m \cos\phi)\, d\theta$$

$$= \frac{1}{4\pi}[V_m I_m \cos\phi \times \theta]_0^{2\pi}$$

$$= \frac{1}{4\pi}(V_m I_m \cos\phi \times 2\pi)$$

$$= \frac{V_m}{\sqrt{2}} \times \frac{I_m}{\sqrt{2}} \times \cos\phi$$

$$= VI \cos\phi \text{ W} \qquad (2.11)$$

The mean power in an a.c. circuit is thus the product of the (r.m.s.) voltage and current multiplied by a constant. This constant is called the *power factor* and is equal to the cosine of the phase angle. When quoting a power factor it is *essential* to state whether it is associated with a leading or a lagging current.

Example 2.6

Calculate the total power dissipated in the circuit of figure 2.6 in example 2.2.

$$\text{Power} = VI \cos\phi \text{ W}$$

$\phi = 21° \, 48'$ leading, therefore the power factor ($\cos\phi$) is 0.9285 leading. Thus

$$\text{Power} = 240 \times 22.3 \times 0.9285$$

$$= 4970 \text{ W}$$

$$= 4.97 \text{ kW}$$

One item of interest is to consider the power dissipated by a pure capacitor or inductor. Section 1.2 has shown that the phase angles in these cases are 90° leading and 90° lagging respectively. This being so, the power factor ($\cos \phi$) is zero in both cases irrespective of the voltage and current values. *Pure inductors and capacitors thus dissipate no power*, being merely energy-storage devices.

This leads to an alternative method of solution for example 2.6. Because L and C dissipate no power, all the power must be dissipated in R. The equation for the power in a resistor is well known

$$P = I_{(\text{r.m.s.})}{}^2 R = (22.3)^2 \times 10$$

$$= 4.97 \text{ kW again}$$

2.3.2 Active and Reactive Current and Power

If the voltage and current in an a.c. circuit are plotted on a phasor diagram as in figure 2.11a, the current may be resolved into two components as shown. The component $I \cos \phi$ in phase with the voltage is called the *active current* whereas the quadrature component $I \sin \phi$ is called the *reactive current*.

If each phasor in figure 2.11a is multipiied by V, we obtain figure 2.11b the angles of which are identical to those of 2.11a, the component $VI \cos \phi$ in phase with the voltage is clearly the power (in watts) whereas the quadrature component $VI \sin \phi$ is called the *reactive power* Q the unit for which is the reactive volt ampere (VAr). The phasor VI is known as the *apparent power* S whose unit is the volt ampere. Clearly

$$\dot{S} = P + jQ$$

from which

$$|S| = \sqrt{(|P|^2 + |Q|^2)}$$

and

$$\phi = \arctan \frac{Q}{P}$$

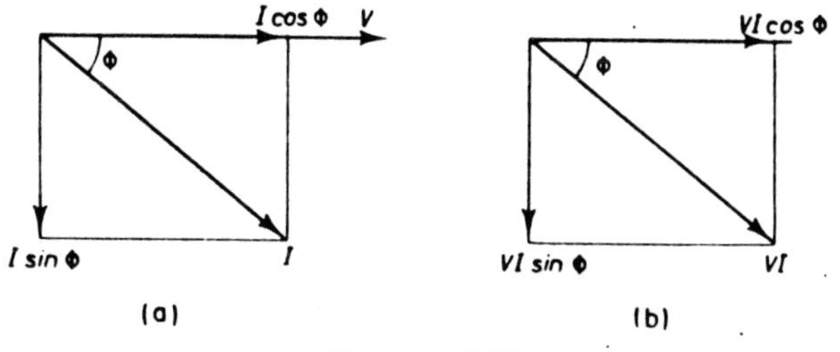

FIGURE 2.11

Active and reactive currents and powers

2.3.3 Power-factor Correction

Because the total power in an a.c. single-phase circuit is $VI \cos \phi$ it is clear that the reactive power $VI \sin \phi$ and, therefore, the reactive current do not contribute to the power of a system. Unfortunately, however, the conductors connecting an electrical source to a load of phase angle ϕ have to be heavy enough to transmit the total current I and not just its active component $I \cos \phi$.

It is economically desirable therefore to ensure that the active component is maximised and the reactive current minimised. This is done by making ϕ as near to zero as possible, that is, approaching unity power-factor. Consider figure 2.12a in which a reactive load is fed from a voltage V. If a capacitor is placed in parallel with it the phasor diagram will be that of figure 2.12b with the capacitor current I leading V by 90°.

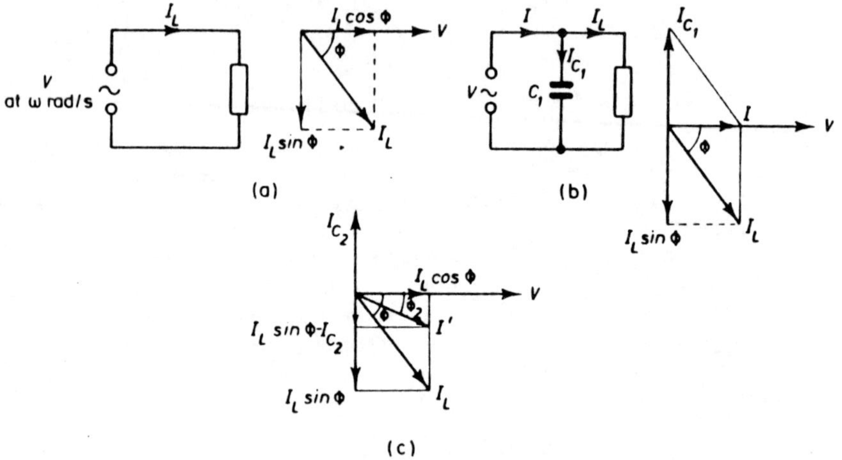

FIGURE 2.12

Full and partial power-factor correction

For full or unity correction C_1 is made such that I_{C_1} is equal in magnitude to $I_L \sin \phi$. This causes the two vertical currents to cancel, making the total current I equal to the active component $I_L \cos \phi$ alone. The conductors carrying energy between the source V and the capacitor-load combination now merely carry the active or useful component of the current, allowing significant savings to be made in material costs. For full correction to unity

$$I_{C_1} = \frac{V}{X_{C_1}} = I_L \sin \phi$$

or

$$V\omega C_1 = I_L \sin \phi$$

$$C_1 = \frac{I_L \sin \phi}{V\omega}$$

Partial Correction

Although power-factor correction capacitors require little maintenance their initial cost may be such as to make full correction unjustifiable. In this case a compromise partial correction to some new power factor $\cos\phi_2$, often approximately 0.9, may be employed. Assume a capacitor of value C_2 is employed in the circuit of figure 2.12b for partial correction. The phasor diagram will be that of figure 2.12c where $I_{C_2} < I_{C_1}$ and thus only cancels a part of $I_L \sin\phi$ leaving a vertical component $I_L \sin\phi - I_{C_2}$.

The new total current is thus I' at a lagging phase-angle ϕ_2 where

$$\phi_2 = \arctan \frac{I_L \sin\phi - I_{C_2}}{I_L \cos\phi}$$

Power-factor correction is chiefly employed to correct the lagging power-factors of the inductive windings of electrical motors and, to a lesser extent, the lagging power-factor produced by discharge-lamp circuitry. In larger installations the physical size of the correction capacitors would be prohibitive and over-excited synchronous motors (which take a leading current) are often employed.

Example 2.7

An a.c. motor takes a current of 80 A at a lagging phase-angle of 53° from a 440 V, 50 Hz supply. Calculate the values and working voltages of the capacitors required to correct the over-all power-factor to (i) unity and (ii) 0.95 lagging. What will be the resulting total current in each case?

Referring to figure 2.13; $\phi = 53^\circ$, then

$$\text{Active motor current} = I_m \cos\phi = 80 \times \cos 53^\circ$$

$$= 80 \times 0.602 = 48.16 \text{ A}$$

$$\text{Reactive motor current} = I_m \sin\phi = 80 \times \sin 53^\circ$$

$$= 80 \times 0.799 = 63.9 \text{ A}$$

(i)

$$I_C = 63.9 = \frac{V}{X_C}$$

therefore

$$C_1 = \frac{63.9}{(440 \times 100\pi)} = 462\ \mu\text{F}$$

$$\text{Working voltage} = V_m \text{ (or } V_{pk}) = \sqrt{(2)}V_{(\text{r.m.s.})}$$

$$= \sqrt{(2)} \times 440 = 622 \text{ V}$$

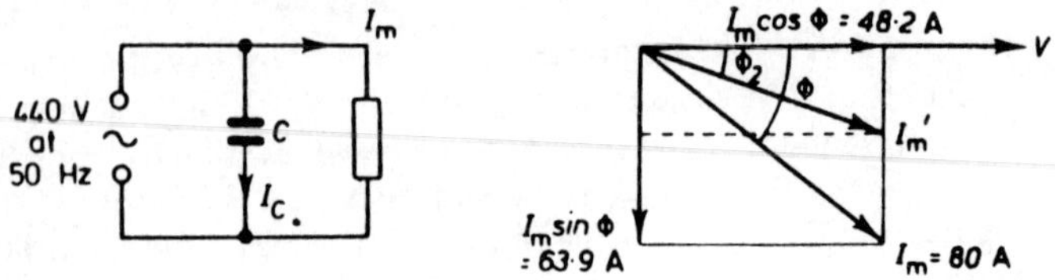

FIGURE 2.13

Example 2.7

(ii)

$$\text{Capacitor current} = I_m \sin\phi - I_m' \sin\phi_2$$

$$= 63.9 - 15.84$$

$$= 48.05 \text{ A}$$

Thus, as before

$$C_2 = \frac{48.05}{440 \times 100\pi} = 347\ \mu\text{F}$$

again

$$\text{Working voltage} = \sqrt{(2)} \times 440 = 622 \text{ V}$$

Total currents

(i) $$I = I_L \cos\phi = 48.16 \text{ A}$$

(ii) $$I = I_m' = I_L \frac{\cos\phi}{\cos\phi_2}$$

$$= \frac{48.16}{0.95} = 50.8 \text{ A}$$

2.3.4 Power from Complex Voltage and Current

Consider the complex voltage and current phasors

$$\dot{V} = a + jb$$

$$\dot{I} = c + jd$$

Both these are represented in figure 2.14 at angles ϕ_1 and ϕ_2 to the x-axis respectively.

$$\text{Power } P = VI \cos(\phi_1 - \phi_2)$$

$$= \sqrt{(a^2 + b^2)}\sqrt{(c^2 + d^2)} \times (\cos\phi_1 \cos\phi_2 + \sin\phi_1 \sin\phi_2)$$

$$= \sqrt{(a^2 + b^2)}\sqrt{(c^2 + d^2)}\left[\frac{a}{\sqrt{(a^2 + b^2)}} \times \frac{c}{\sqrt{(c^2 + d^2)}} + \frac{b}{\sqrt{(a^2 + b^2)}} \times \frac{d}{\sqrt{(c^2 + d^2)}}\right]$$

$$P = ac + bd$$

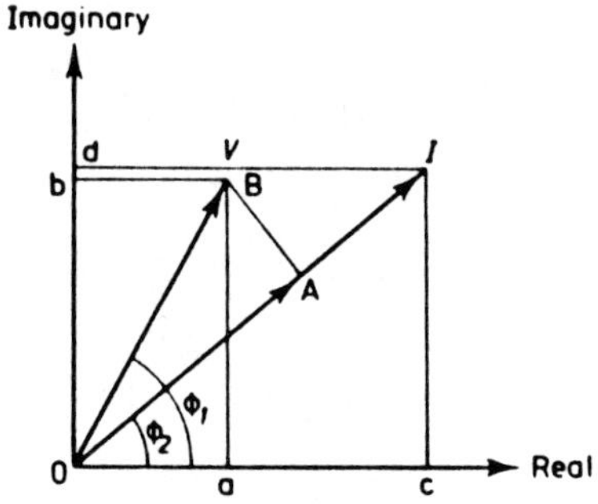

FIGURE 2.14

Power from complex values

Similarly reactive power $Q = VI \sin(\phi_1 - \phi_2)$

$$Q = \sqrt{(a^2 + b^2)}\sqrt{(c^2 + d^2)} \times [\sin\phi_1 \cos\phi_2 - \cos\phi_1 \sin\phi_2]$$

$$= \sqrt{(a^2 + b^2)} \times \sqrt{(c^2 + d^2)}\left[\frac{b}{\sqrt{(a^2 + b^2)}} \times \frac{c}{\sqrt{(c^2 + d^2)}} - \frac{a}{\sqrt{(a^2 + b^2)}} \times \frac{d}{\sqrt{(c^2 + d^2)}}\right]$$

$$Q = bc - ad$$

These two expressions may be obtained more conveniently by multiplying the voltage by the complex conjugate of the current I^*

$$\dot{V}\dot{I}^* = (a + jb)(c - jd)$$

$$= (ac + bd) + j(bc - ad)$$

when the power is given by the real part and the reactive power by the imaginary part, or

$$P = \text{Re}\,(\dot{V}\dot{I}^*)\ \text{W} \qquad (2.12)$$

$$Q = \text{Im}\,(\dot{V}\dot{I}^*)\ \text{VAr} \qquad (2.13)$$

2.3.5 Decibel Notation

In communication and instrumentation practice it is often necessary to compare two powers of widely different magnitudes. The decibel notation has been developed for this purpose. If two powers P_1 and P_2 are dissipated due to voltages V_1 and V_2 across two equal resistors R, then the power ratio may be expressed simply as P_1/P_2.

Alternatively we may use the logarithmic ratio $\log_{10}(P_1/P_2)$ the unit being the bel, B, where

$$\text{Power ratio in bels} = \log_{10}\left(\frac{P_1}{P_2}\right) \text{B} \tag{2.14}$$

The bel is an inconveniently large unit and the decibel, dB, is more commonly used

$$\text{Power ratio in decibels} = 10 \log_{10} (P_1/P_2) \text{ dB}$$

Since

$$\text{power dissipated} = \frac{\text{voltage}^2}{\text{resistance}}$$

$$\text{Power ratio in decibels} = 10 \log_{10} \frac{V_1^2/R}{V_2^2/R}$$

$$= 10 \log_{10} \left(\frac{V_1}{V_2}\right)^2$$

$$= 20 \log_{10} \left(\frac{V_1}{V_2}\right) \tag{2.15}$$

Notice that a power ratio of 2:1 represents

$$10 \log_{10} 2 = 10 \times 0.301 \approx 3 \text{ dB}$$

2.4 Resonance

Resonance is a condition occurring within second-order circuits such that the over-all voltage and current are *in phase*. The two most common examples are the *RLC* series and *RLC* parallel circuits.

2.4.1 Series Resonance

Section 2.2.2 has shown that the impedance and phase angle of the circuit shown in figure 2.15a are given by equations 2.4 and 2.5, namely

$$Z = \sqrt{[R^2 + (X_L - X_C)^2]}$$

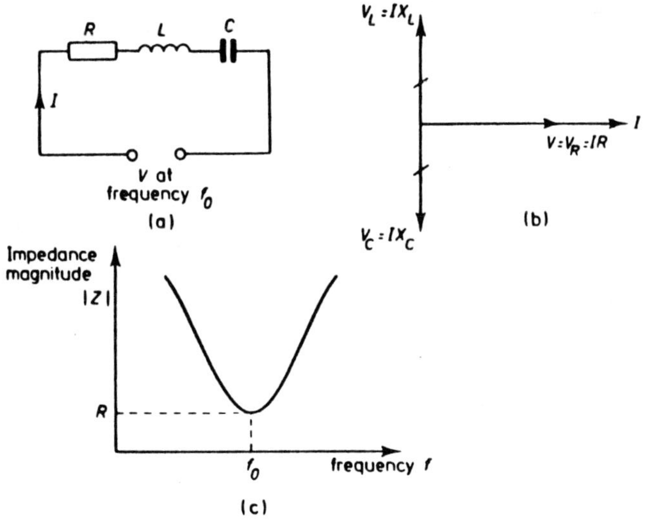

FIGURE 2.15

Series resonance

and

$$\phi = \arctan \frac{X_L - X_C}{R}$$

Consider the situation when the circuit is supplied at a frequency f_0 such that $X_L = X_C$. The vertical phasors being of equal length cancel exactly leaving $V = IR$. The voltage phasor V and the current phasor I are both horizontal in figure 2.15b that is, in phase and therefore the circuit is at resonance. This is confirmed by the expression for the phase angle $\phi = \arctan 0/R = 0$ and a glance at the impedance equation 2.4 shows the impedance in this example to be a minimum of $Z = \sqrt{(R^2)} = R$ at resonance. Plotting a graph of $|Z|$ against frequency reveals this clearly in figure 2.15c.

The *resonant frequency* f_0 is such that $X_L = X_C$, that is

$$2\pi f_0 L = 1/2\pi f_0 C$$

or

$$f_0 = \frac{1}{2\pi\sqrt{(LC)}} \qquad (2.16)$$

while the circuit's impedance at this resonant frequency is

$$Z_D = R$$

and is called the *dynamic impedance*.

Example 2.2 showed that under certain circumstances the capacitor and inductor voltages for such a circuit could exceed the supply voltage. At resonance, in this example, the capacitor and inductor voltages V_C and V_L are equal because since

$$X_L = X_C$$

$$IX_L = IX_C$$

The ratio of the reactance voltages to the supply voltage is thus V_L/V or V_L/V_R since $V = V_R$.

Therefore the *voltage multiplication factor* at the resonant frequency is

$$Q_0 = \frac{V_L}{V_R} = \frac{IX_L}{IR}$$

$$Q_0 = \frac{\omega_0 L}{R} = \frac{2\pi f_0 L}{R} \tag{2.17}$$

2.4.2 Parallel Resonance

Instead of considering the truly parallel circuit of figure 2.7 at resonance we shall examine the behaviour of the practical parallel circuit shown in figure 2.16a. This is because all real inductors exhibit a slight self-resistance R that makes the true parallel circuit unrealistic. The capacitor current clearly leads the applied voltage by 90° and the inductor current I_L will lag the voltage by a phase angle $\phi = \arctan X_L/R$ (see equation 2.5 modified for R and L alone).

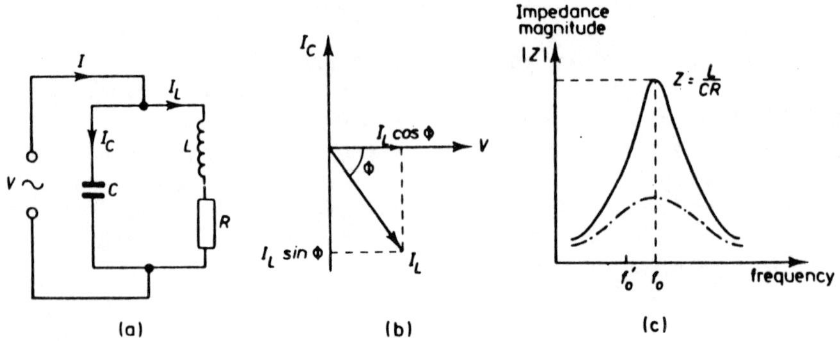

FIGURE 2.16

Parallel resonance

This circuit will resonate at a frequency f_0 such that

$$|I_C| = |I_L \sin\phi|$$

effectively cancelling the vertical phasors and giving a total current

$$I = I_L \cos\phi$$

Since

$$I_C = I_L \sin\phi$$

$$V/X_C = (V/Z)\sin\phi$$

$$V \times 2\pi f_0 C = \left[\frac{V}{\sqrt{(R^2 + X_L^2)}}\right] \sin\phi$$

Because $\phi = \arctan X_L/R$ and $\sin\phi = X_L/\sqrt{(R^2 + X_L^2)}$, we have

$$V \times 2\pi f_0 C = \frac{V \times 2\pi f_0 L}{(R^2 + X_L^2)}$$

or

$$\frac{L}{C} = R^2 + (2\pi f_0 L)^2 \qquad (2.18)$$

$$2\pi f_0 = \sqrt{\left(\frac{1}{LC} - \frac{R^2}{L^2}\right)}$$

resonant frequency

$$f_0 = \frac{1}{2\pi}\sqrt{\left(\frac{1}{LC} - \frac{R^2}{L^2}\right)} \qquad (2.19)$$

The dynamic impedance Z_D (the impedance at resonance) is

$$\frac{V}{I} = \frac{V}{I_L \cos\phi}$$

Since $\phi = \arctan X_L/R$, $\cos\phi = R/\sqrt{(R^2 + X_L^2)}$, we have

$$Z_D = V/\left(\frac{V}{\sqrt{(R^2 + X_L^2)}}\ \frac{R}{\sqrt{(R^2 + X_L^2)}}\right)$$

$$Z_D = \frac{(R^2 + X_L^2)}{R}$$

but from equation 2.18, $(R^2 + X_L^2) = L/C$, therefore

$$Z_D = L/CR \qquad (2.20)$$

Notice that for this circuit, unlike the simple *RLC* series case, the resonant frequency depends on R (equation 2.19), but that for lightly damped circumstances where $R^2/L^2 \ll 1/LC$, equation 2.19 reduces to the series resonance equation 2.16.

Figure 2.16c shows the behaviour of a parallel circuit at frequencies around resonance when the damping is light (low values of R). At higher values of R two effects occur.

(i) The curve becomes flattened towards the shape of the chain-dotted curve shown.

(ii) The frequency for maximum impedance no longer approximates to the resonant frequency ($f_0{}'$) as defined by zero phase-shift.

Example 2.3 showed that a current multiplication effect can occur in parallel circuits. At the resonant frequency, the current multiplication factor Q_0 is given by

$$Q_0 = \frac{I_C}{I} = \frac{I_L \sin\phi}{I_L \cos\phi}$$

$$Q_0 = \tan\phi = \frac{\omega_0 L}{R}$$

Thus the voltage multiplication factor for series resonance and the current multiplication factor for parallel resonance are numerically equal and equal to the *Q-factor*.

2.4.3 Bandwidth of Resonant Filters

Figure 2.17 shows a set of current response curves for an *RLC* series circuit at various values of R when fed with a constant voltage V. Note that in the series case the value of R does not affect the resonant frequency or the frequency of maximum current.

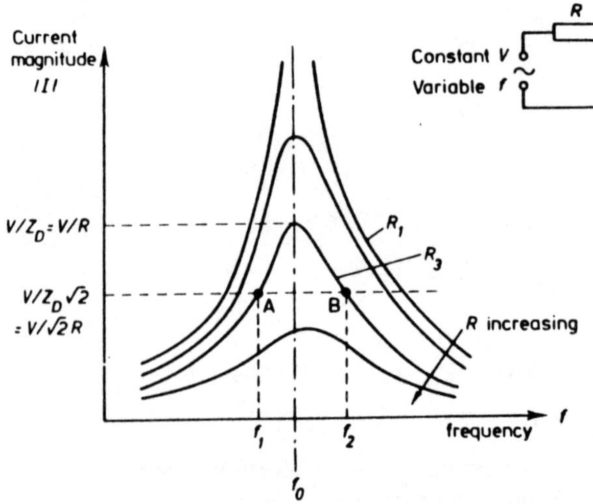

FIGURE 2.17

Tuned filter response

This circuit (and also the parallel circuit) can clearly be used as tuned filters to pass only a narrow band of frequencies around resonance. The *passband* is often defined as that range of frequencies over which the power is at least half its maximum value, that is, the – *3 dB bandwidth*. Half power corresponds to $1/\sqrt{2}$ times the maximum current because power is proportional to current squared In this case the passband extends from f_1 to f_2 when $R = R_3$. The bandwidth is therefore $(f_2 - f_1)$ and at points A and B

$$\frac{V}{R\sqrt{2}} = \frac{V}{\sqrt{[R^2 + (X_L - X_C)^2]}}$$

Therefore

$$2R^2 = R^2 + (X_L - X_C)^2$$

or

$$X_L - X_C \pm R = 0$$

Thus

$$\omega L - \frac{1}{\omega C} \pm R = 0$$

or

$$\omega^2 \pm \frac{\omega R}{L} - \frac{1}{LC} = 0$$

a quadratic equation in ω

$$\omega = \frac{\pm R/L \pm \sqrt{(R^2/L^2 + 4/LC)}}{2}$$

$$\omega = \pm \frac{R}{2L} \pm \sqrt{\left(\frac{R^2}{4L^2} + \frac{1}{LC}\right)}$$

For low damping $R^2/4L^2 \ll 1/LC$, therefore

$$\omega \approx \pm \frac{R}{2L} \pm \omega_0$$

where ω_0 is the resonant angular frequency $\sqrt{(1/LC)}$.
Considering only positive frequencies

$$\omega \approx \omega_0 \pm \frac{R}{2L}$$

giving

$$\omega_1 \approx \omega_0 - \frac{R}{2L}$$

$$\omega_2 \approx \omega_0 + \frac{R}{2L}$$

$$\omega_2 - \omega_1 \approx \frac{R}{L}$$

Thus the bandwidth (Hz) is given by

$$f_2 - f_1 \approx \frac{R}{2\pi L}$$

$$\approx \frac{f_0}{Q_0}$$

(from equation 2.17) or

$$\text{Bandwidth} \approx \frac{\text{centre frequency}}{Q_0} \quad (2.21)$$

Example 2.8

Calculate the resonant frequency and bandwidth of a filter constructed from a coil having 100 mH inductance and 30 Ω series resistance in parallel with a 0.8 μF capacitor.

Calculate the new resonant frequency if a 300 Ω resistor were placed in series with the coil.

The resonant frequency

$$f_0 = \frac{1}{2\pi}\sqrt{\left(\frac{1}{LC} - \frac{R^2}{L^2}\right)}$$

$$= \frac{1}{2\pi}\sqrt{\left(\frac{10^6}{0.1 \times 0.8} - \frac{900}{0.01}\right)}$$

Ignoring the negligible second term in brackets

$$f_0 = \frac{1}{2\pi} \times \frac{10^4}{\sqrt{8}} = 563 \text{ Hz}$$

$$Q_0 = \frac{2\pi f_0 L}{R} = \frac{2\pi \times 563 \times 0.1}{30}$$

$$= 11.8$$

From equation 2.21

$$\text{bandwidth} = \frac{f_0}{Q_0}$$

$$= \frac{563}{11.8} = 47.7 \text{ Hz}$$

If the total resistance were 30 + 300 = 330 Ω

$$f_0' = \frac{1}{2\pi}\sqrt{\left(\frac{10^6}{0.1 \times 0.8} - \frac{330^2}{0.01}\right)}$$

$$= \frac{1}{2\pi}\sqrt{(1.25 \times 10^7 - 1.09 \times 10^7)}$$

$$= \frac{1}{2\pi}\sqrt{(1.6 \times 10^6)} = \frac{1.264 \times 10^3}{2\pi}$$

$$= 201 \text{ Hz}$$

which illustrates the large changes that can occur in resonant frequency when a circuit is heavily damped.

2.5 Network Analysis

Before proceeding to the method of solution of general a.c. networks let us consider a few simplifying theorems.

2.5.1 Thévenin's Theorem

Any two-terminal network may be replaced by a voltage generator equal to the open-circuit output voltage in series with an impedance equal to the input impedance of the network when all voltage and current sources have been replaced by their internal impedances.

This means that any network with two terminals, however complex, may be replaced by a circuit similar to that of figure 2.18b.

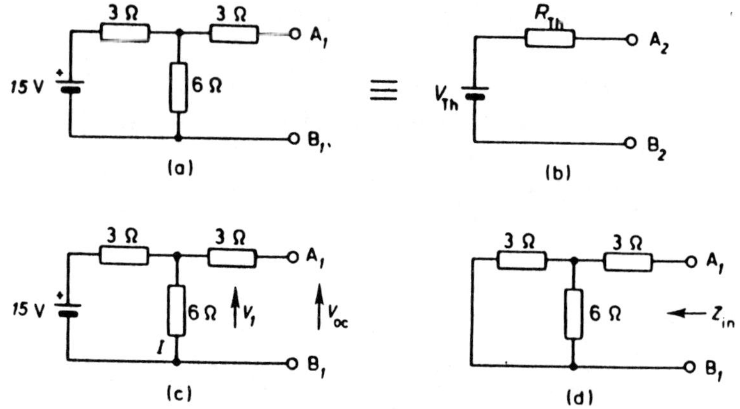

FIGURE 2.18

Illustrating Thévenin's theorem

Example 2.9

Obtain the Thévenin equivalent circuit for figure 2.18a.

The Thévenin voltage (V_{Th}) is equal to the open-circuit output voltage across A_1B_1 that is, the voltage appearing across A_1B_1 when there is no external connection to those terminals (figure 2.18c).

$$I = \frac{15}{(3 + 6)} = \frac{5}{3} \text{ A}$$

therefore

$$V_1 = I \times 6 = 10 \text{ V}$$

Since no current flows through the right-hand 3 Ω resistor

$$V_{oc} = V_{Th} = V_1$$

$$V_{Th} = 10 \text{ V}$$

The Thévenin impedance is obtained by first replacing the voltage source by its internal impedance (in this case zero) which gives figure 2.18d. The input impedance Z_{in} is easily calculated

$$Z_{in} = 3 + \frac{3 \times 6}{3 + 6}$$

$$Z_{Th} = 5 \ \Omega \text{ (resistive)}$$

Therefore the equivalent circuit to figure 2.18a is figure 2.18b where $V_{Th} = 10$ V and $Z_{Th} = 5$ Ω resistive.

Though the above example is a d.c. one, the method is completely general and may be used for a.c. sources and impedances. Another similar theorem allows any network to be simplified in another way.

2.5.2 Norton's Theorem

Any two-terminal network may be replaced by a current generator equal to the short-circuit output current of the network in parallel with an impedance equal to the input impedance of the network when all voltage and current sources have been replaced by their internal impedances.

Example 2.10

Obtain the Norton equivalent circuit for figure 2.19a.

$$\dot{I}_{sc} = \frac{\dot{V}}{\dot{Z}} = \frac{100 + j0}{5 + j0} = (20 + j0) \text{ A}$$

therefore

$$\dot{I}_N = \dot{I}_{sc} = (20 + j0) \text{ A}$$

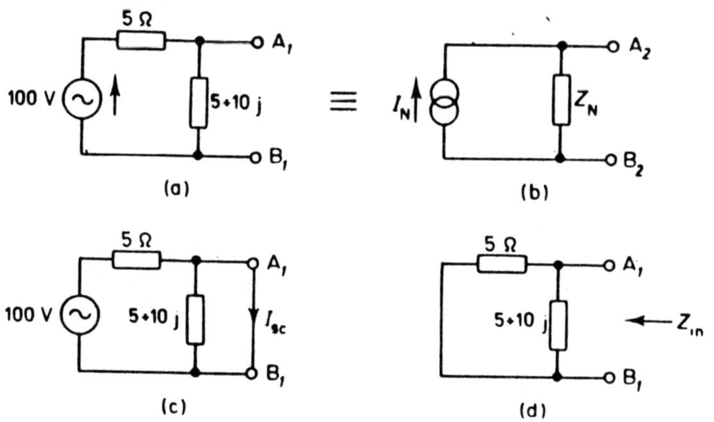

FIGURE 2.19

Illustrating Norton's theorem

Notice that Z_N is defined in the same way as the Thévenin impedance, hence, from figure 2.19d where the voltage generator has been replaced by its internal impedance

$$\dot{Z}_N = \dot{Z}_{in} = \frac{5(5 + 10j)}{5 + (5 + 10j)}$$

$$\dot{Z}_N = \frac{25 + 50j}{10(1 + j)}$$

Rationalising

$$= \frac{(2.5 + 5j)(1 - j)}{1^2 + 1^2}$$

$$= (1.25 + 2.5j)(1 - j)$$

$$= 1.25 - 1.25j + 2.5j + 2.5$$

$$\dot{Z}_N = (3.75 + 1.25j)\,\Omega$$

The concept of a current generator may be new to the reader who may find difficulty in imagining a source that will deliver a constant current into whatever load is placed across it. Approximations to a constant-voltage generator are occasionally met. For instance, a large accumulator having negligible internal impedance will present an almost constant voltage to wide ranges of load.

A constant-current generator is best imagined by a constant-voltage source in series with a very large resistance (figure 2.20b) where applying either load A or load B will result in an approximately constant current of 10^{-4} A passing through the load. Note that whereas a constant-voltage generator has negligible internal impedance, a constant-current generator should have infinite internal impedance ideally.

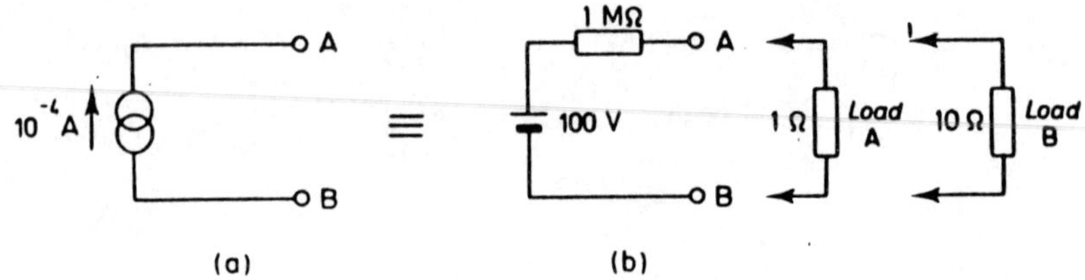

FIGURE 2.20

Visualising the constant-current generator

2.5.3 Current Division Law

While many students often apply the voltage division rule (figure 2.21a) unconsciously as a matter of 'common sense', the current division rule appears less easily understood. Consider the voltage V, common to both resistors in figure 2.21b.

$$V = I\left(\frac{R_1 R_2}{R_1 + R_2}\right)$$

but

$$I_1 = \frac{V}{R_1} = I\left(\frac{R_2}{R_1 + R_2}\right)$$

Similarly

$$I_2 = I\left(\frac{R_1}{R_1 + R_2}\right) \qquad (2.22)$$

This is the current division rule. Compare its similarities and differences with the voltage rule equation of figure 2.21a.

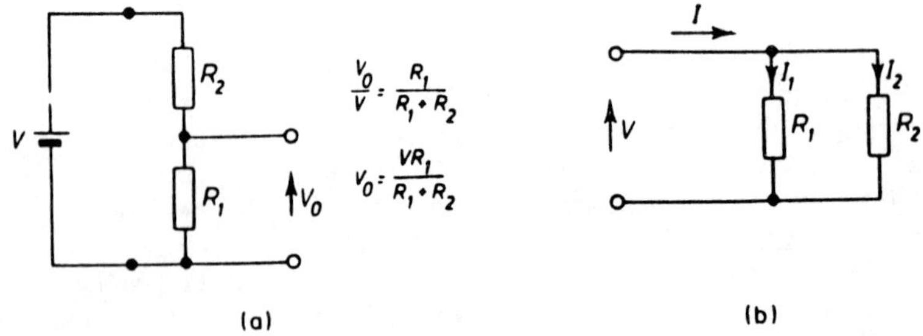

FIGURE 2.21

The voltage and current division rules

2.5.4 General Mesh Analysis

A general method will be given by which the reader may write down the simultaneous equations for the solution of any linear network systematically, thus minimising the possibility of errors. A determinant method for conveniently solving these equations will be shown.

Nearly any network may be redrawn in the manner shown in figure 2.22a as a series of *meshes* or circuit loops. If we number the meshes 1 to n and insert a hypothetical clockwise circulating-current (I_1 to I_n) in each mesh, we may write down Kirchhoff's voltage equation for each mesh assuming clockwise currents positive and voltages positive if they induce clockwise currents.

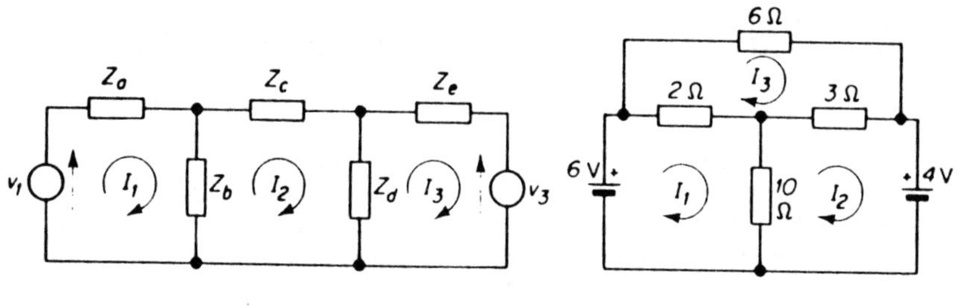

(a) (b)

FIGURE 2.22

General mesh analysis

For mesh 1 $\quad +V_1 = +I_1 Z_a + (I_1 - I_2) Z_b$

For mesh 2 $\quad 0 = +(I_2 - I_1) Z_b + I_2 Z_c + (I_2 - I_3) Z_d$

For mesh 3 $\quad -V_3 = +(I_3 - I_2) Z_d + I_3 Z_e$

(Note the sign of V_3 since it opposes clockwise currents.) Rearranging

$$\left.\begin{aligned} V_1 &= I_1 (Z_a + Z_b) - I_2 (Z_b) \\ 0 &= -I_1 (Z_b) + I_2 (Z_b + Z_c + Z_d) - I_3 (Z_d) \\ -V_3 &= -I_2 (Z_d) + I_3 (Z_d + Z_e) \end{aligned}\right\} \quad (2.23)$$

Equations 2.23 may be rewritten

$$\left.\begin{aligned} \Sigma V_1 &= +I_1 Z_{11} - I_2 Z_{12} - I_3 Z_{13} \\ \Sigma V_2 &= -I_1 Z_{21} + I_2 Z_{22} - I_3 Z_{23} \\ \Sigma V_3 &= -I_1 Z_{31} - I_2 Z_{32} + I_3 Z_{33} \end{aligned}\right\} \quad (2.24)$$

where Z_{nn} is the *self* impedance of mesh n (in this example $Z_a + Z_b$ for mesh 1), Z_{nm} is the *mutual* or shared impedance between meshes n and m (in this example Z_b between meshes 1 and 2), and ΣV_n is the algebraic sum of the voltage sources in mesh n. (Note that the signs of these mutual impedances are negative since they link current flowing in opposite directions.)

After a little practice equations 2.24 may be written directly from the circuit diagram with little chance of error.

Example 2.11

Write down the mesh equations for the circuit shown in figure 2.22b.

Number the meshes and insert clockwise circulating currents as shown.

For mesh 1 $\quad +6 = I_1(2 + 10) - I_2(10) - I_3(2)$

For mesh 2 $\quad -4 = -I_1(10) + I_2(10 + 3) - I_3(3)$

For mesh 3 $\quad 0 = \quad I_1(2) \quad I_2(3) + I_3(2 + 3 + 6)$

We may check these equations for errors in two ways

(i) The positive signs should proceed in one diagonal line, all other terms on the right-hand side being negative.

(ii) The numerical values of the coefficients should be symmetrical about the top-left and bottom-right corners, in this case

$$\left.\begin{matrix} 12 & -10 & -2 \\ -10 & 13 & -3 \\ -2 & -3 & 11 \end{matrix}\right\} \qquad (2.25)$$

A convenient method for solution of the mesh equations 2.24 is by the use of determinants. Letting the coefficients of the right-hand side be written in a frame as a determinant, we may write

$$\frac{1}{\begin{vmatrix} Z_{11} & -Z_{12} & -Z_{13} \\ -Z_{21} & Z_{22} & -Z_{23} \\ -Z_{31} & -Z_{32} & Z_{33} \end{vmatrix}} = \frac{I_1}{\begin{vmatrix} \Sigma V_1 & -Z_{12} & -Z_{13} \\ \Sigma V_2 & Z_{22} & -Z_{23} \\ \Sigma V_3 & -Z_{32} & Z_{33} \end{vmatrix}}$$

$$= \frac{I_2}{\begin{vmatrix} Z_{11} & \Sigma V_1 & -Z_{13} \\ -Z_{21} & \Sigma V_2 & -Z_{23} \\ -Z_{31} & \Sigma V_3 & Z_{33} \end{vmatrix}} = \frac{I_3}{\begin{vmatrix} Z_{11} & -Z_{12} & \Sigma V_1 \\ -Z_{21} & Z_{22} & \Sigma V_2 \\ -Z_{31} & -Z_{32} & \Sigma V_3 \end{vmatrix}}$$

or more conveniently

$$\frac{1}{\Delta_Z} = \frac{I_1}{\Delta_1} = \frac{I_2}{\Delta_2} = \frac{I_3}{\Delta_3} \qquad (2.26)$$

where Δ_Z is the impedance determinant (2.25) and Δ_n is Δ_Z with the n th column replaced with the mesh voltages

Example 2.12

Using determinants calculate the current through the 10 Ω resistor in figure 2.22b and state its direction

The impedance determinant (Δ_Z) is

$$\begin{vmatrix} 12 & -10 & -2 \\ -10 & 13 & -3 \\ -2 & -3 & 11 \end{vmatrix}$$

which may be evaluated as

$$\Delta_Z = 12(143 - 9) + 10(-110 - 6) - 2(30 + 26)$$

$$= 1608 - 1160 - 112 = 336$$

Similarly

$$\Delta_1 = \begin{vmatrix} +6 & -10 & -2 \\ -4 & 13 & -3 \\ 0 & -3 & 11 \end{vmatrix}$$

$$= 6(143 - 9) + 10(-44 + 0) - 2(12 - 0)$$

$$= 804 - 440 - 24 = 340$$

From equation 2.26

$$I_1 = \frac{\Delta_1}{\Delta_Z} = \frac{340}{336}$$

$$= 1.012 \text{ A}$$

Also

$$\Delta_2 = \begin{vmatrix} 12 & +6 & -2 \\ -10 & -4 & -3 \\ -2 & 0 & 11 \end{vmatrix}$$

$$= 12(-44 + 0) - 6(-110 - 6) - 2(0 - 8)$$

$$= -528 + 696 + 16 = 184$$

From equation 2.26

$$I_2 = \frac{\Delta_2}{\Delta_Z} = \frac{184}{336}$$

$$= 0.548 \text{ A}$$

Therefore the current through the 10 Ω resistor is $I_1 - I_2$ downwards

$$I_1 - I_2 = 1.012 - 0.548$$

$$= 0.464 \text{ A downwards}$$

This somewhat laborious calculation is nevertheless systematic and therefore less prone to error than a random attempt at solution of the simultaneous equations 2.23 or 2.24. When arranged in this systematic way they are eminently suitable, of course, for computer solution.

2.5.5 *RC* Phase-shifting Network

Figure 2.23a shows the circuit of a popular phase-shifter for obtaining firing voltages for controlled rectifiers (section 5.3.1). Figure 2.23b (the phasor diagram) shows the current I leading the transformer secondary voltage $2V$ by some leading phase-angle ϕ the magnitude of which will depend on the settings of C and R. The output voltage is taken between the midpoint of the transformer (point A) and the junction between V_R and V_C (point B). As C and R are varied the locus of point B will be a semicircle since V_R and V_C are always at right angles to each other.

In triangle OAB, OA = OB since they are both radii of the semicircle. Thus

$$\hat{AOB} = \phi = \hat{ABO} \text{ (isosceles triangle)}$$

therefore

$$\hat{OAB} = 180^\circ - 2\phi$$

and

$$\hat{BAD} = 2\phi$$

But $\hat{BAD} = 2\phi$ is the phase angle between $2V$ and V_{out}, that is, θ. Therefore as C and R vary, V_{out} is always constant in magnitude V and at an angle $\theta = 2\phi$ to the supply voltage $2V$. The phase angle $\phi = \arctan (X_C/R)$ and hence the phase shift $\theta = 2 \arctan (X_C/R)$ which may vary from 0 to 180° as X_C varies from 0 to infinity.

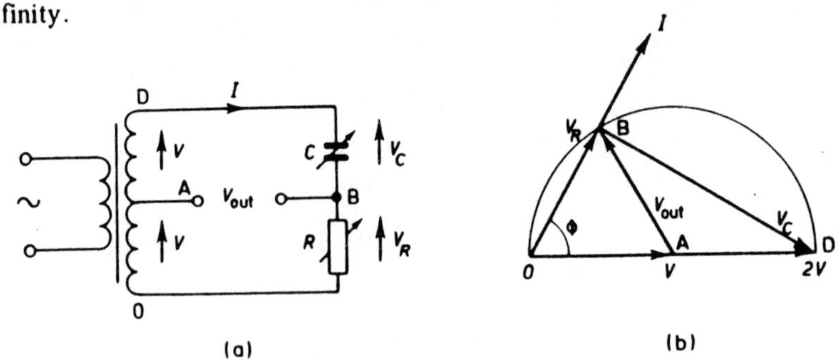

FIGURE 2.23

The *RC* phase-shifter

2.5.6 Maximum Power Transfer Theorem

In many situations when using instrumentation equipment it is necessary to transfer the maximum signal power from a source, for example a transducer, to its load (amplifier or display equipment). Consider figure 2.24 in which a source having an internal impedance $R + \mathrm{j}X$ is connected to its load of impedance $R_L + \mathrm{j}X_L$. The magnitude of the current

$$|I| = \frac{V}{Z} = \frac{V}{\sqrt{[(R + R_L)^2 + (X + X_L)^2]}}$$

hence the load power is $I^2 R_L$, that is

$$P_L = \frac{V^2 R_L}{(R + R_L)^2 + (X + X_L)^2} \qquad (2.27)$$

(i) If R_L is fixed and X_L can be varied, maximum load power can be obtained when $(X + X_L) = 0$, by inspection of equation 2.27. That is, when

$$X_L = -X$$

(ii) Assuming $(X + X_L = 0)$ and R_L is variable, maximum power will be transferred when

$$\frac{\mathrm{d}P}{\mathrm{d}R_L} = \frac{V^2 (R + R_L)^2 - 2V^2 R_L (R + R_L)}{(R + R_L)^4} = 0$$

that is, when

$$(R + R_L)^2 - 2R_L (R + R_L) = 0$$

or

$$R_L = R$$

(iii) If both X_L and R_L can be varied, the condition for maximum power from (i) and (ii) together is

$$\dot{Z}_L = R - \mathrm{j}X$$

or that the load impedance must be the *complex conjugate* of the source impedance.

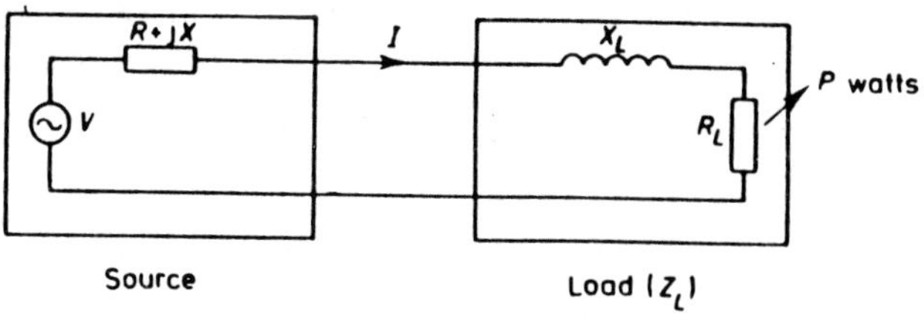

FIGURE 2.24

Maximum power transfer

Various methods of matching loads to their sources have been evolved. One of these, transformer matching is considered in section 4.2.

2.6 Three-phase Systems

2.6.1 Principles of Three-phase Generation

It is well known that a single-phase alternating voltage of the form $v_r = V_m \sin\theta$ is generated when a single coil is rotated in a stationary magnetic field (figure 2.25a and b). If, however, three coils, separated mechanically by 120° of arc, are revolved together the voltages produced by each will be as shown in figure 2.25d. The voltage produced by coil R_1R_2 will be as before

$$v_r = V_m \sin\theta$$

but

$$v_y = V_m \sin(\theta - 120°) \text{ for coil } Y_1Y_2$$

and

$$v_b = V_m \sin(\theta - 240°) \text{ for coil } B_1B_2$$

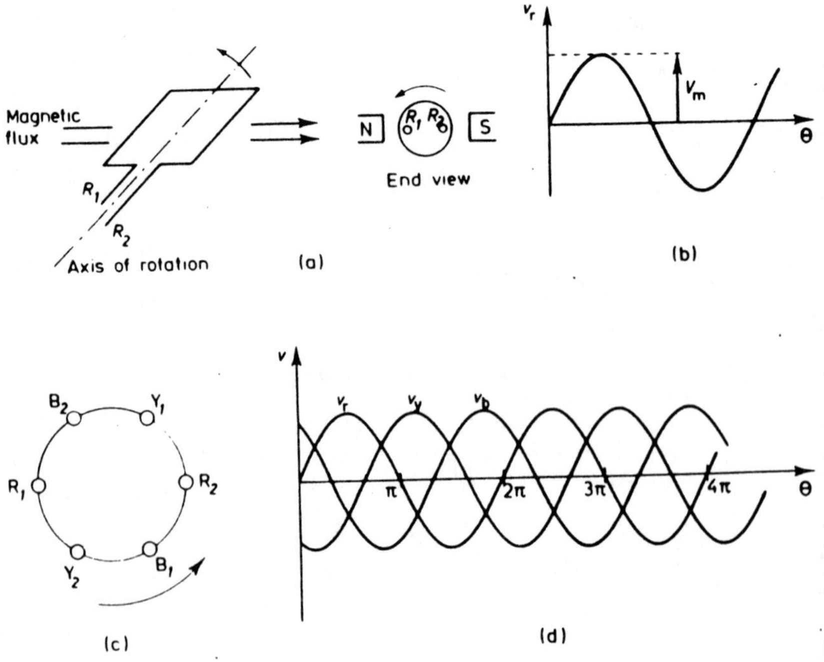

FIGURE 2.25

Principles of three-phase generation

2.6.2 Star Connection

Clearly the three separate generator coils $R_1 R_2$, $Y_1 Y_2$ and $B_1 B_2$ may be connected to three separate loads by a total of six conductors. Consider however, if they were connected to three identical resistive loads as shown in figure 2.26a. The voltage between any two interconnecting *lines* say red R and yellow Y would be given by V_{RY} in the phasor diagram of figure 2.26b. By the laws of phasor addition

$$\dot{V}_{RY} = \dot{V}_{RN} + \dot{V}_{NY}$$

and

$$\dot{V}_{NY} = -\dot{V}_{YN}$$

It is clear by geometry that OABC is a parallelogram where OB bisects angle AOC. Hence angle AOD = 60°/2 = 30° and OD is V_{RN} cos 30° in length. Hence

$$OB = |V_{RY}| = 2|V_{RN}| \cos 30^\circ$$

or more generally, for this so-called *star connection*

$$\left.\begin{aligned} \text{line voltage} &= \sqrt{3} \times \text{phase voltage} \\ V_L &= \sqrt{3} \times V_P \\ \text{Clearly also} & \\ \dot{I}_L \text{ (line current)} &= \dot{I}_P \text{ (phase current)} \end{aligned}\right\} \qquad (2.28)$$

where the *phase voltage* is the r.m.s. voltage between any line and neutral and the line voltage is the r.m.s. voltage between any pair of lines.

From the symmetry of figure 2.26a it is clear that for a *balanced load* (equal loads in each phase), points N and N′ are at the same potential hence a neutral wire connecting the generator and load star points would carry no current.

In practice this connection is usually included because of the difficulty of ensuring exact equality in the three loads.

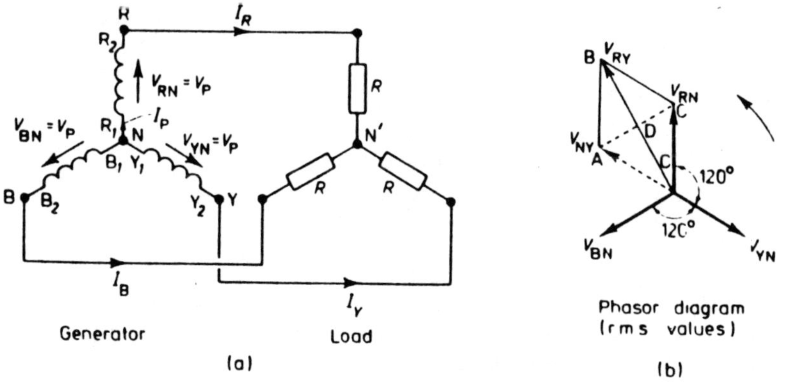

FIGURE 2.26

Three-phase star connection

2.6.3 Delta Connection

In figure 2.26b the sum of the three phase voltages $\dot{V}_{RN}$, $\dot{V}_{YN}$ and $\dot{V}_{BN}$ is clearly zero. This allows an alternative delta connection for the windings as in figure 2.27a. Before connection to the load there will be no circulating currents in the generator coils because, being connected end-to-end, we have effectively summed the three voltages. In this delta (Δ) connection the line voltages and phase voltages are obviously equal

$$\dot{V}_L = \dot{V}_P$$

For a balanced load $R_1 = R_2 = R_3$ and therefore the phasors representing the load currents I_{R_L}, I_{Y_L} and I_{B_L} will be equal in magnitude and in phase with $\dot{V}_{RN}$, $\dot{V}_{YN}$ and $\dot{V}_{BN}$ respectively. Drawing this gives figure 2.27b. Now the line current is

$$\dot{I}_L = \dot{I}_{R_L} - \dot{I}_{Y_L}$$

and thus by a similar construction and argument on figure 2.27b it can be seen that, for delta connection the line current equals $\sqrt{3}$ times the phase current.

$$\left.\begin{aligned} I_L &= \sqrt{3} \times I_P \\ \dot{V}_L &= \dot{V}_P \end{aligned}\right\} \qquad (2.29)$$

Diagrams 2.26 and 2.27 show the generator and load connected in like manner in each case but this need not be so; star – delta and delta – star interconnections are very common

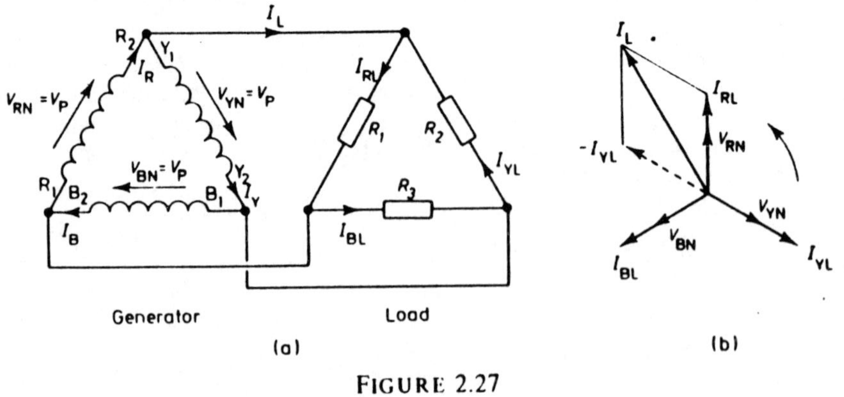

FIGURE 2.27

Three-phase delta connection

Example 2.13

A three-phase star-connected generator generates 240 V per phase as shown in figure 2.28a. It is attached to three identical delta-connected load coils each of which has 20 Ω resistance and 20 Ω reactance. Calculate

(a) the line voltage
(b) the load phase voltage and current and
(c) the line current.

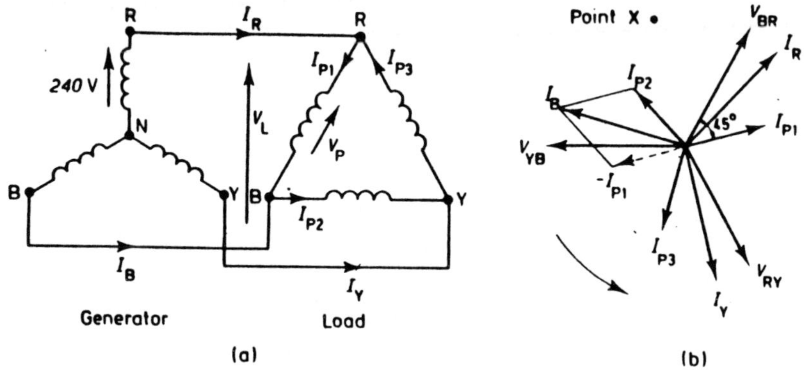

FIGURE 2.28

Example 2.13

Indicate the load voltages and currents and the line currents on a phasor diagram.

(a) *For the generator*

$$V_L = \sqrt{3}\,V_P = \sqrt{3} \times 240 = 416 \text{ V}$$

(b) *For the load*, impedance per phase

$$Z_L = \sqrt{(R_L{}^2 + X_L{}^2)} = \sqrt{(400 + 400)} = 28.3\ \Omega$$

and phase angle

$$\phi = \arctan X_L/R = \arctan 1 = 45^\circ \text{ lagging}$$

Phase voltage

$$\dot{V}_P = \dot{V}_L = 416 \text{ V}$$

Phase current

$$I_{P_1} = \frac{V_P}{Z_L} = \frac{416}{28.3} = 14.7 \text{ A}$$

(c) *Line current*

$$I_L = \sqrt{3}\,I_P = \sqrt{3} \times 14.7 = 25.4 \text{ A}$$

The phasor diagram of figure 2.28b is drawn by indicating any one of the line voltages, say V_{RY}, in some arbitrary direction and then sketching the others at 120° intervals. Note how the sequence of lettering is maintained: V_{RY}, V_{YB} and V_{BR}, the subscripts being RY – YB – BR.

One of the phase currents in the load is then sketched 45° behind (lagging) its appropriate subscript voltage. (For example V_{BR} gives rise to I_{P_1}.) I_{P_3} and I_{P_2} are then included at 120° intervals always 45° behind.the voltage. Selecting one of the line currents, say I_B

$$\dot{I}_B = \dot{I}_{P_2} - \dot{I}_{P_1}$$

Hence sketching $-I_{P_1}$ and summing with I_{P_2} gives I_B. Then I_R and I_Y may be included at 120° intervals. Always check that the phase sequence is correct. For example, an observer standing at X would see I_R, I_Y and I_B come past in that order as the diagram rotated. This is the sequence RYB that was first adopted in figure 2.26a.

2.6.4 Power and Reactive Power

The total power dissipated in a three-phase load is always the sum of the three individual phase powers. Similarly the total reactive power is given by the sum of the individual reactive powers for each phase. This may easily be seen from figure 2.29 in which all three active powers P_1, P_2 and P_3 for the unbalanced load shown are in the same direction. Similarly all three reactive powers, Q_1, Q_2 and Q_3 are in the vertical direction. The quantities which must not be added arithmetically are the apparent powers S_1, S_2 and S_3 because they differ in direction.

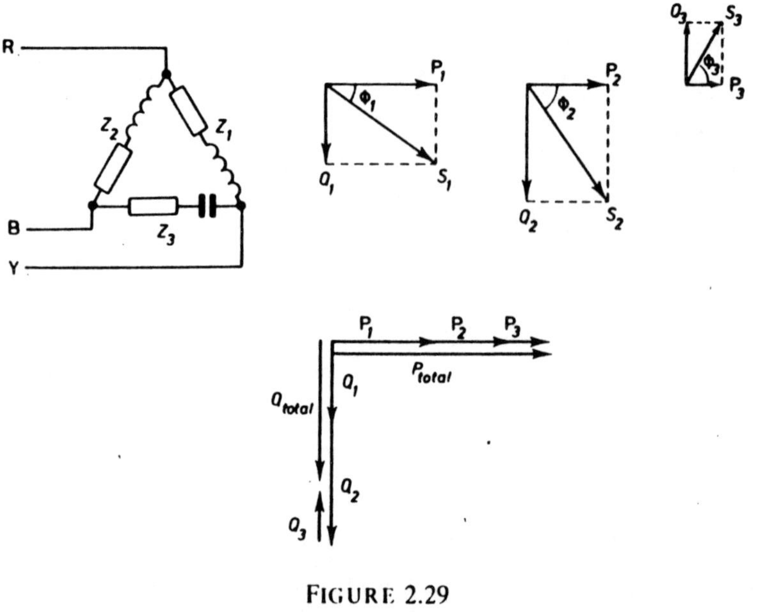

FIGURE 2.29

The addition of powers and reactive powers

For a *balanced load* in which $Z_1 = Z_2 = Z_3$, the powers and reactive powers per phase will be equal hence

$$P_{total} = 3P_1 = 3V_P I_P \cos\phi$$

and

$$Q_{total} = 3Q_1 = 3V_P I_P \sin\phi$$

Star load	Delta load
$V_P = \dfrac{V_L}{\sqrt{3}}$	$V_P = V_L$
$I_P = I_L$	$I_P = \dfrac{I_L}{\sqrt{3}}$
$V_P I_P = \dfrac{V_L I_L}{\sqrt{3}}$	$V_P I_P = \dfrac{V_L I_L}{\sqrt{3}}$

therefore for star- or delta-connected *balanced* loads

$$\left.\begin{aligned} P_{total} &= \frac{3V_L I_L}{\sqrt{3}}\cos\phi = \sqrt{3}\, V_L I_L \cos\phi \\ Q_{total} &= \frac{3V_L I_L}{\sqrt{3}}\sin\phi = \sqrt{3}\, V_L I_L \sin\phi \end{aligned}\right\} \qquad (2.30)$$

For *unbalanced loads* the total power and reactive power must be obtained from addition of the phase powers and reactive powers calculated separately (figure 2.29).

Note that, unless stated otherwise, three-phase voltages and currents always refer to the *line* values.

Example 2.14

Calculate the total power and reactive power dissipated by the load in example 2.13 (figure 2.28). Calculate the values of the three capacitors required to correct the over-all power-factor to unity when the capacitors are connected (i) in star across the load and (ii) in delta across the load. The supply frequency is 50 Hz.

From the previous example, $V_L = 416$ V, $I_L = 25.4$ A and $\phi = 45°$ lagging. Hence the total power and reactive power are

$$\begin{aligned} P &= \sqrt{3}\, V_L I_L \cos\phi \\ &= \sqrt{3} \times 416 \times 25.4 \times 0.707 \\ &= 12950 \text{ W} \end{aligned}$$

$$Q = \sqrt{3}\, V_L I_L \sin\phi$$

since $\sin\phi = \cos\phi$ at 45°

$$Q = 12950 \text{ VAr}$$

For correction to unity power-factor, all this reactive power must be cancelled by the capacitors. Thus reactive power per capacitor is 12 950/3 = 4317 VAr.

(i) If the capacitors are connected in delta, each receives the full line voltage of 416 V. Therefore the capacitor current, I_C is the reactive power/capacitor voltage

$$I_C = \frac{4317}{416} = 10.4 \text{ A}$$

$$I_C = \frac{V}{X_C} = V \times 2\pi f C$$

$$C = \frac{I_C}{2\pi f V}$$

$$= \frac{10.4}{314 \times 416}$$

$$= 79.6\ \mu\text{F}$$

(ii) When the capacitors are star-connected each receives only $1/\sqrt{3}$ of the line voltage or $416/\sqrt{3}$ = 240 V. Therefore as before the capacitor current is

$$I_C = \frac{4317}{240} = 18 \text{ A}$$

$$C = \frac{I_C}{2\pi f V} = \frac{18}{(314 \times 240)}$$

$$= 239\ \mu\text{F}$$

Comparison of the answers to parts (i) and (ii) shows that the delta connection only requires one-third of the capacitance needed for star connection. Delta connection is therefore more usual although the working voltage of each capacitor must be $\sqrt{3}$ times higher than for star connection.

2.7 Power Measurement[5]

2.7.1 Electrodynamic Instruments

These instruments (figure 2.30a) consist of a moving coil carrying an instantaneous current i_1 within a magnetic field produced by a further pair of fixed coils carrying a current i_2. Current is fed to the moving coil via hair-springs at top and bottom that also provide the restoring torque to control the deflection.

Because the magnetic path consists mainly of air, the flux density B between the two fixed coils is proportional to i_2. If the distance between the fixed coils is not greater than their diameter the flux density is fairly uniform between them. Since the torque produced on the moving coil is proportional to $B \times i_1$ and B is itself proportional to i_2, both torque and final deflection depend on $i_1 \times i_2$.

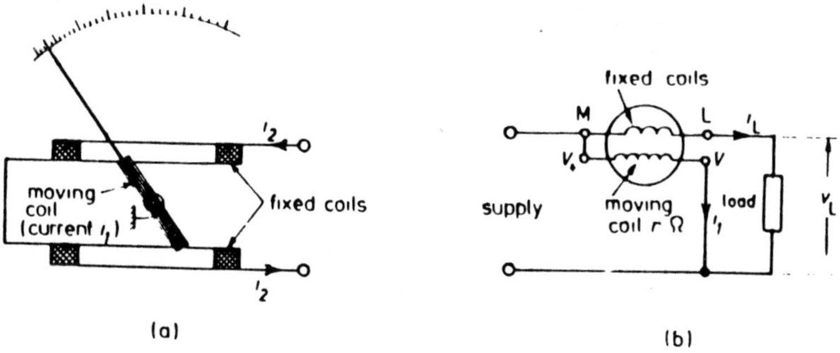

FIGURE 2.30

The construction and use of the electrodynamic wattmeter

If the fixed and moving coils are placed in series, $i_1 = i_2$ and therefore the deflection is proportional to $i_1{}^2$. When used to measure an alternating current or voltage the deflection is proportional to the mean value of the current or voltage squared because of the mechanical inertia of the movement. We thus have a true r.m.s. instrument. It is rarely used in this form because of its high cost.

2.7.2 Single-phase Measurements

If the fixed and moving coils are connected between a source and its load as shown in figure 2.30b, the deflection will be proportional to the instantaneous power

$$\text{Deflection} \propto i_1 \times i_2$$

$$\propto \frac{v_L}{r} \times i_L$$

$$\propto v_L i_L$$

$$\propto \text{instantaneous load power}$$

If the voltage drop across the fixed coil is neglected, the p.d. across the voltage-coil circuit is approximately v_L since the fixed coils are designed to have low resistance.

This instrument will measure the *mean* electrical power over a cycle because its mechanical inertia prevents the pointer from following the variations in instantaneous power throughout the cycle.

It is possible to correct for the power lost in the current coils -- the instrument will read high by an amount $i_L{}^2 R_C$ where R_C is the resistance of the current coils. The value of R_C is stated by the manufacturer.

The letters against the four instrument terminals in figure 2.30b are those conventionally used by the manufacturers.

2.7.3 Three-phase Measurements

If a three-phase load is balanced and its neutral or star point is accessible, a single-phase measurement of power may be made and multiplied by three using the circuit of figure 2.31a. If, however, the load is unbalanced, whether delta- or star-connected, two wattmeters must be used as in figure 2.31b.

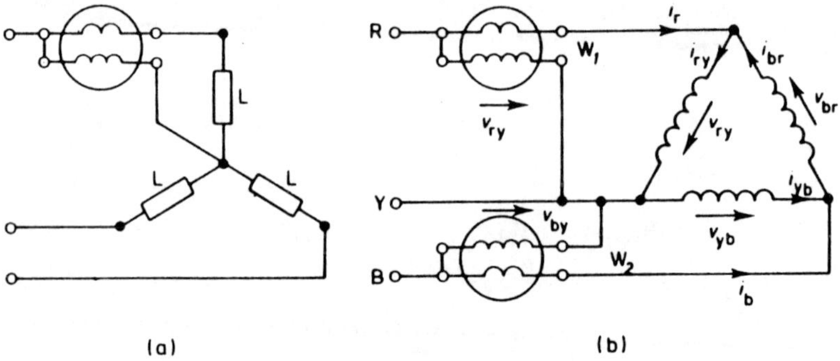

FIGURE 2.31

Three-phase power measurement

$$\text{Instantaneous load power } p = v_{ry}i_{ry} + v_{yb}i_{yb} + v_{br}i_{br}$$
$$= v_{ry}i_{ry} + v_{yb}i_{yb} - i_{br}(v_{ry} + v_{yb})$$

since

$$v_{br} = -v_{rb} = -(v_{ry} + v_{yb})$$

Thus

$$p = v_{ry}(i_{ry} - i_{br}) - v_{yb}(i_{br} - i_{yb})$$
$$= v_{ry}i_r + v_{by}i_b$$
$$= \text{wattmeter } W_1 \text{ reading} + \text{wattmeter } W_2 \text{ reading}$$

Because of instrument inertia the mean power is the *algebraic sum* of the wattmeter readings

$$P = W_1 + W_2 \tag{2.31}$$

The word algebraic is included because for certain power factors one of the wattmeters gives negative readings and this change of sign must be included in the summation.

Power Factor from Wattmeter Readings

If when using the above method of measuring three-phase power the load is balanced, the load power-factor may also be deduced from W_1 and W_2.

Figure 2.32 shows the line voltage phasors V_{BR}, V_{RY} and V_{YB}. The load phase-currents I_{br}, I_{ry} and I_{yb} lag their respective voltages by ϕ, the load phase-

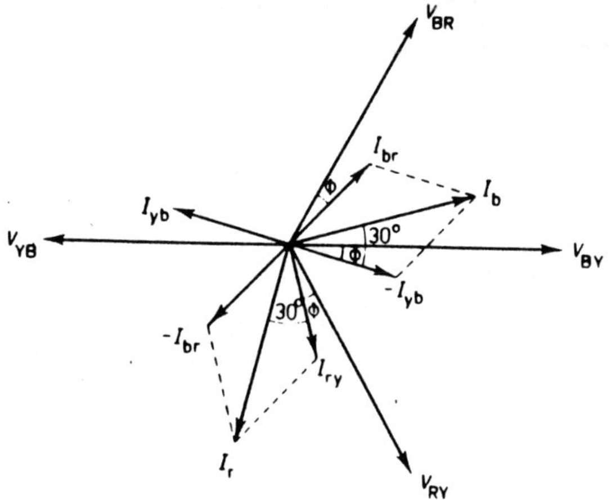

FIGURE 2.32

Phasor diagram of figure 2.31

angle. The two wattmeter coil currents I_b and I_r are constructed as shown (figure 2.32)

$$I_b = I_{br} - I_{yb}$$
$$I_r = I_{ry} - I_{br}$$

W_1 is fed by V_{RY} and I_r, the angle between them being $(30° + \phi)$ from figure 2.32. Similarly W_2 is fed by V_{BY} and I_b, the angle between them being $(30° - \phi)$.

Since $V_{RY} = V_{BY} = V_L$ and $I_r = I_b = I_L$

$$W_1 = V_L I_L \cos(30° + \phi)$$
$$W_2 = V_L I_L \cos(30° - \phi)$$
$$W_1 + W_2 = V_L I_L [\cos(30° + \phi) + \cos(30° - \phi)]$$
$$= V_L I_L [\cos 30° \cos\phi - \sin 30° \sin\phi + \cos 30° \cos\phi + \sin 30° \sin\phi]$$
$$W_1 + W_2 = V_L I_L (2 \cos 30° \cos\phi)$$
$$= \sqrt{3}\, V_L I_L \cos\phi$$

Similarly

$$W_2 - W_1 = V_L I_L \cos(30° - \phi) - V_L I_L \cos(30° + \phi)$$
$$= V_L I_L [(\cos 30° \cos\phi + \sin 30° \sin\phi - \cos 30° \cos\phi + \sin 30° \sin\phi]$$
$$= V_L I_L \sin\phi$$

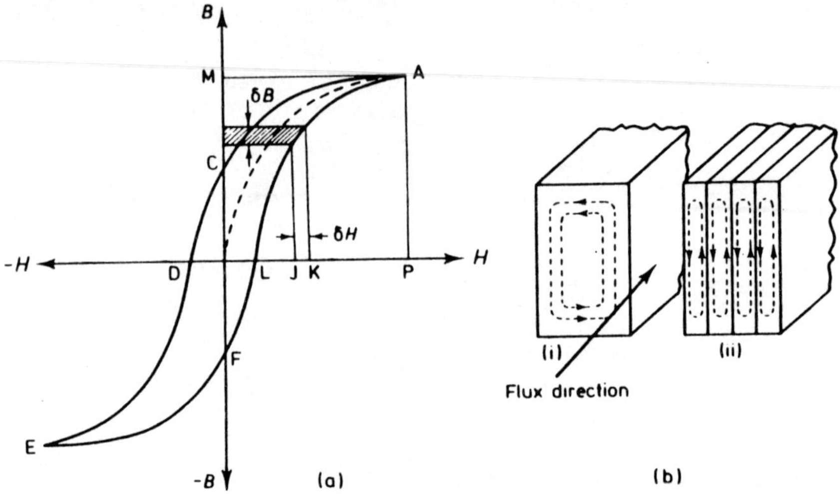

FIGURE 2.33

Hysteresis and eddy-current losses

Therefore

$$\tan\phi = \frac{\sin\phi}{\cos\phi} = \sqrt{3}\left(\frac{W_2 - W_1}{W_2 + W_1}\right)$$

or the phase angle

$$\phi = \arctan\sqrt{3}\left(\frac{W_2 - W_1}{W_2 + W_1}\right) \tag{2.32}$$

2.8 Magnetic Losses[6]

Magnetic or iron losses occur whenever an alternating current is used to magnetise magnetic materials causing magnetic reversals. These losses may be conveniently divided into hysteresis and eddy-current losses.

2.8.1 Hysteresis Loop

The reader will be aware that when a coil of N turns carrying a current i is used to magnetise a previously demagnetised specimen, the magnetising force H is given by

$$H = Ni/l \text{ AT/m} \tag{2.33}$$

where l is the path length of the magnetic circuit.

As i is increased from zero, H increases and a plot of H against the resulting magnetic flux density B resembles the dashed line in figure 2.33a. If i and therefore H is increased indefinitely the curve will level off indicating magnetic *saturation*. If H is now decreased to zero, the curve does not retrace its path but follows the line AC. The *remanent flux-density*, OC when the specimen has been previously saturated is called the *retentivity*, or remanence, of the material. This property of retaining some flux density after the removal of H is clearly of the utmost importance to designers of permanent magnets and electrical machines. This remanent magnetism may be destroyed by mechanical or thermal shock or by the application of a demagnetising force in the opposite direction OD. The value of this *coercive force* required just to destroy the retentivity is called the *coercivity* of the material. The coercivity of permanent-magnet material should clearly be high if the magnet's properties are to be durable.

If H continues to be increased in the reverse direction negative saturation occurs at the point E. The effect of making H more positive is to return the curve to positive saturation via points F and L. The figure ACDEFLA is the *hysteresis loop* of the material which contains most of the information on the material's magnetic properties. The word hysteresis means 'lagging behind', referring to the fact that changes in flux density B lag behind the value of magnetising force H.

2.8.2 Magnitude of Hysteresis Losses

Let an increase δi cause a change δH from OJ to OK in time δt. From equation 2.33

$$\delta H = N\frac{\delta i}{l}$$

The corresponding flux change $\delta\Phi$ will be $a\,\delta B$ where a is the cross-sectional area of the material. The flux change produces a back e.m.f. in the coil of $-N\delta\Phi/\delta t = -aN\delta B/\delta t$. This back e.m.f. must be neutralised by the supply voltage

$$v = +aN\delta B/\delta t$$

The power flowing in the circuit during this change is

$$p = vi = iaN\frac{\delta B}{\delta t}$$

and the energy applied to the specimen in time δt is therefore

$$\delta W = iaN\delta B$$

$$= \text{OJ} \times l \times a \times \delta B$$

$$= \text{OJ} \times \text{volume} \times \delta B$$

Therefore the energy stored per unit volume in time δt is

$$\delta W = \text{area of shaded element}$$

As H rises from zero to OP the energy/unit volume given to the material is equal to the area within the figure FLAMCF. As H falls from OP to zero the energy/unit volume *returned* by the material is equal to the area within the figure CAMC.

In the complete half-cycle the energy absorbed therefore is

$$\text{FLAMCF} - \text{CAMC} = \text{FLACF}$$

That is, the area within the right-hand side of the hysteresis loop. Because the figure is symmetrical about the y-axis, the area within the complete loop represents the energy absorbed/unit volume per cycle.

$$\begin{pmatrix}\text{Hysteresis energy loss}\\ \text{per unit volume}\end{pmatrix} = \begin{pmatrix}\text{Area within figure}\\ \text{EFLACDE}\end{pmatrix} \text{ J per cycle}$$

The area must of course be converted to BH units in the following way by multiplication with the graph-axis scales.

Area in BH units = area (m^2) × flux-density scale (T/m) × magnetising force scale (AT/m per m)

Example 2.15

The area within the hysteresis loop of a given material is 36 cm^2 when drawn to the following scales: 1 cm = 100 AT/m and 1 cm = 0.2 T. Calculate the hysteresis loss in watts per cubic metre at a frequency of 60 Hz.

$$\begin{aligned}\text{Energy loss per m}^3 \text{ per cycle} &= \text{area of loop in } BH \text{ units}\\ &= 36 \times 10^{-4} \times 10^4 \times 20\\ &= 720 \text{ J}\end{aligned}$$

Since there are 60 complete cycles per second

$$\begin{aligned}\text{Power loss/m}^3 &= 720 \times 60 \text{ J/s}\\ &= 43\ 200 \text{ W}\end{aligned}$$

This method of area measurement, though applicable to isolated specimens, is often difficult practically. The above example shows the power loss to be *proportional to the frequency*. Steinmetz has provided a useful empirical equation for hysteresis losses within the typical engineering flux-density range

$$0.1 \text{ T} < B < 1.5 \text{ T}$$

$$W = kvfB_{max}^{1.6} \tag{2.34}$$

where k is the Steinmetz constant for the material (table 2.1), v is the volume, f the frequency and B_{max} is the maximum flux density in the cycle.

Materials with the lowest coefficients (for example silicon steel) are those used for alternating magnetic devices such as transformer cores.

2.8.3 Eddy-current Losses

When a varying flux passes through an electrically conducting magnetic material e.m.f.s are induced which produce eddy currents within the material in the directions shown in figure 2.33b (i). These eddy currents heat the material, resulting in energy loss. These induced e.m.f.s obey the e.m.f. equation (section

TABLE 2.1
Steinmetz coefficients

Material	Coefficient k
Hard cast steel	7034
Cast steel	754 - 3014
Cast iron	2763 - 4019
Soft iron	502
Dynamo sheet steel	502
0.2 per cent silicon iron	528
4.8 per cent silicon iron	191

4.2) and are thus proportional to the flux frequency. Because power is proportional to the voltage squared

$$\text{eddy-current power loss} \propto (\text{frequency})^2$$

Eddy-current losses may be minimised in two ways

(i) The use of high resistivity core material to minimise the eddy currents from a particular value of induced e.m.f. and
(ii) Constructing the core from thin mutually insulated laminations as shown in figure 2.33b (ii).

Imagine the core divided into four laminations as shown. Because the area enclosed by each strip cross-section has fallen by four, the enclosed flux and therefore the e.m.f. will have fallen by four. As there are four such strips this alone would make no difference. However laminating the core has also increased the path resistance for the eddy currents by a factor of four so that

$$\frac{\text{power loss per lamination}}{\text{power loss for solid core}} = \frac{V_L^2}{V_S^2} \times \frac{R_S}{R_L} = \frac{1^2}{4^2} \times \frac{1}{4} = \frac{1}{4^3}$$

where V_S and R_S are the induced voltage and resistance in the solid core and V_L and R_L are those for one lamination; but there are four laminations so

$$\frac{\text{power loss in laminated core}}{\text{power loss in solid core}} = \frac{1}{4^3} \times 4 = \left(\frac{1}{4}\right)^2$$

In words, the eddy-current power losses are reduced by the square of the number of laminations. It is uneconomic to reduce the lamination thickness below 0.5 mm because of production difficulties.

.9 Problems

.1 Find the average and r.m.s. values together with the form and peak factors of a repetitive sawtooth voltage waveform. The voltage rises linearly from zero to 100 V in 2 seconds and drops instantaneously to zero to repeat

2.2 A coil of wire takes 10 A when connected to a 100 V d.c. supply. When it is transferred to a 100 V, 50 Hz a.c. supply it only takes 5 A. Account for this and calculate the inductance of the coil.

2.3 An *RLC* series circuit exhibits the following component voltages when carrying a current of 0.5 A. Resistor voltage 10 V, inductor voltage 50 V and capacitor voltage 70 V. If *R* and *L* are both pure parameters and *C* has a value of 10 μF, calculate (i) the total applied voltage, (ii) the supply frequency, (iii) the value of *L* and (iv) the total circuit power-factor.

2.4 A coil of 10 Ω resistance and 0.0382 H inductance is connected in parallel with a 100 μF capacitor. Calculate the individual and total currents drawn from a 240 V 50 Hz supply and determine the power and power factor with the aid of a phasor diagram.

2.5 A circuit consists of a parallel combination of a resistor *R* and a capacitor *C* in series with another parallel combination consisting of a further resistor *R* and an inductor *L*. Using complex impedance methods show that the behaviour of the complete circuit is independent of frequency if $R = \sqrt{[L/C]}$

2.6 A 50 Hz 240 V single-phase power source has the following loads placed across it in parallel: 4 kW at a power factor of 0.8 lagging, 6 kVA at a power factor of 0.6 lagging and 5 kVA which contains 1.2 kVAr of leading reactive power. Determine the over-all power factor of the system and the value of a capacitor which, if connected across the system would restore the over-all power factor to unity.

2.7 A 240 V supply feeds a load whose complex impedance is (10 – j12) Ω Using the j operator calculate the power and reactive power in the load.

2.8 A coil consists of 2 mH inductance and 30 Ω of series resistance. Together with a parallel capacitor it forms a resonant circuit at 500 kHz. If this circuit is fed with 1 V at the resonant frequency calculate (i) the value of capacitor required for resonance, (ii) the *Q* value of the circuit, (iii) its 3 dB bandwidth and (iv) the supply and capacitor currents at resonance.

2.9 A 240 V source having an internal resistance of 2 Ω feeds a composite circuit consisting of a coil of impedance (10 + j12) Ω in parallel with a capacitor of – j12 Ω. Using Thévenin's theorem calculate the current in the coil.

2.10 Solve problem 2.9 using Norton's theorem and the current division rule

2.11 Calculate I_3 in figure 2.22a (p. 45) if $\dot{Z}_a = (5 - j2)\ \Omega$, $\dot{Z}_b = 3\ \Omega$, $\dot{Z}_c = j5\ \Omega$, $\dot{Z}_d = 5\ \Omega$ and $\dot{Z}_e = (2 - j2)\ \Omega$. V_1 is a sinusoidal source of $10\angle 30°$ V and V_3 is zero.

2.12 A 500 V, three-phase motor presents a balanced load to the supply. The readings of two wattmeters measuring the input power are 28.15 kW and

13.35 kW. Calculate the power factor, line current and reactive power taken by the motor.

2.13 Calculate the value of each of three identical delta-connected capacitors required to bring the power factor of example 2.12 to unity at a frequency of 50 Hz.

2.14 Measurements of the total iron (eddy and hysteresis) losses in a small a.c. induction motor yield the following results. At operating frequencies of 45 Hz and 55 Hz the total iron losses are 3.44 W and 4.59 W respectively. Calculate the magnitude of these losses at the normal operating frequency of 50 Hz.

3 Physics of Devices

3.1 The Bohr Atom[7,8]

The reader will be aware from previous study of the Bohr model of the atom that each element consists of identical atoms. An atom contains a relatively massive nucleus of protons and neutrons orbited by electrons moving in certain permitted paths or shells. The proton carries a positive charge of $+1.602 \times 10^{-19}$ and an electron carries an exactly equal negative charge. In the normal atom the number of electrons in the shells equals the number of protons in the nucleus, hence the atom is electrically neutral.

Almost all the properties of differing elements may be explained by the disposition of the electrons within their shells. Elements whose outermost shell is full (the rare gases helium, neon, argon, krypton and xenon) are very stable. Conversely, elements whose outermost shell has only one electron (hydrogen, sodium, and potassium) or, alternatively, has one electron short of a full complement (fluorine, chlorine, bromine), are intensely chemically reactive. This is because the atom wishes to achieve a more stable form in which its outermost shell (valence shell) is complete. It can do this by sharing its excess electrons or borrowing more electrons from neighbouring atoms of appropriate elements to form compounds by covalent sharing as in figure 3.1.

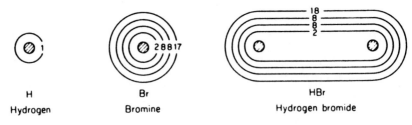

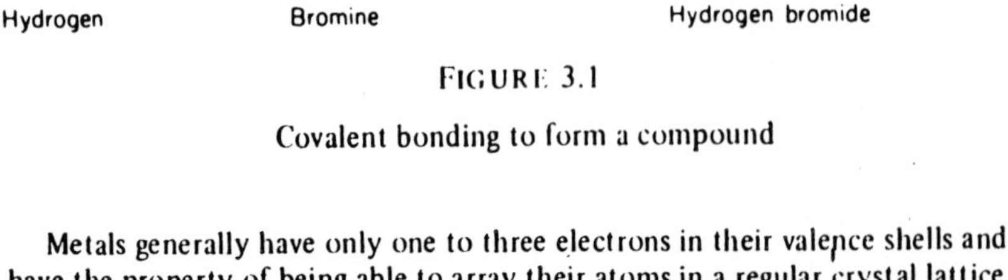

FIGURE 3.1

Covalent bonding to form a compound

Metals generally have only one to three electrons in their valence shells and have the property of being able to array their atoms in a regular crystal lattice. The valence electrons are only loosely bound to their parent atoms and are able to move through the lattice in the form of an electron cloud or gas. There are approximately 10^{29} of these 'free' conduction electrons in a cubic metre of copper.

An important class of elements that we shall consider later are the semiconductors such as carbon, germanium and silicon in which the valence level is half-filled. These materials are capable of forming a regular crystal lattice by covalent sharing of valence electrons with surrounding atoms of the same element (see figure 3.2).

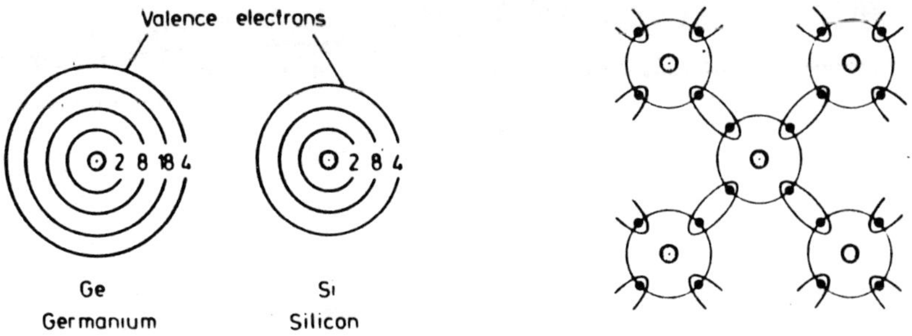

FIGURE 3.2

Typical semiconductor atoms and the sharing of their valence electrons to form a crystal lattice

3.2 Electron Emission[7, 8]

Many devices rely for their action on the emission of free electrons from a metallic surface. Figure 3.3 shows the electrical and gravitational forces between electrons and their surrounding atoms at four positions near the surface of a metal. Inside the metal at (a) the net forces are zero whereas at b and c there are considerable forces attempting to prevent the electron's escape. As the distance x from the surface increases, these forces weaken since both operate on the inverse-square law. When x is about 10^{-6} m these forces are negligible and the electron is 'free' of the surface.

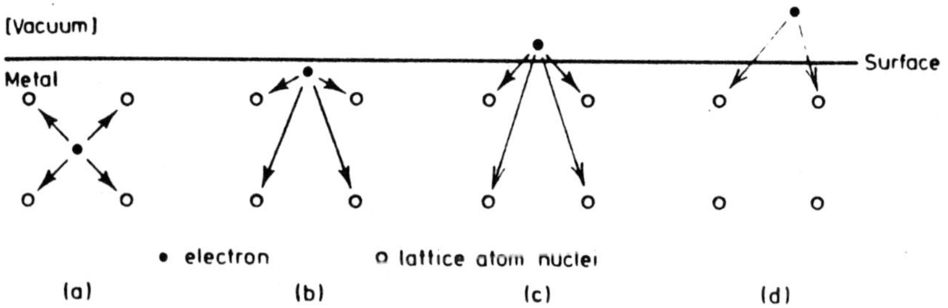

FIGURE 3.3

Restoring forces in electron emission

3.2.1 Work Function

Clearly external work must be done on the surface against these restoring forces to cause electrons to become free (be emitted). At normal temperatures, random thermal lattice vibrations will cause some electrons to have a velocity component towards the surface. As external energy is applied it is these electrons which will be emitted first. *The work function* ϕ is defined as that external energy which must be applied just to cause electron emission. These atomic energies are so small compared to the joule that they are more conveniently expressed in electron volts (eV).

One electron volt is that energy acquired by an electron when it is accelerated through a potential difference of one volt, hence

$$\text{energy} = \text{power} \times \text{time} = VIt$$

$$= \text{voltage} \times \text{charge}$$

therefore

$$1 \text{ eV} = 1 \times \text{electronic charge}$$

$$1 \text{ eV} \doteq 1.6 \times 10^{-19} \text{ J}$$

Note that the work function in joules = $e\phi$ if ϕ is in electron volts. Typical work functions are given in table 3.1. The work-function energy for emission may be applied externally in several ways.

TABLE 3.1

Work functions of common emitting materials

Material	Work function ϕ (eV)
Tungsten	4.52
Caesium	1.81
Barium - tungsten	1.6
Barium - strontium oxide	1.3

3.2.2 Thermionic Emission

This occurs when the surface is heated externally, causing random thermal agitation which endows some electrons with sufficient energy to escape. Heating is usually achieved by forming the surface on a thin hollow cylinder inside which is a current-carrying wire at temperatures between 1000 and 2500 K. This form of indirectly heated cathode was used as an electron source in electronic valves but survives today in the cathode-ray tube (figure 3.4).

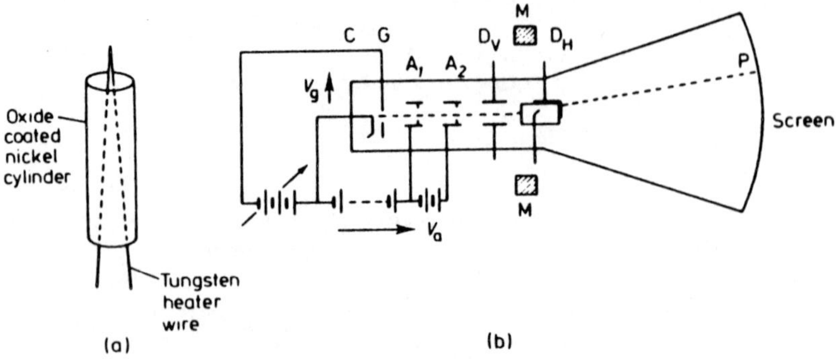

FIGURE 3.4

An indirectly heated cathode and its use in the cathode-ray tube

A cathode C constructed as in figure 3.4a emits electrons which are accelerated towards the screen by a positive potential of more than a thousand volts on anode A_2. A variable positive potential on A_1 allows the operator to focus the beam into a spot just where it hits the phosphor-coated screen causing a glowing spot to appear. A variable negative potential on the grid plate G allows varying numbers of electrons to be accelerated, thus controlling the spot brightness. Two sets of electrostatic deflector plates move the spot in the vertical (D_V) and horizontal (D_H) directions.

Because of problems of tube geometry the electrostatic deflection is limited to screen diameters up to about 15 cm. The larger tubes used in television receivers and display oscilloscopes are usually electromagnetically deflected by toroidal coils M placed around the tube neck.

3.2.3 Photoemission

Electron emission occurs when a surface is bombarded with electromagnetic radiation (X-rays, ultraviolet, visible or infrared light). Because the electromagnetic energy arrives in discrete quantities or *quanta*, no emission occurs until the quantum energy exceeds the surface work-function. The quantum energy is hf joules where h is Planck's constant (6.625×10^{-34} J s) and f is the radiation frequency. Remember also that $c = f\lambda$ for any wave motion where c is the velocity and λ the wavelength. Thus there will be a threshold wavelength λ_0 above which no photoemission occurs

$$\lambda_0 = \frac{c}{f_0}$$

and

$$hf_0 = e\phi$$

therefore

$$\lambda_0 = \frac{ch}{e\phi}$$

where c is the velocity of electromagnetic radiation (3×10^8 m/s). Wavelengths are still often quoted in Ångstrom units instead of metres (1 Å = 10^{-10} m).

If the wavelength is such that the quantum energy hf exceeds the work function $e\phi$ of the surface, the surplus energy is transferred to the emitted electron as kinetic energy.

$$hf - e\phi = \frac{mv^2}{2}$$

This is Einstein's equation and care must be taken to use consistent energy units (joules) throughout, therefore if e = electronic charge in C and ϕ = work function in eV, then m = electronic rest mass (9.11×10^{-31} kg) and v = escape velocity (m/s).

Example 3.1

Calculate the threshold frequency of a photocathode coated with caesium and determine the electron escape velocity if bombarded with radiation at a wavelength of 5000 Å. From table 3.1, $\phi = 1.81$ eV.

$$hf_0 = e\phi$$

$$f_0 = \frac{(1.6 \times 10^{-19} \times 1.81)}{6.63 \times 10^{-34}}$$

$$= 4.37 \times 10^{14} \text{ Hz}$$

$$\lambda_0 = c/f_0 = \frac{3 \times 10^8}{4.37 \times 10^{14}}$$

$$= 6.88 \times 10^{-7} \text{ m}$$

If $\lambda = 5000$ Å $= 5 \times 10^{-7}$ m, then $f = c/\lambda = 3 \times 10^8/(5 \times 10^{-7})$ Therefore the quantum energy

$$hf = \frac{6.63 \times 10^{-34} \times 3 \times 10^8}{5 \times 10^{-7}}$$

$$= 3.98 \times 10^{-19} \text{ J}$$

Work function

$$e\phi = 1.6 \times 10^{-19} \times 1.81 = 2.89 \times 10^{-19} \text{ J}$$

therefore the kinetic energy of the electrons is

$$mv^2/2 = hf - e\phi$$

$$= (3.98 - 2.89) \times 10^{-19}$$

therefore

$$v = \sqrt{\left(\frac{2 \times 1.09 \times 10^{-19}}{9.11 \times 10^{-31}}\right)}$$

$$= \sqrt{(23.9 \times 10^{10})}$$

$$= 4.89 \times 10^5 \text{ m/s}$$

A popular type of photocell is the photoemissive tube whose construction is shown in figure 3.5a. The photocathode is usually a nickel semicylinder the inside surface being coated with low work-function photoemissive material. It might be thought that the spectral response curve of such a device would resemble figure 3.5b. We must remember, however, that the tube current depends on the *number* of emitted electrons and not on their kinetic energy. Both of these factors are obviously zero below the threshold frequency f_0 or above the threshold wavelength λ_0. Nevertheless at higher frequencies (lower wavelengths), the preferential absorption of certain wavelengths by each cathode material together with the inability of the glass envelope to transmit them gives spectral response curves shaped as in figure 3.5c.

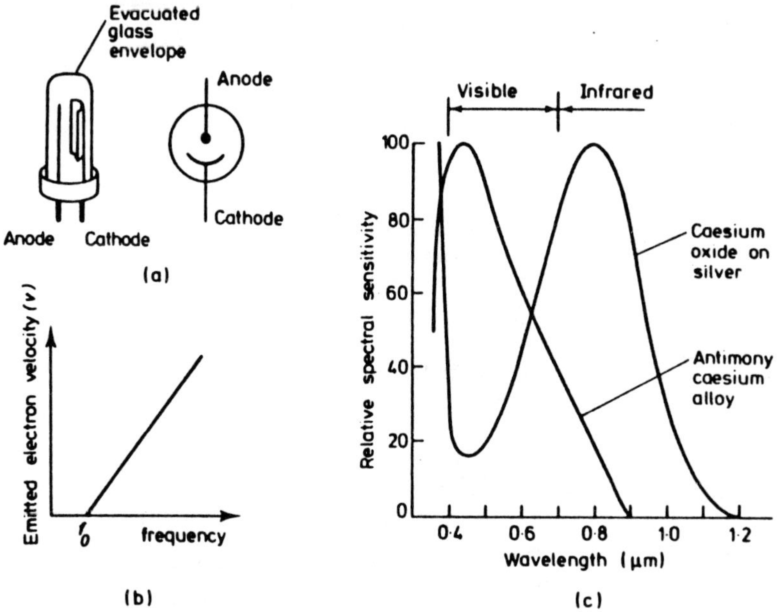

FIGURE 3.5

(a) A photoemissive cell and its circuit symbol;
(b) electron-velocity – frequency graph;
(c) spectral-response curves (not to same vertical scale)

The emitted electrons are collected on a positive wire anode and the anode current indicates their number. For any given wavelength the number of emitted electrons depends on the intensity of the incident radiation. The speed of response of such photocells is practically instantaneous.

3.2.4 Secondary Emission

In secondary emission the external work to liberate electrons is provided by bombardment of the surface with a beam of incident electrons called *primaries*. The surface disturbance produced can cause emission of *secondary* electrons but not necessarily in proportion to the energy (or velocity) of the primaries. Figure 3.6a shows the effect of primary electrons having various incident energies. Primaries whose individual kinetic energy is less than the work function can cause no emission, merely heating the surface (i). At higher energies secondary emission occurs first singly (ii) and then by multiple emission (iii). At the highest energies, however, (iv) the electron release takes place so far inside the surface that there is little possibility of emission because the secondaries' escape is hindered by their collision with lattice atoms.

The number of secondaries emitted per incident primary electron is called the *secondary emission coefficient* δ for that surface at that primary energy.

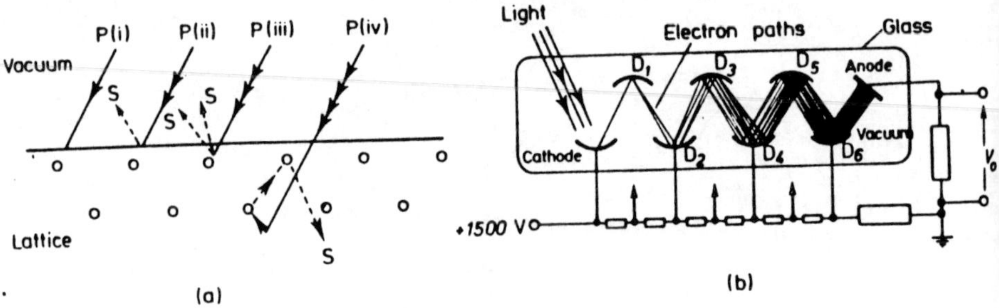

FIGURE 3.6

(a) Secondary emission at various incident energies;
(b) its use in the photomultiplier tube

Some materials such as caesium oxide on silver exhibit values of δ as high as 10 for primary energies of 500 eV. Such materials are employed in the photomultiplier or electron-multiplier tube (figure 3.6b) in which a series of n dynodes, each at an increasingly positive voltage, multiply repeatedly the small number of electrons produced at a photocathode. The current gain δ^n can exceed one million giving an extremely sensitive photoelectric device suitable for applications such as the astronomical measurement of star intensities.

3.3 Gas Discharge Devices

Conduction, that is, the motion of charged particles, is difficult in gases at normal pressure because of collisions with the relatively densely packed gas molecules. Certain discharge lamps do operate at these pressures but the majority of gas-filled devices are partially evacuated down to a pressure below 1 torr (mm of mercury).

3.3.1 Ionisation and Excitation

At these lower pressures electrons are relatively free to move under the influence of an applied electrostatic field with only occasional collisions with gas molecules. If they are accelerated by the influence of an external electrostatic field they can easily acquire sufficient kinetic energy between collisions to *ionise* gas molecules with which they collide.

Ionisation is the process in which the outermost electrons of an atom become detached from the remainder of the atom. The atom, previously electrically neutral, is now deficient in negative charge; it is thus positively charged. These positively charged atoms are called *ions* (from the Greek – 'wanderer') because, although charged and therefore capable of acceleration in an electric field, they are much more massive and therefore more sluggish than an electron. Because of this mass they have great momentum and are therefore liable to damage any negatively charged electrode to which they are attracted.

A very definite amount of energy is required to ionise an atom – its *ionisation energy* (joules). If this energy is expressed in electron volts (eV) it is commonly known as the *ionisation potential.* Energies less than the above are sometimes capable of raising an orbital electron to a higher energy state (or shell) without detaching it from the parent atom. The atom is now said to be 'excited'. This process is usually revealed by a glow in the gas caused by excited electrons falling back to their original levels and releasing energy δE in the form of quanta of light as they do so. The frequency f of this radiation may be obtained from the energy equation

$$hf = \delta E$$

It is these radiation frequencies corresponding to transitions between strictly defined atomic energy levels that constitute the emission spectra of various atoms.

3.3.2 Gas-filled Photocells

Perhaps the simplest electronic application of ionisation effects is the gas amplification effect used in some photocells. Such photocells are constructed in a similar manner to that in figure 3.5a and a small quantity of inert gas is inserted after evacuation. The electrons released from the photocathode by radiation are accelerated by the positive anode and collide with gas molecules causing multiple ionisation. That is, more than one electron escapes from the gas molecule per collision. This process increases the number of free electrons available for conduction and therefore increases the anode current for a given incident radiation. This *gas amplification factor* can reach as high as 10 but although the sensitivity of the tube is increased it becomes more electrically fragile. A glance at figure 3.7 shows that whereas the vacuum phototube has an anode current largely unaffected by the anode voltage, the gas-tube current rises appreciably at higher voltages. The gas-tube anode current is, however, not linearly proportional to the incident luminous flux as in the case of the vacuum photocell. This renders it unsuitable for many measurement applications but it can successfully be employed as a sensitive on–off light detector. Its electrical fragility derives

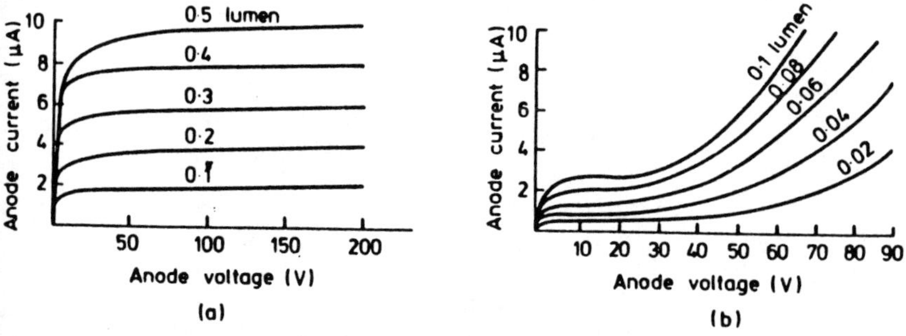

FIGURE 3.7

A comparison of the electrical characteristics of photocells from the possibility of photocathode damage by the massive positively charged gas ions. The permitted anode voltage is normally restricted to 100 V to minimise this damage. Because of the finite time (approximately 10^{-4} s) required for gas ionisation to occur after the commencement of electron motion these devices are much slower than their vacuum counterparts. The upper limit of response rarely exceeds 10^4 Hz.

3.3.3 Cold-cathode Discharge Tubes

These are constructionally the simplest gas devices. Figure 3.8a shows a pair of metal electrodes sealed into a glass envelope containing inert gas at a low pressure. Application of an increasing potential between the electrodes results in the electrical characteristic of figure 3.8b. At normal temperatures and with no applied voltage there will still be a few ion - electron pairs formed by ambient radiation. At low voltages the electrons move towards the anode and the ions towards the cathode as shown. The number of these charged particles is largely unaffected by voltage, causing the first part of the characteristic OA to be nearly horizontal. After point A the electron velocities become great enough to cause some ionisation until at V_b sufficient charged particles are moving quickly enough to ionise nearly all the gas and breakdown occurs. V_b is the *breakdown voltage.*

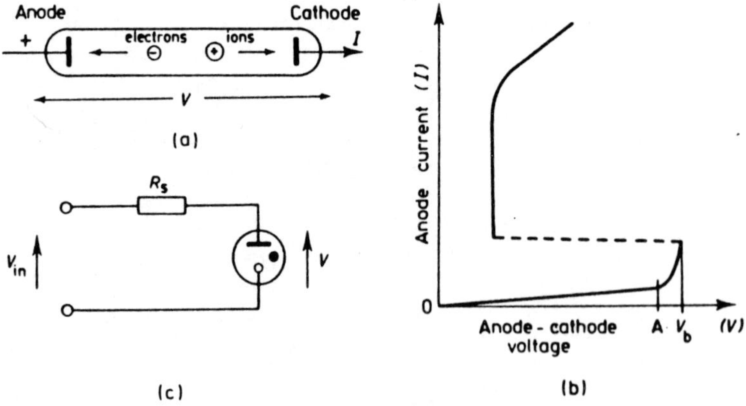

FIGURE 3.8

The cold-cathode discharge tube, its characteristics and use

The curve now passes through an unstable (dashed) region. The current rises and the voltage falls to approximately the ionisation potential where it remains over a considerable current range (a tenfold increase). This is the *glow discharge region* in which neon-indicator and voltage-stabiliser tubes operate. Figure 3.8c shows how such tubes may be operated with a series resistor to limit the current to safe values once breakdown and discharge occur. Care must be taken that V_{in} exceeds the tube's breakdown voltage and that R_s is only just large enough to limit the current to a value in the centre of the discharge region during normal operation. Such tubes may act as indicator lamps when operated on either d.c. or a.c. supplies. When neon gas is used the familiar red glow occurs but it is generally of too low an intensity for general illumination. The fluorescent tubes used for illumination have their inside walls coated with a phosphor which glows brightly in the presence of the ion – electron stream.

Alternatively gas tubes may be used for d.c. *voltage stabilisation* because the tube voltage remains sensibly constant over a wide current range. Although V_{in} in figure 3.8c may vary considerably, V will be almost constant.

3.3.4 Power Rectifiers

Because of the low voltage-drop across a gas tube during discharge it is a much more efficient rectifier at high current levels than its vacuum counterpart. At high currents the ionic bombardment of the cathode is severe. A mercury pool has been found to be almost indestructible in this respect; hence many devices use mercury vapour as a gas.

Earlier mercury-arc rectifiers used a glass envelope shaped as in figure 3.9a. Because of its fragility, modern designs now have steel containers.

The *ignitron* (figure 3.9b and c) is typical of such gas-discharge rectifiers. A quantity of mercury is sealed into an evacuated steel container which may be double-walled to permit circulation of a liquid coolant. A graphite anode is supported within the vessel by a steel rod which provides both thermal and electrical conduction to the exterior. This rod passes through the vessel walls via a glass – metal seal and may be fitted with cooling fins in large examples. On applying a potential between anode and cathode no conduction can occur until ionisation of the gas vapour has been initiated by a third electrode – the igniter. This latter consists of a cone of boron carbide which dips into, but is not wetted by, the mercury cathode. Application of a voltage between igniter and cathode causes sparking to occur between their surfaces with consequent ionisation of the surrounding gas. Almost immediately, conduction occurs between the anode and cathode and the igniter no longer has any effect since the cathode hot-spot, produced by the main arc at the cathode, sustains ionisation.

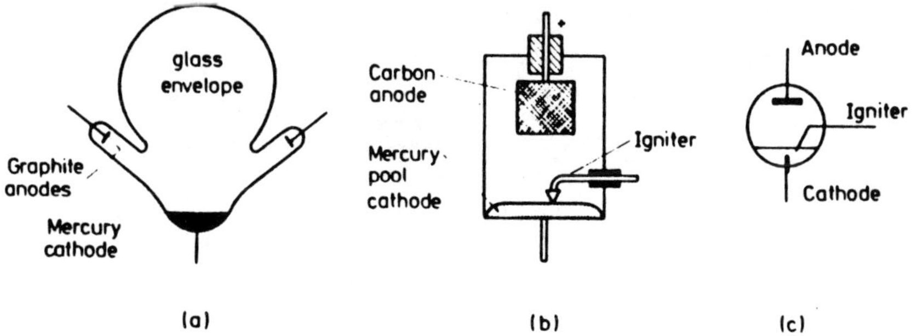

FIGURE 3.9

The mercury-arc rectifier; (a) its traditional form; (b) the ignitron; and (c) the ignitron circuit symbol

The discharge can only be extinguished by removing the anode to cathode voltage, which would automatically occur at the end of each positive half-cycle if the ignitron were fed from an a.c. source. Since conduction may only occur when the anode is positive with respect to the cathode the device may be used as a rectifier to handle currents up to 10000 A at high efficiencies. The anode to cathode potential drop is almost constant at approximately 15 V once conduction starts.

The discharge has to be restarted by the igniter on each half-cycle but a short (few microseconds) pulse is sufficient to initiate conduction. The precise positioning of this pulse within the cycle will allow accurate control of the mean anode current using the methods of section 5.3.

3.4 Semiconductor Devices[8, 9]

As mentioned in section 3.1 semiconductors are elements or compounds capable of forming a regular crystal lattice with a covalent bonding structure. The two most commonly used substances, silicon and germanium, are both quadravalent (four valence electrons) and form lattices as in figure 3.2.

3.4.1 Intrinsic Conduction

At absolute zero the lattice is stationary and conduction cannot occur because there are no free charges for conduction. At ambient temperatures, however, there is considerable thermal lattice vibration and many covalent bonds are ruptured. This means that an electron shared between two adjacent atoms leaves their sphere of influence and can wander through the lattice. The area from which it originated is now deficient in negative charge since the region was originally neutral (see figure 3.10a). This positive region is called a *positive hole* because a wandering electron may later recombine with it to fill it and resume the unbroken bond structure. Notice that the numbers of thermally produced free electrons and holes are always equal. We speak of thermally generated *hole-electron pairs*. If an electric field is applied across the material, electrostatic

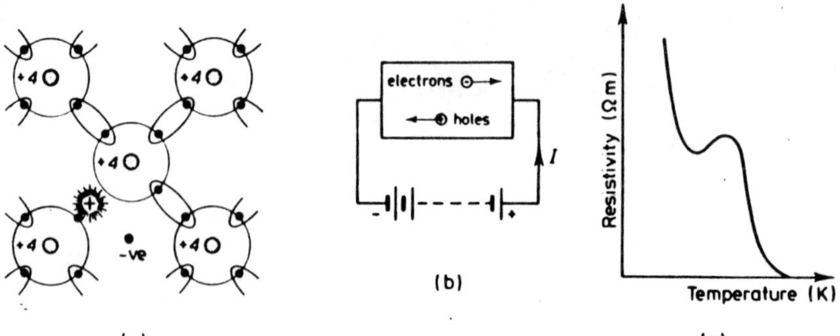

FIGURE 3.10

Intrinsic semiconduction

forces will cause the electrons to migrate or drift towards the positive end and holes to migrate towards the negative end (figure 3.10b). A little thought will reveal that these opposite motions of holes and electrons result in conventional current flow in the *same* direction.

The reader may be finding difficulty with the concept of a positive hole behaving as a particle for migration purposes. The following analogy can prove helpful. Imagine the front row of seats in the balcony as seen from the stage of a theatre. Only one of the seats is empty – a positive hole – somewhere at the left-hand end. A member of the audience leaves his seat in the centre of the row and exits via the right-hand end; on returning he does not wish to disturb the others and so returns not to his own seat but to the other vacant one at the left. Two casual observations from the stage before and after the above events could lead an intoxicated actor to believe that an empty seat had moved from the left to the centre!

Conduction caused by the drift of thermal hole-electron pairs in pure semiconductor materials is called *intrinsic conduction*. It is clearly very temperature-dependent, becoming easier as the rising temperature accelerates the rate at which hole-electron pairs are generated and recombine. The lifetime of a pair between generation and recombination is a mere 0.1 ms at room temperatures! Unlike metals, therefore, the resistivity of semiconductors decreases with temperature as in figure 3.10c – a potentially unstable situation. As the temperature rises, any current from an applied voltage will also rise, causing further heating and temperature elevation. This situation is self-maintaining and will result in *thermal runaway* which can cause the material to melt and become unusable. This great temperature-dependence of intrinsic materials is employed in the *thermistor* for temperature measurement. Beads or rods of metallic oxide mixtures exhibit resistance changes of the order of $10^3 : 1$ up to 300 °C. They are frequently used in bridge circuits for measurement purposes.

3.4.2 Extrinsic Conduction

The previous phenomena all occur in pure semiconductor materials in which the impurity levels are less than 1 part in 10^{10}. The majority of electronic devices, however, use semiconducting properties caused by controlled concentrations of selected impurity atoms. These impurity atoms are deliberately introduced during manufacture in concentrations as low as 1 part in 10^8; nevertheless they produce a marked effect on the electrical properties. The impurity materials are chosen for their crystalline similarity to the pure semiconductor so that they can lie neatly in the crystal lattice. They are, however divided into two classes: those such as phosphorus, antimony or arsenic, having five valence electrons and those like boron, aluminium, gallium or indium, which have only three valence electrons. Their effect when in concentrations as low as 1 part in 10^8 is shown in figure 3.11a and b. The arsenic (As) atom has one spare electron loosely bound to it and available for conduction. The indium (In) atom contributes one spare positive hole to the material. The lower diagrams show the extrinsic materials at ambient temperatures when there are also a few intrinsic thermal electron-hole pairs present. Clearly, in the case of arsenic doping, electrons are the majority carriers and holes the minority ones. This is *n*-type material having a preponderance of free negative charges. Conversely

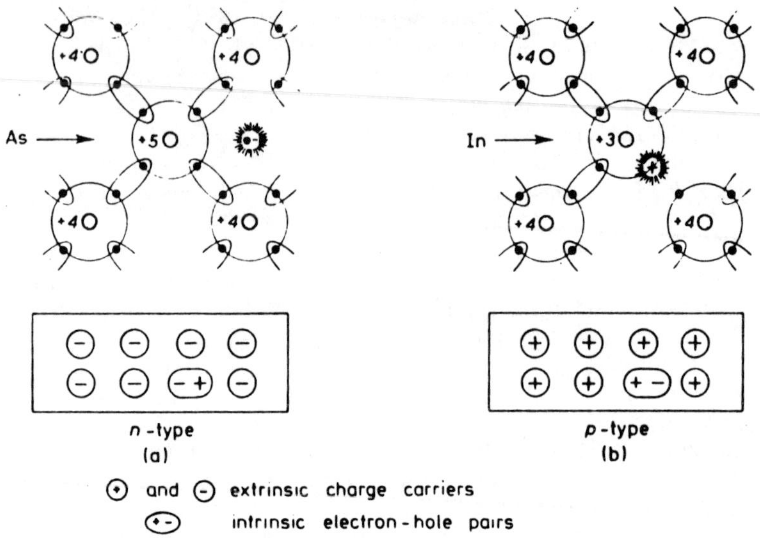

FIGURE 3.11

p- and *n*-type extrinsic semiconductors

with indium doping, positive holes are in the majority, the material is *p*-type possessing many free positive charges. Great care should be taken to remember that although blocks of these materials are conventionally drawn as shown, they are both electrically *neutral*. The surplus holes or electrons are exactly balanced by the proton charges in the impurity atom nuclei.

In practice, the impurity concentration is adjusted so that the extrinsic carriers (one per impurity atom) outnumber the intrinsic carriers at room temperatures – the material is thus overwhelmingly extrinsic. At temperatures of 80 °C for germanium and 200 °C for silicon the number of thermal intrinsic carriers begins to approach the fixed concentration of extrinsic carriers and the materials begin to lose their *p*- and *n*-type properties, reverting to intrinsic behaviour. This effect limits permissible container (or case) temperature for semiconductor devices to approximately 60 ° and 120 °C for germanium and silicon devices respectively (see section 5.1).

3.4.3 The *p*-*n* Diode

Consider two pieces of semiconductor material arranged as in figure 3.12a. They are both electrically neutral and possess majority and minority carriers as shown. If they could be brought together as in figure 3.12b so that the crystal lattice was continuous at their junction, diffusion effects would cause charge motion as shown. Diffusion is the process whereby local concentrations of material are dispersed to give a homogeneous environment. It is the process whereby a drop of ink disperses itself evenly throughout a bucket of water. The excess holes migrate left into the *n*-type making it positively charged whereas the excess electrons migrate to the right into the *p*-type causing it to become negatively

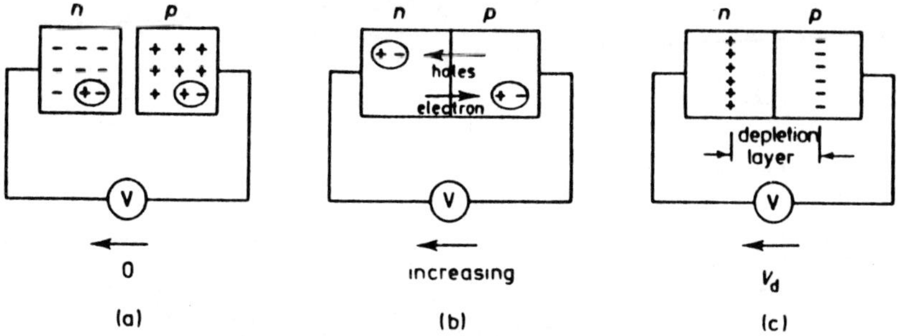

FIGURE 3.12

The $p-n$ junction

charged. This gradual acquisition of charges by the two regions sets up a potential across the junction as shown. It is of such a polarity as gradually to oppose further motion of the majority carriers until, at figure 3.12c, a dynamic equilibrium is established in which there is no net motion of carriers across the junction. The potential V_d across the junction is called the *barrier potential*. It is not, however, a barrier to the minority thermal carriers on each side. After some moments the majority carriers which have crossed the junction tend to arrange themselves as shown in figure 3.12c leaving a depletion layer between 10^{-6} and 10^{-8} m wide on either side of the junction in which there are very few charge carriers. As may be imagined, an almost insulating region flanked by charges of opposite polarities resembles a capacitor electrically and this will be important in our later study of transistors (sections 3.4.5 and 8.2.1). In practice, these junctions are never formed between two separate pieces of material. Lattice continuity can only be ensured by using one long piece of pure semiconductor and forming a junction at its centre by melting-in n-type impurities from one end and p-type impurities from the other. .

Suppose that an external voltage of the polarity shown in figure 3.13a is applied to the $p-n$ junction, the direction is such as to reinforce the barrier

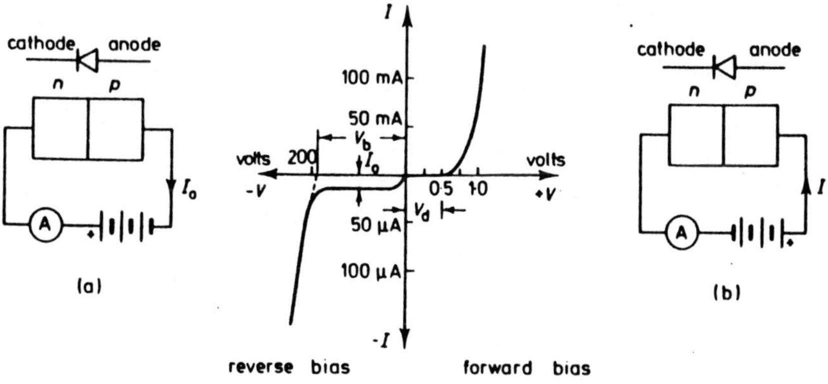

FIGURE 3.13

Typical silicon-diode characteristics (note change of scales)

potential making majority charge motion even more unlikely. This is called *reverse bias*, the only current I_0 is that caused by thermal minority carriers. This reverse leakage current is therefore very temperature sensitive, being typically 10 μA for a silicon device and 4 mA for a germanium one at room temperature and at a reverse potential of 100 V. At a sufficiently high reverse voltage V_b breakdown begins to occur by charges being pulled through the barrier by the intense electrostatic field. Generally silicon devices can withstand two or three times the reverse voltage possible in germanium.

In the forward-biased direction (figure 3.13b) the current rises very rapidly once the forward voltage drop V_d has been reached. This point corresponds to the barrier potential being neutralised by the external forward bias which is applied in the opposite direction. Once this potential barrier is destroyed, high forward currents occur by majority carrier motion. This forward voltage drop for silicon (0.6 V) is roughly twice that for germanium.

One can see therefore that the $p-n$ junction is a rectifying device, allowing current to flow much more easily in the forward direction which is indicated by the direction of the triangle in the circuit symbol shown.

3.4.4 The Bipolar Transistor

Consider either of the arrangements of figure 3.14a or b in which a thin (approximately 20 μm) slice of p- or n-type semiconductor called the base is flanked by sections of the opposite type. External voltages are applied so that in both cases junction J_1 is forward-biased and J_2 is reverse-biased. Majority carriers (electrons) in the $n-p-n$ case and vice versa) will easily pass across J_1 from the left-hand *emitter* regions into the bases as shown by the broad arrows. Because the base region is thin and has a low concentration of its own majority carriers, little recombination occurs in the base. The greater part (0.98 to 0.995) of the carrier stream passes into the right-hand or *collector* region since J_2 does not present a barrier to those carriers which have the sign of minority carriers in the base material. The remainder of the original emitter carriers leave the device via the base terminal. Because electrons move in the opposite sense to a conventional

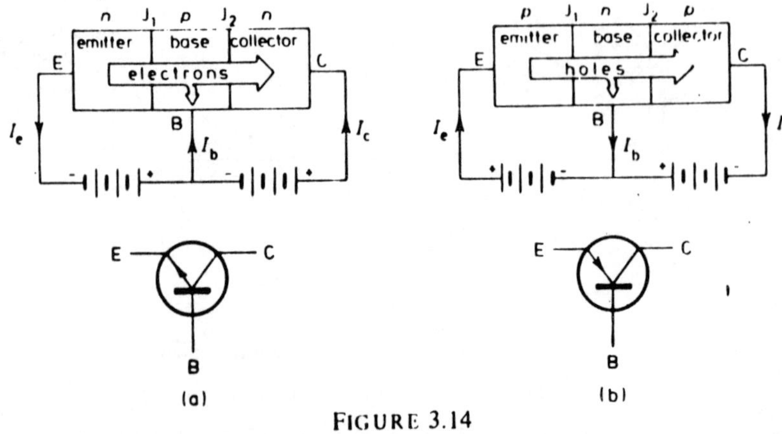

FIGURE 3.14

The bipolar transistor and its circuit symbols (a) *npn*, (b) *pnp* current and because holes move in the same sense as a conventional current, the directions of emitter, base and collector currents are as shown.

In both cases, by Kirchhoff's law

$$I_e = I_b + I_c$$

and if, for example

$$I_c = 0.99\, I_e$$

then

$$\frac{I_c}{I_b} = \frac{0.99\, I_e}{(1 - 0.99)\, I_e}$$

$$\frac{I_c}{I_b} = 99$$

or

$$\Delta I_c = 99\, \Delta I_b$$

This equation shows that any variation of base current ΔI_b will appear, magnified considerably, as a variation in collector current ΔI_c. The transistor is thus a device for the amplification of current changes. The manner in which this ability is applied practically is discussed fully in section 8.1.

Both *npn* and *pnp* transistors are employed in modern practice using both silicon and germanium although, because of production considerations, silicon *npn* and germanium *pnp* devices tend to be most common. *npn* and *pnp* types require supply voltages of opposite polarities and this fact is put to advantage in some electronic circuit designs.

In *phototransistors* incident radiation is allowed to fall upon the emitter-base junction J_1. The radiant energy produces free charges which pass across the emitter-base junction and are amplified by the transistor action. Germanium phototransistors are more sensitive to infrared radiation than other photodevices since a quantum energy of 0.7 eV is all that is required for operation.

3.4.5 The Field-effect Transistor (FET)

In the bipolar transistor, the controlling current I_b is produced by an input voltage across J_1. This junction is forward-biased and has a low impedance. In some applications this is a disadvantage because it produces electrical loading of the previous circuits.

In the field-effect transistor (figure 3.15) however the input voltage is applied across a reverse-biased junction which has a much higher input impedance. The device consists essentially of a channel of *n*- or *p*-type material. Majority carriers flow along this from the source end to the drain end under the influence of the applied supply voltage V_{DS}. Two gate regions of opposite type material are encountered half-way down the channel. These regions have their junctions with the channel reverse-biased by the application of V_{GS} between gate and source. The magnitude of this applied voltage V_{GS} determines the width of the depletion layer surrounding the gate-channel junctions. As V_{GS} is increased and the depletion layers increase in width they tend to pinch-off the flow of carriers from source to drain. In the FET therefore it is an *input voltage* which controls the soutce - drain current. The electrical characteristic curves of figure 3.15c clearly show that over the operating region the drain current is almost independent of V_{DS}. It is almost linearly related to V_{GS} until breakdown occurs at the reverse-biased gate - drain junction.

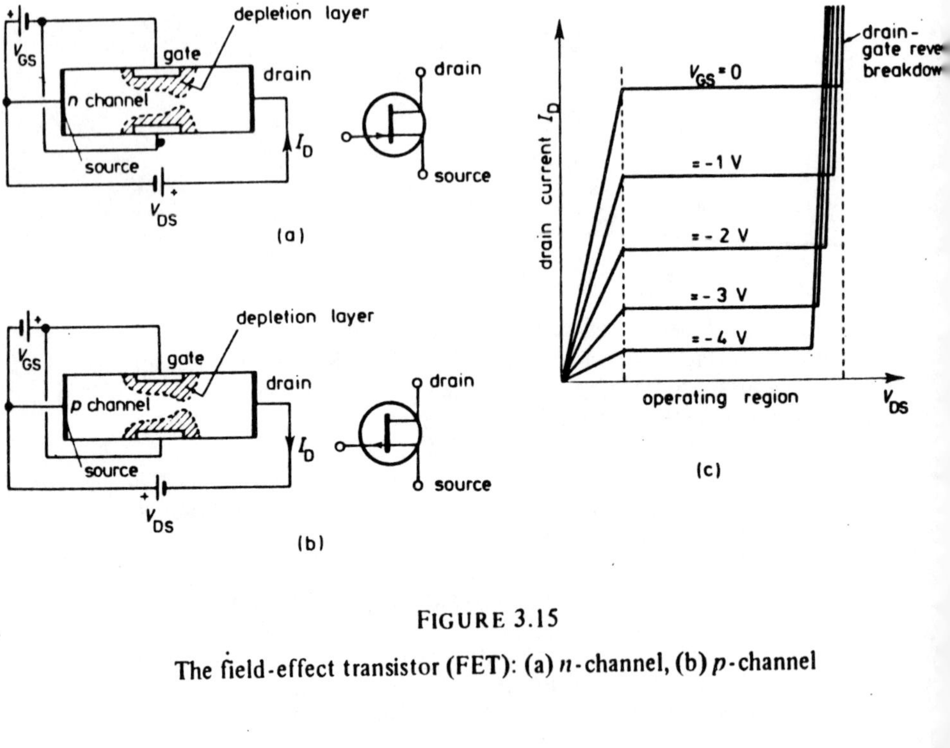

FIGURE 3.15

The field-effect transistor (FET): (a) *n*-channel, (b) *p*-channel

3.4.6 The Thyristor

The thyristor is essentially an electronic relay in which variations of a small input or gate current I_g set the level at which the device will become conducting. In addition, the thyristor is unidirectional, that is it rectifies any a.c. supply applied to it. A non-rectifying version, the triac, is discussed in section 5.4.2.

Figure 3.16a shows the basic thyristor construction comprising four alternating regions of *p*- and *n*-type material. Voltages applied between anode and cathode in either direction will not normally produce conduction because either J_2 or J_1 and J_3 will be reverse-biased. If however an external voltage V is applied in such a direction that only J_2 is reverse-biased, J_2 will eventually break down at some voltage V_b. If a small current I_g is applied into the gate terminal the voltage required for breakdown decreases.

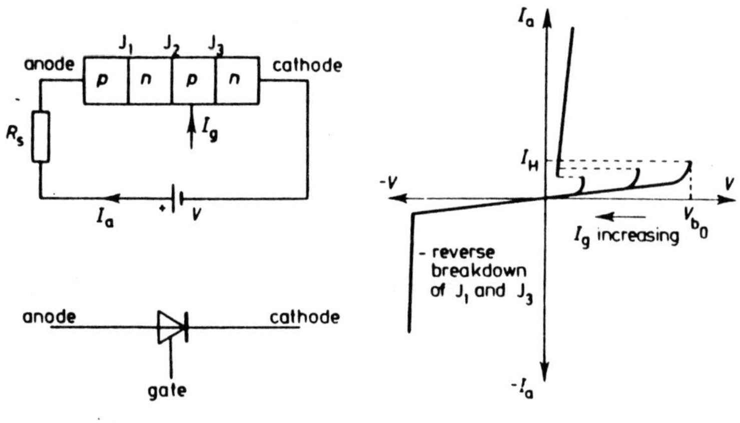

(a) (b)

FIGURE 3.16

The thyristor, its circuit symbol and electrical characteristics

Once breakdown has occurred the current through the device will not stop until V is reduced so that I_a falls below the holding current I_H. In other words, after breakdown, the gate current has no effect on the value of I_a until the non-conducting state is regained. If V is reversed the device merely acts as a reverse-biased diode, breaking down at some fixed reverse voltage which is unaffected by gate current.

3.5 Problems

3.1 Calculate the energy carried by photons of red light (λ = 6439 Å) and ultraviolet light (λ = 3302 Å).

Each of the above wavelengths strikes a surface of work function 2 eV. If electron emission occurs, calculate the mean velocity of the emitted electrons.

3.2 A surface has a secondary emission coefficient of 9 when it is bombarded with 300 eV primary electrons. Calculate the emission current in mA per W of primary electron power.

3.3 The vacuum and gas photocells whose characteristics are given in figure 3.7 are both used with supply voltages of 80 V and load resistors of 4 MΩ. Using load-line techniques, estimate the photocell current when both cells are illuminated with fluxes of 0.1 lumen. Hence, if they are of identical construction, calculate the gas amplification factor.

4 Power Distribution and Machines

In this chapter the main emphasis will be directed to the distribution and conversion of electrical power into mechanical energy. The subject of electrical generation is mentioned briefly to contribute to the reader's understanding of the nature of constraints placed on the consumer's end of the distribution network.

4.1 A.C. versus D.C.

One of the chief reasons for the adoption of alternating as distinct from direct current electrical distribution systems in the national power network ('the Grid system') is the ease with which its potential may be selected as required by means of *transformers*.

In certain very specialised applications a return is being made to d.c. distribution where interconnections between a.c. systems are required. These are being adopted because instability problems may occur where the length of a.c. transmission lines becomes comparable to the wavelength of the frequency used (6000 km for 50 Hz). Switching-current surges arise and the whole system may be closed down by its safety-control system. A.C. cables also suffer a rise in temperature in the insulation due to capacitive effects (see dielectric heating, section 5.6) and the insulation has to withstand the peak instantaneous voltage V_m whereas the effective equivalent d.c. voltage would only be the r.m.s. value $0.707\ V_m$. Where interconnections between separate a.c. systems are required (Denmark – Sweden, France – Britain and North Island – South Island in New Zealand) d.c. links eliminate the need for synchronisation (ensuring that the frequency and phase relationship of the two systems are identical). The delay in adoption of d.c. transmission techniques was caused by the late development of an efficient and simple form of inverter to reconstitute a.c. voltages from a d.c. supply (section 5.5). Orders have recently been placed for high-voltage d.c. transmission links within the National Grid, but in the foreseeable future a.c. systems will be used without exception for consumer supply.

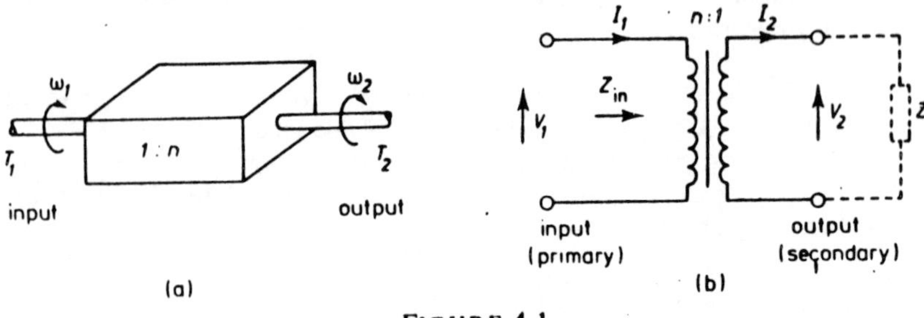

FIGURE 4.1

(a) Idealised step-up gearbox and (b) step-down transformer of turns-ratio *n*

4.2 The Transformer

The transformer is merely an electromagnetic device for altering the ratios between voltage and current in the two electrical subsystems which it couples; its action is analogous to that of a mechanical gearbox (figure 4.1a).

$$\text{Torque ratio} = \frac{T_2}{T_1} = \frac{1}{n} \qquad \text{Voltage ratio} = \frac{V_2}{V_1} = \frac{1}{n}$$

$$\text{Speed ratio} = \frac{\omega_2}{\omega_1} = n \qquad \text{Current ratio} = \frac{I_2}{I_1} = n$$

These equations show that idealised gearboxes and transformers with no internal losses transmit power unchanged

$$T_2\omega_2 = T_1\omega_1 \qquad V_2 I_2 (\cos\phi) = V_1 I_1 (\cos\phi)$$

The equations for both devices are similar, but with voltage V substituted for torque T and current I substituted for angular velocity ω. It is well known that a viscous frictional coefficient (N m per rad/s) applied to the output shaft of a gearbox will be reflected at the input shaft multiplied by the square of the gearbox step-up ratio

$$\frac{T_1}{\omega_1} = n^2 \frac{T_2}{\omega_2}$$

The equivalent electrical equation for the transformer would be

$$\frac{V_1}{I_1} = n^2 \frac{V_2}{I_2}$$

or

$$Z_{\text{in}} = n^2 Z_{\text{L}}$$

V_1/I_1 is the impedance looking into the transformer primary terminals and V_2/I_2 is the value of any load impedance placed across the transformer output (Z_{L} in figure 4.1b). Transformers are frequently used as impedance-changing devices in measurement and instrumentation circuits.

In neither the mechanical nor the electrical case will the device be ideal (have unity efficiency) and therefore the power output will be less than the total power input by an amount equal to the internal losses within the device. In a gearbox these are mainly drag from the lubricant and friction at gear-teeth and bearings. Transformer losses are of two types: first, magnetic (or iron) losses within the core caused by eddy-current and hysteresis effects and second, thermal losses caused by the currents passing through the winding conductors (copper losses).

Example 4.1

An ideal transformer connects a 240 V supply to a resistive load of 10 Ω which requires a supply at 12 V. Calculate the transformer turns-ratio, the primary and secondary currents together with the load impedance 'seen' by looking into the primary terminals of the transformer.

$$I_2 = \frac{\text{load voltage } (V_2)}{\text{load resistance}} = \frac{12}{10} = 1.2 \text{ A}$$

$$\frac{V_1}{V_2} = n = \frac{240}{12} = 20:1$$

$$\frac{I_1}{I_2} = \frac{1}{n}$$

$$I_1 = \frac{I_2}{n} = \frac{1.2}{10} = 120 \text{ mA}$$

$$Z_{\text{in}} = n^2 Z_{\text{L}} = 20^2 \times 10 = 4 \text{ k}\Omega$$

4.2.1 Method of Operation of Transformers

A transformer makes use of the phenomenon of *mutual inductance* in which the input voltage is applied to a winding (usually referred to as the 'primary') causing a current to flow which produces a flux which links with a 'secondary' output winding. Changes in flux will only induce voltages within the secondary when the primary applied voltage and therefore the current is varied. This condition is usually fulfilled by applying a sinusoidal alternating voltage to the primary. It is essential that almost all of this flux should link with the secondary winding (low leakage) and so the windings are mounted on a high-permeability path of metal – the core (figure 4.2). In this way the leakage flux is kept to negligible proportions (less than 0.1 per cent in a good design). Since the flux is alternating the core temperature will rise from eddy-current and hysteresis effects (see section 2.8) but these are minimised by the use of 'stalloy' (2 - 5 per cent silicon steel) laminated cores. The narrow hysteresis loop of this material ensures minimal hysteresis losses and eddy-current losses are limited by the high resistivity and by the use of laminations.

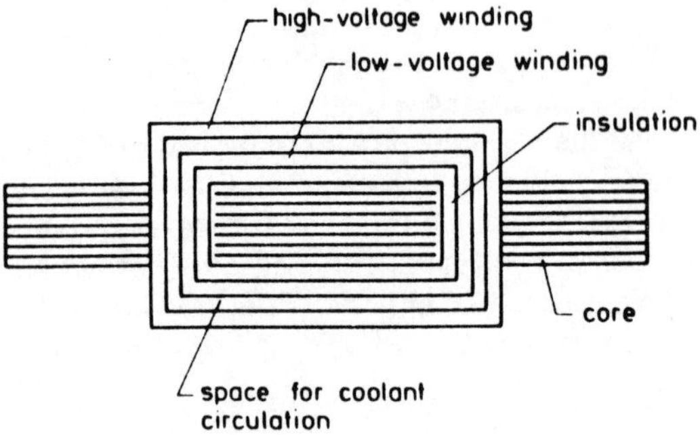

FIGURE 4.2

Cross-section of a concentrically wound higher power transformer

The combination of eddy-current and hysteresis losses, known as the *iron losses,* should just equal the copper or heating losses (caused by the primary and secondary currents flowing in their respective windings) for maximum long-term efficiency. The detailed operation of a transformer on load and the development of a phasor diagram for this situation are covered in many standard texts[4] and are too complex for discussion here; referring to figure 4.3b a simple explanaation of operation will be attempted. With the secondary open-circuited (that is, on no load) the primary winding draws sufficient current I_m to magnetise the core fully; this magnetisation current I_m may only be 0.05 of the full-load primary current. When the secondary circuit is closed a secondary current I_2 flows, dependent in value on the secondary voltage and load impedance. By Lenz's law this current will flow in a direction such that it opposes the flux producing it. This fall in core flux decreases the back e.m.f. in the primary winding allowing an increase in primary current to restore the flux to its original value. Thus the secondary and primary currents rise up to their steady-state values, the relationship between them being $I_2 = n \times I_1$.

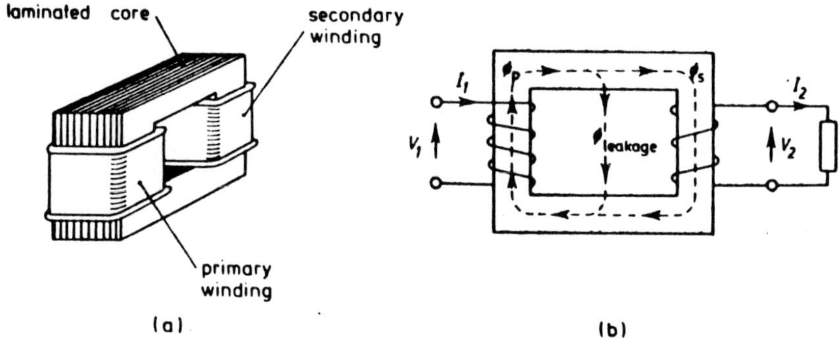

FIGURE 4.3

(a) An arrangement of two magnetically coupled windings on a core to produce a low-power transformer; (b) a schematic diagram of a single-phase transformer

For a well-designed transformer with negligible leakage the primary and secondary fluxes will be approximately equal, hence

$$\phi_p = \phi_s = \phi$$

The voltages on the two windings will be given by

$$v_1 = N_1 \frac{d\phi}{dt} \qquad v_2 = N_2 \frac{d\phi}{dt}$$

Hence

$$\frac{v_1}{v_2} = \frac{N_1}{N_2} = n \text{ (the turns ratio)}$$

and thus V_1/V_2 (the ratio of the r.m.s. voltages) is equal to n.

The E.M.F. Equation

Assume that the flux in a transformer core varies sinusoidally as in the expression $\phi = \Phi_m \sin \omega t$. The voltage induced in a winding of N turns around this core will have an instantaneous value

$$v = N \frac{d\phi}{dt} = N\Phi_m \omega \cos \omega t$$

This voltage is obviously sinusoidal (though not in phase with the flux) therefore its r.m.s. and maximum values (V and V_m) are related by $V = V_m/\sqrt{2}$.

By inspection the maximum value of $v = N\Phi_m$ when $\cos \omega t = 1$. Thus

$$V = \frac{\omega}{\sqrt{2}} N\Phi_m$$

$$= \frac{2\pi f}{\sqrt{2}} N\Phi_m$$

$$= 4.44\Phi_m Nf \text{ (the e.m.f. equation)}$$

4.2.2 Transformer Construction

The upper limit of core-flux density (approximately 1 T) is determined by the onset of nonlinearity in the B/H curve as it approaches saturation; thus the core has to be of sufficient cross-sectional area to support the required total value of flux Φ_m. This means bulk and weight for high voltages. If this is not possible – for example, in aeronautical practice, for weight reasons – the frequency must be raised, often to 80 or 400 Hz to allow a smaller core to be employed. Care must be taken not to operate such transformers on a 50 Hz supply since their cores are incapable of sustaining the high flux-density required.

Example 4.2

Calculate the theoretical r.m.s. voltage obtainable from a 200-turn coil wound on a transformer core of square cross-section and 1 cm side if a 50 Hz alternating flux-density of maximum value 0.5 T (Wb/m^2) exists. Calculate the per-unit saving in core weight possible if this transformer could be operated at 80 Hz; all other factors being equal.

$$\text{flux} = \text{flux-density} \times \text{cross-sectional area}$$

$$\Phi_m = B_m \times a = 0.5 \times 10^{-4} \text{ Wb}$$

therefore

$$V = 4.44 \times N \times f \times \Phi_m$$

$$= 4.44 \times 200 \times 50 \times 0.5 \times 10^{-4}$$

$$= 2.22 \text{ V}$$

For a frequency of 80 Hz the maximum flux is given by

$$\Phi_{m_2} = \frac{V}{4.44\, nf} = \frac{2.22}{4.44 \times 200 \times 80} = 0.31 \times 10^{-4} \text{ Wb}$$

$$\frac{B_{m_1}}{B_{m_2}} \times \frac{a_1}{a_2} = \frac{\Phi_{m_1}}{\Phi_{m_2}}$$

For the same value of B_m

$$a_2 = \frac{a_1 \Phi_{m_2}}{\Phi_{m_1}} = 10^{-4} \times \frac{0.31}{0.5} = 0.62 \times 10^{-4} \text{ m}^2$$

For a given core length the weight must be proportional to the cross-sectional area.

$$\text{percentage area reduction} = \frac{(1 - 0.62) \times 10^{-4}}{10^{-4}}$$

$$\text{percentage weight reduction} = 0.38$$

The actual method of construction depends on the apparent power (VA) rating of the transformer. The winding arrangement as shown in figure 4.3a is unsuitable for all but the lowest power devices since it allows considerable magnetic leakage. More usually the windings are concentrically wound on the centre-limb of a three-leg core (figure 4.2). In high-power transformers one winding at least is usually at high voltage and requires careful insulation and high-current windings often have their conductors carefully shaped from rectangular metal bar instead of from circular wire.

The heat from leakage, core and winding (copper) losses is dissipated by normal air cooling in ratings up to approximately 10 kVA and by immersing the whole transformer in a mineral oil of high thermal- and low electrical-conductivity above this rating. Oil immersion has the advantage that it protects the winding insulation from the ingress of water. The high thermal-conductivity oil transfers the heat to the tank walls whence it is radiated into the atmosphere. Often the tank walls are blackened or ribbed to improve radiation but a major improvement can be made by fitting the tank with external pipes, often double-banked through which the oil circulates by natural convection.

More efficient convection cooling methods include the use of external cooling tanks fitted to the outside of the transformer. Size limitations imposed by road and rail transportation limit the maximum heat dissipation by convection means only to about 5000 kW.

Needless to say, the manufacturer's recommendations on siting must be strictly adhered to; these usually require the transformer to be free-standing and away from buildings to ensure an uninterrupted air-flow. Even with rigid construction some sound from lamination vibrations at 50 Hz will be inevitable and this may prove objectionable close to buildings.

It should be borne in mind that the 'negligible' leakage flux may be very considerable (for a transformer carrying megawatts of power) and severe magnetic pick-up may occur in sensitive electronic equipment in the vicinity; careful magnetic screening of this equipment may be needed.

When a transformer of medium or large size has to be enclosed within a building, the use of pumped oil-cooling to external radiators must be considered. If the complete installation must be enclosed, water-cooling coils may be placed within the oil tank. A water-pipe leak in such circumstances would prove disastrous since the transformer oil's electrical insulating properties would be drastically reduced, so this method is reserved for extreme conditions. Even in air-cooled transformers care must be taken to ensure that no contamination is caused by moisture absorption from the air. A conservator drum is mounted on the top of the oil tank and this allows for expansion of the oil on heating. The air above it is kept dry by silica-gel crystals in the breather pipe. A drain cock is fitted so that any contaminated oil which, being heavier than dry oil, sinks to the conservator floor, may be run off at intervals. This is usually the only routine servicing needed and since the efficiency of transformers used for power-supply purposes is usually above 0.98 they are a very useful and trouble-free item of plant.

4.2.3 Summary of Transformer Uses

(1) By far the most common is conversion of power system voltages between 11 kV or 33 kV at which most turbo-alternators generate to the 132 kV or 400 kV at which the National Grid operates, and subsequent reduction in voltage to 11 kV or 6.6 kV for area distribution, thence to the 415 V three-phase distribution employed for small industrial and domestic (240 V single-phase) use, (see section 4.3).

(2) The transformation of voltages at low power levels within such items of equipment as the power supplies for electronic equipment, battery-chargers, converters and inverters (see section 5.5).

(3) For impedance-matching within a measurement system to ensure maximum power-transfer of the signal between sub-systems (figures 6.13 and 6.14).

(4) For electrical isolation between circuits. Energy is transferred between the two windings by a magnetic field, and the absence of direct electrical connection can be used for safety purposes. In figure 4.4a a possibly fatal shock might be sustained by contact with the live side of the mains supply, but since the right-hand side of the circuit in figure 4.4b is 'floating' or isolated from earth, a shock would not be sustained unless contact were to be made with conductors b and c simultaneously.

Another application of the isolation property is the removal of d.c. components from complex waveforms. Since only flux *changes* induce secondary voltages, any d.c. component contained in an electrical signal will not appear in the secondary circuit.

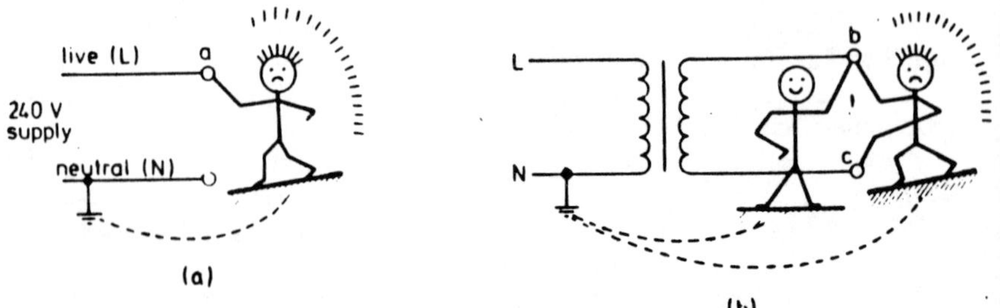

FIGURE 4.4

Showing how, without isolation, it is possible to die 'single-handed'

4.2.4 The Auto-transformer

Where a.c. voltage transformation without electrical isolation between the primary and secondary is sufficient, a simpler arrangement known as the auto-transformer may be employed because of its lower cost. This has only one winding as shown in figure 4.5a with a tapping at some point B. The total winding forms the primary while only section BC forms the secondary. Thus

$$\frac{V_2}{V_1} = \frac{N_{BC}}{N_{AC}}$$

Variable-ratio auto-transformers are available in which the tapping point is a movable carbon brush which enables the output voltage to be varied. One trade name under which these are marketed is the 'Variac'; they are commonly encountered in laboratories in current ratings from 2 to 30 A.

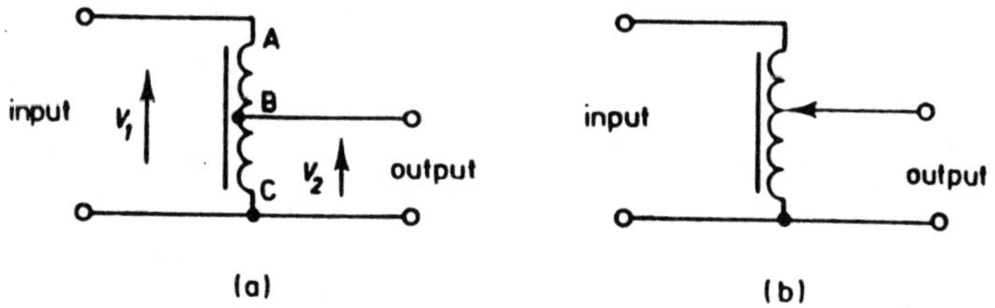

FIGURE 4.5

The auto-transformer (a) in its simplest form and (b) as a variable-ratio transformer

4.3 The Distribution of Electrical Energy

Since electrical power is proportional to the product of voltage and current, the transmission of high energy across country involves sophisticated techniques adapted to high current and voltage levels. Difficulties encountered with high current are the rising costs of conducting materials, especially copper. Aluminium is increasingly used since it is cheaper than copper for the cables of large cross-section required to minimise cable resistance and therefore distribution loss. The design of switches to break circuits carrying large currents to inductive loads is also difficult. The cost of energy transmission at high voltage is not so acute since the only problems that arise are those of insulation. If transmission towers (pylons) may be used, advantage can be taken of the fact that air is a moderately good insulator and voltages up to 400 kV and possibly to 750 and 1000 kV may be transmitted across rural areas without the interconductor spacing becoming unmanageable. If, however, the presence of National Parks or other aesthetic considerations precludes the use of open transmission towers, underground cables must be employed which use sophisticated insulation techniques and cost approximately ten times as much as open-wire systems.

At high voltages the cost of conductors for a given power transfer is of paramount importance and it can be shown[10] that the amount of copper required

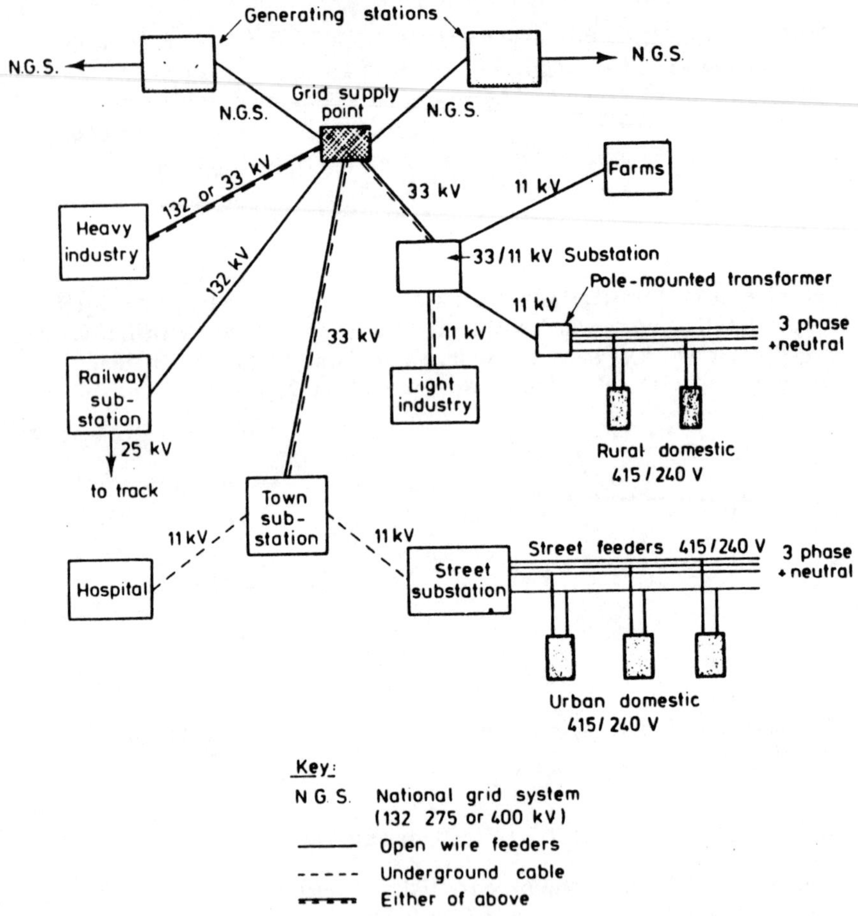

FIGURE 4.6

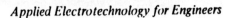

A greatly simplified diagram of the national electrical transmission and distribution system

for single-phase two-wire systems is 1.33 times greater than that for three-phase three-wire systems, other factors being equal. All distribution systems of higher voltage than the 240 V local system are therefore of the three-phase three-wire type. The single-phase 240 V system is itself obtained by adding a fourth neutral line for domestic purposes in order that the statutory limit of 250 V for lighting is not exceeded. Connection between one of the three-phase lines and neutral wire thus gives a line-neutral voltage of $415/\sqrt{3} = 240$ V (see section 2.6). Connections to individual domestic premises are arranged so that the load presented to the local 11 kV/415 V substation secondary circuit is as nearly balanced between the phases as possible, thus minimising the neutral current.

Potentials of 400 kV to 1 MV are clearly unsuitable for any consumers because of safety considerations and insulation problems, hence these potentials together with the 132 kV system are reserved for interconnection of generating stations via the National Grid. Supplies to heavy industry (steel, shipbuilding, etc.) are made at 33 kV or sometimes at 132 kV. A Grid supply point is provided on the consumers' premises to lower the voltage to 11 kV and 415 V at transformer substations. Supplies to light industry, farms and remote rural domestic consumers at 11 kV are provided from a 33 kV/11 kV substation fed from the output of a Grid supply-point. Energy at 11 kV is not allowed to enter domestic premises for safety reasons; it is transformed down to 415/240 V, usually by a pole-mounted transformer at the last transmission pole. Supplies to urban and suburban areas are usually via underground 11 kV cables. Hospitals requiring high power are directly supplied at 11 kV; street transformer-substations feed the 415 V three-phase street distributor from which individual houses receive 240 V single-phase spurs. Electric railways are separately supplied at 25 kV produced at special railway substations fed directly from the National Grid.

4.4 Principles of Electro-mechanical Energy Conversion

An electro-mechanical transducer is a device for converting electrical energy into mechanical energy or vice versa. Transfer between the two forms takes place via electric and magnetic fields as follows

Electrical energy	–	Energy storage in electric and magnetic fields	–	Mechanical energy

→ Motors

Generators ←

In transducers for power systems, as opposed to measurement devices, the electric field is negligible and the conversion takes place via the magnetic field.

If a mechanical member is moved by a force produced from an electrical source, mechanical work is done and the energy for this work is obtained from the magnetic field. A reverse process occurs during electrical generation, although it should be noted that a magnetic field must exist before an e.m.f. may be induced causing some transducers (for example, the induction motor) to be incapable of sustaining a reverse energy flow.

As in all systems losses occur which are dissipated to the surroundings as heat, yielding the following balance equation in terms of energy

$$\text{Net mechanical output} = \text{Electrical input} - \text{Electrical and mechanical losses} - \text{Energy stored}$$

$$\underset{(W_\mathrm{m})}{\text{Output}} = \underset{(W_\mathrm{e})}{\text{Input}} - \underset{(W_\mathrm{l})}{\text{Losses}} - \underset{(W_\mathrm{s})}{\text{Stored}}$$

A serious problem is the inability of present methods to store large quantities of energy in electrical form. A similarity in construction of electrical motors and generators mentioned above allows indirect methods of storage. An example is a pumped-water storage scheme where water is pumped to a high level reservoir during periods when surplus energy is available within the National Grid. During peak energy-demand periods this water is used to power, as turbines, the pumps previously used to raise it. The electrical motors attached to the pump are then employed in their generator mode to feed energy back into the National Grid.

4.5 Alternating-current Machines

It has been shown in section 2.6 that a three-phase supply may be generated by a rotating magnetic field. Conversely, a polyphase supply may be reconstituted into a rotating magnetic field by a similar geometric arrangement of windings. Let the three-phase current waveform of figure 4.7b be applied to the three star-connected coils shown in figure 4.7a.

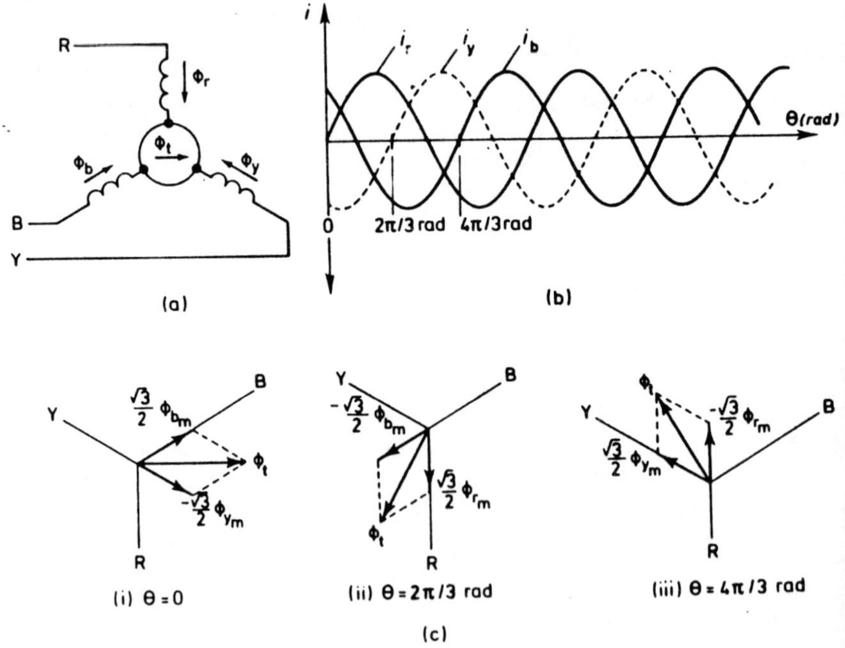

FIGURE 4.7

(a) The winding arrangement, (b) the current waveforms and (c) the flux directions encountered in one method of producing a rotating flux

Assuming that the flux magnitude produced in each of the three windings is proportional to the instantaneous value of the current in that winding, the magnitude of each component flux ϕ_r, ϕ_y and ϕ_b, together with the magnitude and direction of their resultant ϕ_t is shown at angle 0, $2\pi/3$ and $4\pi/3$ rad in figure 4.7c (i), (ii) and (iii) respectively. Note that the phase sequence is red, yellow, blue. This represents successive shifts of $2\pi/3$ radians anticlockwise

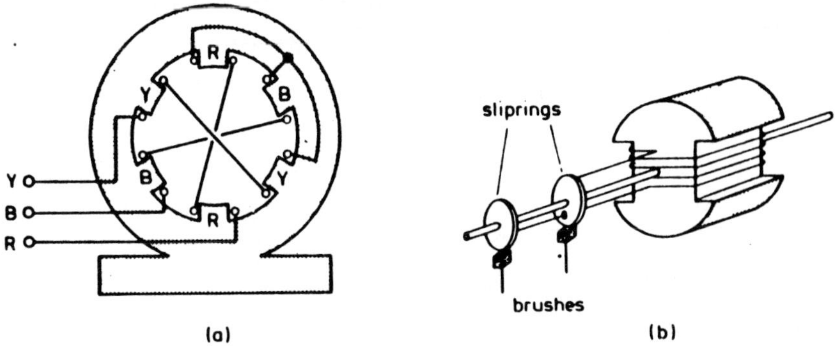

FIGURE 4.8

(a) A possible star-connection of a two-pole three-phase stator and
(b) a primitive two-pole rotor for a synchronous machine

(by convention) in the phasor diagram. The resultant flux has rotated by an equal angle clockwise. If connections between any *two* of the supply lines and the coils are interchanged, the flux can be shown to reverse its direction of rotation. Figure 4.7a shows only a theoretical winding arrangement; the actual configuration of a three-phase, two-pole machine is depicted in figure 4.8a. In the above example the flux rotates at the supply frequency, *the synchronous speed*. If, however, the machine were a four-pole design the flux would now only rotate at half the previous speed (figure 4.9 b), since in following one electrical cycle it would only have rotated through 180°. Thus if f is the supply frequency, p the number of pole pairs and n_s the synchronous speed of flux rotation

$$n_s = \frac{f}{p} \text{ rev/s}$$

The magnitude of the rotating flux in figure 4.7 is constant and may be shown by simple geometry to be 1.5 times the maximum value of any individual phase flux.

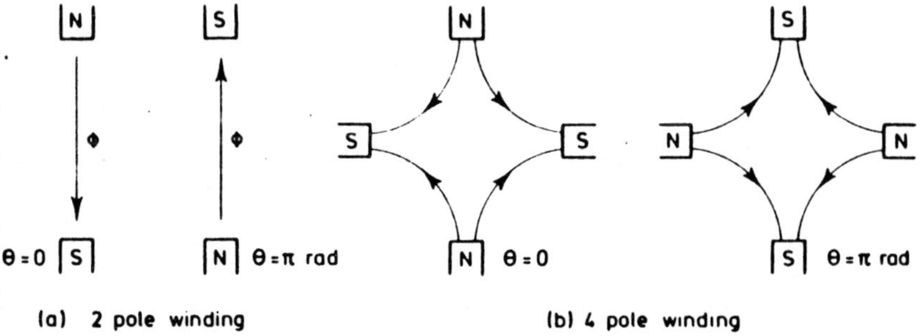

FIGURE 4.9

Showing that in half an electrical cycle (π radians) the flux pattern of a four-pole machine has only rotated through half the angle of that in a two-pole machine

Example 4.3

Calculate the speed of rotation of flux in a three-phase 50 Hz four-pole motor driving an extractor fan. What is the simplest method of converting it to a fresh-air intake fan?

$$\text{Speed of flux rotation } n = \frac{\text{supply frequency } f}{\text{no. of pole pairs } p}$$

$$= \frac{50}{2} = 25 \text{ rev/s}$$

or

$$N = 25 \times 60 = 1500 \text{ rev/min}$$

To reverse the direction of rotor rotation, the flux rotation must be reversed. This is easily achieved by interchanging any two of the three supply-leads to the motor.

4.5.1 The Synchronous Machine

It was shown in section 2.6.1 that a primitive generator using the rotor of figure 4.8b could produce a sinusoidal e.m.f. of the same frequency as its rotational speed. This is a very elementary form of synchronous machine used as a generator. If a compass needle were placed within the rotating field (figure 4.7c) produced by a polyphase supply, it would (frictional forces being minimal) rotate with the field. If the magnet were to be mounted on a shaft, this rotation could be utilised to do external work, the magnet lagging with increasing mechanical load-torque until the *pull-out* torque is reached when synchronous speed cannot be maintained and motor action ceases. The field-coil arrangement is called the *stator* in electrical machines and the rotating bar-magnet would form the *rotor* in the above example.

In practice, a permanent magnet is not used since the flux densities obtainable are too low and the heating and vibration would soon result in a loss of remanent magnetism. An electromagnet fed with a d.c. excitation current via slip-rings as in figure 4.8b is employed

The need for a separate d.c. supply and the fixed speed of operation severely limit the application of this type of *motor*, though of course all a.c. *generators* fall into the synchronous-machine category. Synchronous speed for the standard 50 Hz mains frequency is $n = f/p = 50/1$ rev/s or $50 \times 60 = 3000$ rev/min for the simplest two-pole machine, falling to 1500, 1000, 750 rev/min, etc., for four-, six- and eight-pole machines respectively. This does allow some *discontinuous* motor-speed control because the windings of a multipole machine may be externally switched to effect a variation in the number of poles employed, albeit with a loss in efficiency.

The chief use of synchronous motors is to effect power-factor improvement since if the excitation current is made large for a given mechanical load it will draw a leading current from the supply to offset the lagging reactive current of other plant. In this application a synchronous machine is often referred to as a *synchronous capacitor*. This may appear an absurd term with which to describe the machine; however, it derives from another method of power-factor improvement, namely the parallel connection of a capacitor to the load (see section 2.3.3).

4.5.2 The Three-phase Induction Motor

The most widely used a.c. motor, because of its inherent simplicity, is the induction motor, shown in its primitive form in figure 4.10. A closed coil of wire rotating about an axis parallel to its sides AB and CD is inserted into a rotating magnetic field produced by three-phase or other means. The magnetic field moves with respect to the rotor conductors which are initially stationary causing the moving flux to cut the stationary conductors AB and CD. An e.m.f. will be induced in the rotor conductors which, according to Lenz's law, will cause rotor currents to flow giving rise to a rotor flux opposing the stator flux (action and reaction).

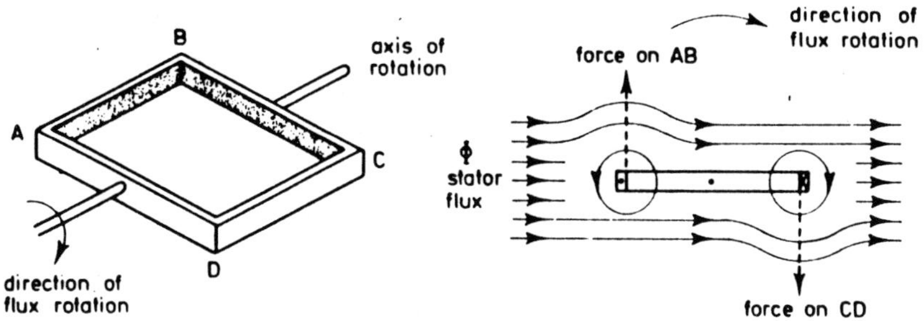

FIGURE 4.10

A primitive rotor for an induction motor and the forces acting upon it

Application of Lenz's law and the screw rule to figure 4.10 shows that the coil will experience a force in a direction tending to accelerate it in the direction of rotation of the flux. When the rotor is initially stationary, the machine is essentially a transformer with a short-circuited secondary. The low resistance of the rotor conductors permits large rotor currents, causing large stator currents to be drawn in order to maintain the field-flux magnitude constant (see section 4.2.1, p. 87). As the rotor accelerates there is progressively less relative angular velocity between rotor and stator, producing reduced rotor currents and hence reduced stator currents. Consider the situation that would arise if it were possible for the rotor to accelerate until it revolved at the synchronous speed. There would be no relative motion between field and rotor therefore no induced rotor e.m.f.s or currents and hence no rotor torque. The rotor would thus decelerate because of frictional and other losses until the relative motion of rotor and field were such that the magnetic torque produced balanced the mechanical torque required to overcome the losses. The machine would thus rotate at a speed slightly less than synchronous speed (contrast the synchronous machine). The induction motor relies for its action on this relative motion between *rotor* and *field flux*. This difference between the rotor and field-flux speeds is called the *slip speed*.

Slip is usually expressed in percentage or per unit values as a proportion of the synchronous speed. Thus

$$s \text{ (per unit slip)} = \frac{\text{synchronous speed} - \text{rotor speed}}{\text{synchronous speed}}$$

$$s = \frac{n_s - n_r}{n_s}$$

At standstill the slip is thus unity and at synchronous speed it is zero. Thus the rotor speed

$$n_r = n_s(1 - s)$$

and the relative speed of rotation between the flux and the moving rotor is

$$n_s - n_r = sn_s$$

The synchronous speed is of course the speed of flux rotation. From section 4.5 this is

$$n_s = \frac{\text{supply frequency}}{\text{stator pole-pairs}} = \frac{f}{p}$$

In practice many conductors are used, laid in slots running axially along the surface of a laminated rotor-drum. The purpose of the drum is to provide a low permeability path for the magnetic field, giving maximum field strength for a given magnetomotive force. The conductors have their extremities connected together at each end of the rotor. Since the shape of the conductors and their shorting rings resembles a cage, the term *squirrel cage* is used to describe this form of induction motor (figure 4.11)

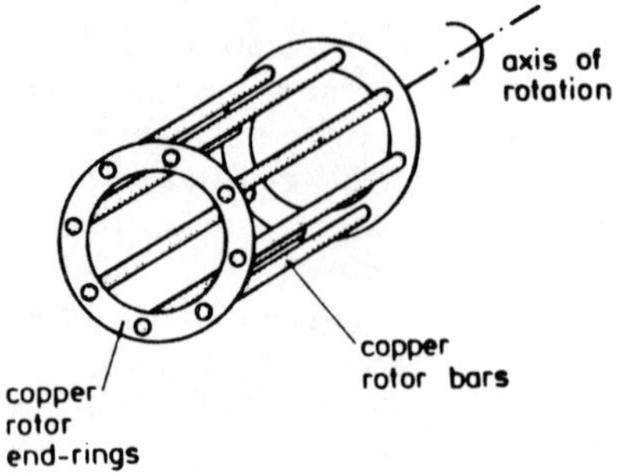

FIGURE 4.11

The arrangement of the conductors in a squirrel-cage rotor; the rotor drum is omitted for clarity

The Rotor E.M.F.

At standstill the stator and the rotor windings may be considered as a simple transformer, the rotor e.m.f. V_0 will thus be sinusoidal at the supply frequency. The rotor current will have a similar form but will be limited in magnitude by the rotor impedance Z_r. This impedance is formed by the rotor winding self-resistance R_r and the standstill rotor reactance X_0.

Reactance is present because the rotor current produces an m.m.f. and flux associated with the rotor itself, as distinct from the flux produced by the stator. Thus the standstill rotor current I_{r0} is given by

$$\frac{V_0}{Z_r} = \frac{V_0}{\sqrt{(R_r^2 + X_0^2)}}$$

As the rotor accelerates the relative speed of rotation between the rotor and the stator flux decreases from n_s at standstill to sn_s. This causes two effects: (i) the induced rotor e.m.f. decreases from V_0 to a value $V_r = sV_0$ since induced e.m.f. is proportional to the rate of flux cutting, and (ii) the frequency f_r of the rotor e.m.f. is similarly reduced to sf where f is the stator supply frequency. The second of these effects lowers the rotor reactance (previously $X_0 = 2\pi fL$) to $X_r = 2\pi sfL$, thus rotor reactance $X_r = sX_0$. The rotor resistance is substantially constant depending as it does solely upon the conductor sizes and materials and temperature.

Once the rotor revolves, the above expression for rotor current is thus modified to

$$I_r = \frac{V_r}{\sqrt{(R_r^2 + X_r^2)}} = \frac{sV_0}{\sqrt{[R_r^2 + (sX_0)^2]}}$$

The Torque Characteristics

These may be derived qualitatively by setting up an energy-balanced equation between the gross rotor power, the mechanical load on the rotor and the rotor losses, neglecting friction and windage losses.

$$\text{Mechanical output} = \text{Gross rotor-input from field} - \text{Rotor losses}$$

$$\omega_r T = \omega_s T - \text{Rotor losses}$$

$$T(\omega_s - \omega_r) = \text{Rotor losses}$$

where ω_s and ω_r are the *angular* synchronous and rotor speeds (rad/s)

Therefore

$$T \times 2\pi(n_s - n_r) = V_r \times I_r \times \cos\phi_r$$

$$T \times 2\pi \times s \times n_s = V_r \times \frac{V_r}{Z_r} \times \cos\phi_r$$

$$= \frac{V_r^2}{Z_r} \times \frac{R_r}{Z_r} = \frac{V_r^2 \times R_r}{(R_r^2 + X_r^2)}$$

$$= \frac{s^2 V_0^2 \times R_r}{R_r^2 + (sX_0)^2}$$

Therefore torque is proportional to

$$\frac{sV_0^2 \times R_r}{R_r^2 + (sX_0)^2}$$

after division throughout by s and noting that n_s is constant for any one machine. The shape of the speed – torque curve may be obtained from the above expression by inspection. Typically $X_0 > R_r$ so that $(sX_0)^2 >> R_r^2$ and at low speeds therefore torque $\propto 1/s$. At high speeds where s is low, $R_r^2 >> (sX_0)^2$ therefore torque $\propto s$. At some intermediate speed we can show by differentiating with respect to s that the torque is a maximum when $sX_0 = R_r$. This is left to be verified by the reader. Figure 4.12 (where $R_r = X_0/8$), is a typical torque curve for an induction machine.

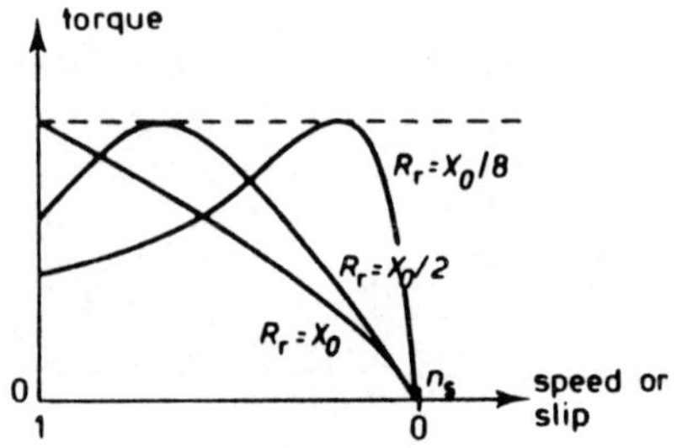

FIGURE 4.12

Induction motor-torque – speed curves for various value of rotor resistance

Starting Arrangements

Even though, in a particular application, the low starting torque may be acceptable, the high starting currents taken from the supply (typically three to six times the full-load current) may produce unacceptable voltage drops in supply cables when motors are started by direct connection to the supply ('direct-on'). For motors whose rating exceeds about 1 kW it is customary to employ some device for limiting the motor starting current.

For any particular value of s (say standstill) T can be increased by artificially increasing R_r. This may be appreciated by looking again at the torque expression given above, this time treating s as a constant and differentiating with respect to R_r. Again it can be shown that maximum starting torque may be obtained by making $R_r = X_0$. This can be achieved either by inserting an external resistor in series with the rotor-current path via slip-rings or by using a motor with two rotor windings, one of high resistance that provides the starting torque and the main low-resistance winding providing the major torque at operating speeds. The external resistance has to be decreased to zero after starting since the high rotor-circuit power losses produced will impair the over-all machine efficiency; a diagram of a typical arrangement is given in figure 4.13 while figure 4.12 shows the modification of the torque – speed curves which ensues. The disadvantage of the slip-ring induction motor is that some of the inherent simplicity of construction is lost and therefore both cost and maintenance are increased.

Alternative methods of limiting the starting current initially apply less than the full working voltage to the stator windings. This may be done either by an auto-transformer to reduce the supply voltage or by making use of the fact that two voltages are obtainable when the stator windings are connected in star or delta.

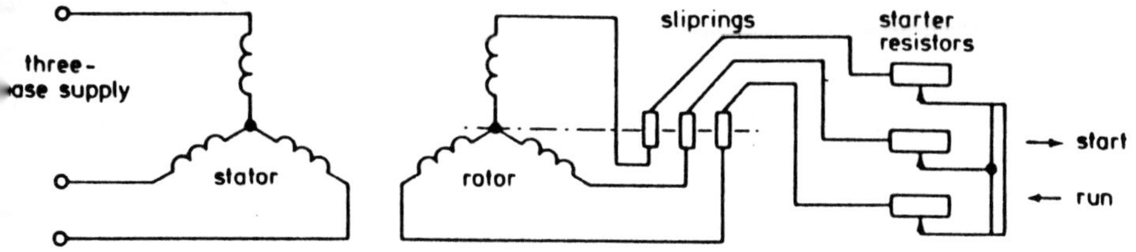

FIGURE 4.13

External resistors inserted in the rotor circuit of a wound-rotor induction motor to improve starting torque

In the star - delta starter (figure 4.14) both ends of each of the stator windings must be brought out of the machine on terminals to allow the machine to be started in star connection where each winding will receive only $1/\sqrt{3}$ of the supply voltage on initial connection. This limits the starting current per phase to $1/\sqrt{3}$ of its value when switched direct-on or the line current to $1/3$ of its former value. Within a few seconds, when the speed is steady, the starter switch is quickly moved to the running (delta) connection where each winding now receives its full rated voltage.

Because full voltage is not applied until the machine is rotating, the starting current is limited to a safe value.

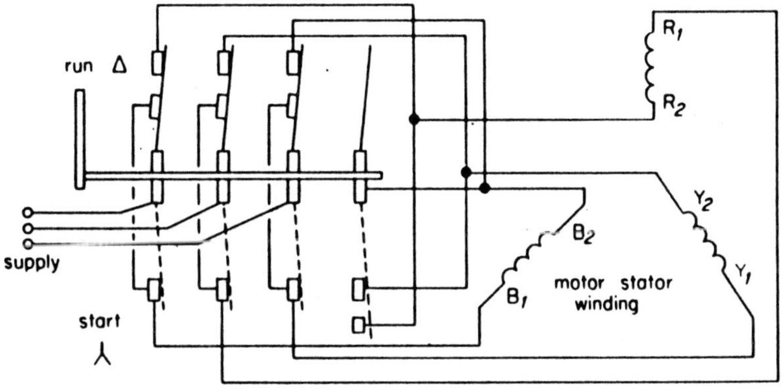

FIGURE 4.14

The star - delta starter

Example 4.4

All the stator flux in a star-connected, three-phase, two-pole, slip-ring induction motor may be assumed to link with the rotor windings. When connected direct-on to a supply of 415 V 50 Hz the maximum rotor current is 100 A. The stand-still values of rotor reactance and resistance are 1.2 Ω/phase and 0.5 Ω/phase respectively. Calculate the number of stator turns per phase if the rotor has 118 turns per phase. At what motor speed will maximum torque occur?

Rotor e.m.f./phase = Rotor current/phase × impedance/phase

Standstill impedance/phase $= \sqrt{(R_r^2 + X_0^2)}$

$= \sqrt{(0.5^2 + 1.2^2)}$

$\doteq 1.3\ \Omega$

Rotor e.m.f./phase $= 100 \times 1.3 = 130$ V

Three-phase supply voltage = 415 V (line to line)

Therefore supply voltage/phase $= 415/\sqrt{3} = 240$ V

Ratio of stator turns/phase to rotor turns/phase = ratio of stator voltage to rotor voltage = 240 : 130

therefore

$$\text{stator turns} = \frac{240}{130} \times 118 = 218 \text{ (to nearest integer)}$$

Maximum torque will occur when rotor reactance equals rotor resistance that is, when $X_r = R_r$ or $sX_0 = R_r$, thus at a slip s

$$s = \frac{R_r}{X_0} = \frac{0.5}{1.2} = 0.417$$

Synchronous speed for a two-pole 50 Hz machine is

$$n_s = \frac{f}{p} = \frac{50}{1} \text{ rev/s}$$

Therefore

$$\text{slip speed} = sn_s = 50 \times 0.417 = 20.8 \text{ rev/s}$$

$$= 20.8 \times 60 = 1250 \text{ rev/min}$$

$$\text{Rotor speed} = \text{synchronous speed} - \text{slip speed}$$

$$= 3000 - 1250 = 1750 \text{ rev/min}$$

4.5.3 The Capacitor-start and Run (Split-phase) Motor

This machine operates in a similar manner to the three-phase squirrel-cage motor. However, the rotating field is produced by two coils instead of three, which are placed at right-angles to each other. Figures 4.15a and b show that if the currents through the two coils are displaced in phase by 90° with respect to each other a rotating field is produced. This phase shift is obtained by a capacitor in series with one winding as shown in figure 4.15c. If the capacitor's reactance is high compared with the impedance of the coil in series with it, the current in this coil will be displaced by about 90° with respect to the other

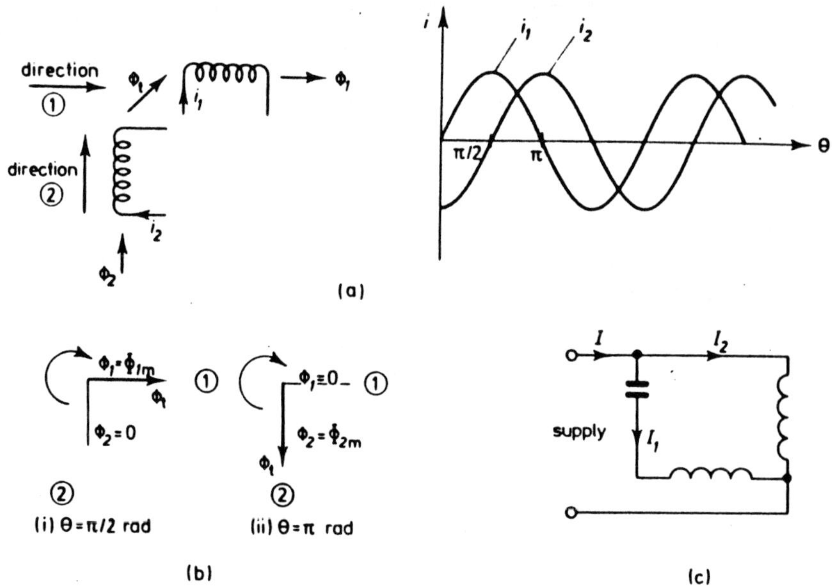

FIGURE 4.15

The split-phase rotor; (a) coil arrangement and current relationships, (b) resultant flux directions and (c) circuit diagram

coil current. This enables both windings to be supplied from one single-phase voltage source.

Because of the poor efficiency of these motors compared to three-phase machines they rarely exceed 4 kW output and hence the expense of a wound rotor is not justified; a squirrel-cage rotor is thus employed. These motors are commonly used in domestic washing-machines.

4.5.4 The Single-phase Induction Motor

This is the simplest possible type of a.c. machine consisting of a single-phase winding together with a cage rotor. It is used in profusion for small-power constant-speed applications such as chart motors in instruments, tape recorders and record players. The upper power limit is approximately 800 W since the torque applied to the rotor is not constant as in three-phase motors, but is impulsive throughout the period of one revolution. The field produced by the stator does not rotate but is fixed in space, only its magnitude varies throughout the cycle sinusoidally.

The Two Rotating Fluxes Theory shows (see figure 4.16) that although there is one stationary flux, it may be considered as being composed of two constant-magnitude rotating fluxes. These fluxes are considered to be superimposed and to rotate at the same speed in opposite directions. Hence, after a

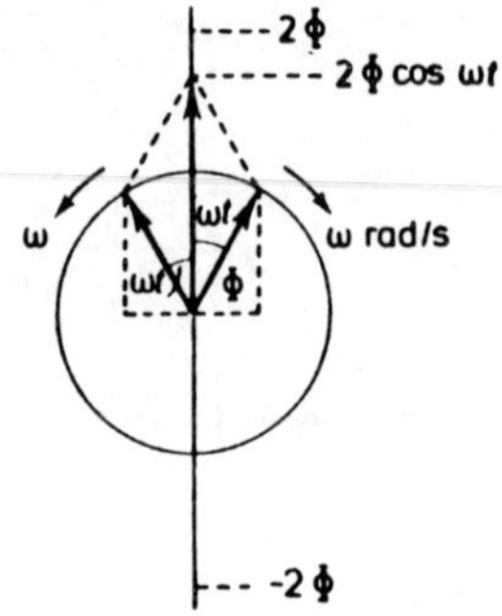

FIGURE 4.16

Two-rotating-flux theory

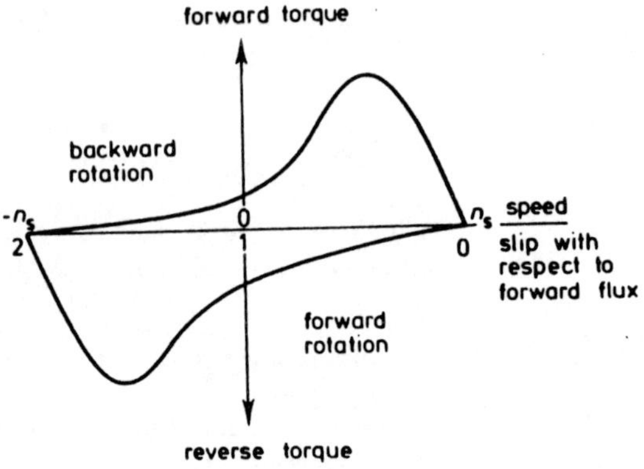

FIGURE 4.17

Torque - speed curves for a single-phase induction motor

time t, the angle described by each will be ωt as shown. The horizontal components of each flux will cancel one another at all values of ωt, but the vertical components will vary between 2Φ at $\omega t = 0$, through zero at $\omega t = \pi/2$ and thence to a minimum of -2Φ at $\omega t = \pi$. Thus their resultant is a stationary flux of maximum value 2Φ and sinusoidal in form. The converse, therefore, is that any sinusoidal stationary flux may be resolved into two counter-rotating fluxes and it is this concept that helps to explain the physical behaviour of a single-phase induction motor. Figure 4.17 shows a torque - slip diagram for each of the fluxes. In the case of the backward-rotating flux the diagram is reversed and inverted since positive slip to the forward flux is negative slip to the backward flux. At standstill there is no starting torque, but if the rotor is externally rotated in either direction the asymmetry of the diagram shows an accelerating torque in that direction. Thus the motor is bidirectional, which may not always be an advantage!

In-built Starting Arrangements

Because there is no starting torque and since the direction of rotation depends on the initial conditions, some form of starting arrangement is essential. All the methods make use of flux phase-shift to simulate a two-phase field (as used in the split-phase motor of section 4.5.3) for starting purposes.

All but the smallest machines employ an auxiliary winding at right-angles to the main winding, the current through it being phase-displaced with respect to the main winding current by one of two methods. Figure 4.18a shows an auxiliary winding of high-resistance material giving a current I_{aux} which lags the supply voltage less than the highly reactive starting current through the main winding I_m. A lower total current I and better power-factor at starting may be obtained by supplying the auxiliary winding via a capacitor as in figure 4.18b.

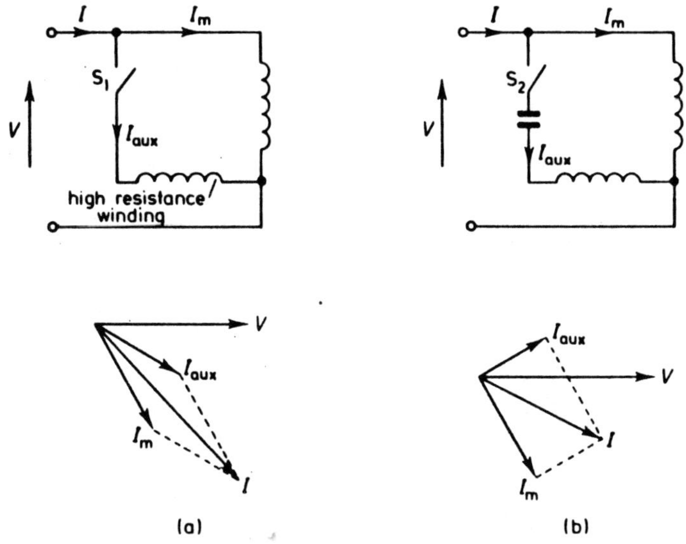

FIGURE 4.18

Two methods of providing phase-shift for the starter winding
S_1 and S_2 are both closed when the rotor is stationary

The capacitor and auxiliary windings in both these methods need only be short-term rated since a centrifugal switch can be arranged to disconnect the starting circuit once a pre-determined speed has been obtained (contrast with the split-phase motor of section 4.5.3).

Low-power motors rely on field asymmetry induced by 'shading' the poles as shown in figure 4.19. The copper bands act as short-circuited turns, in which e.m.f.s are induced which phase-shift the flux Φ_s within the shaded region compared with the unshaded flux Φ. These two phase-displaced fluxes can be seen to be at an angle to each other and will therefore produce a starting torque as explained previously.

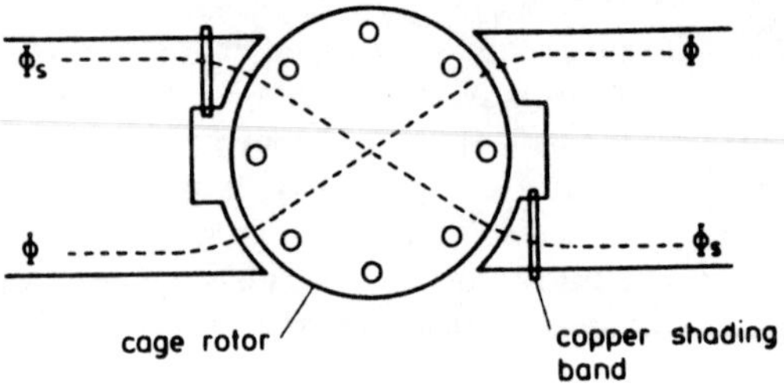

FIGURE 4.19

The shaded-pole method of starting low-power single-phase motors

4.6 Direct-current machines

Sections 4.5.1 to 4.5.4 show that the speed control of the simpler a.c. machines considered is very limited, being confined to speeds not far removed from the synchronous speed. In the past, a.c. commutator machines have been evolved in which speed control over wide ranges was possible (Schrage motors, etc.) but their complexity, together with the advent of cheap efficient semiconductor rectifiers, has rendered them less common. Advantage is now taken of the wide range of speed control inherently available in d.c. machines; the d.c. supply often being obtained locally by rectification.

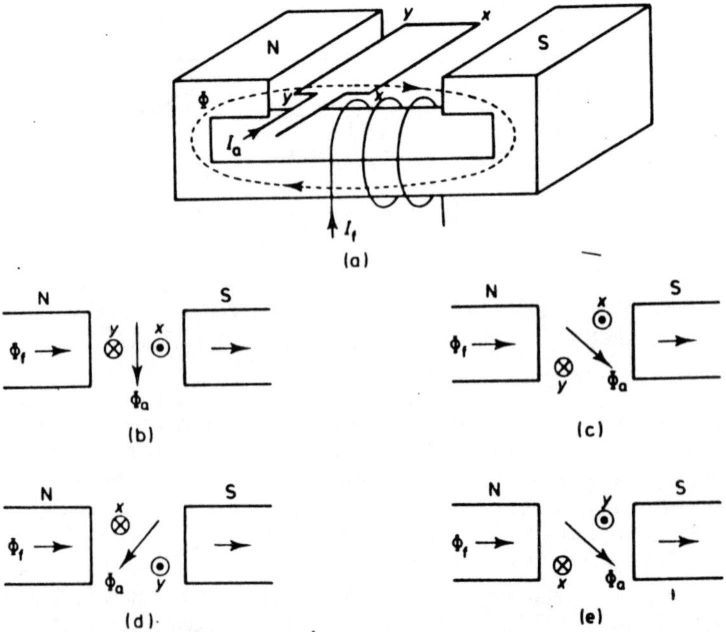

FIGURE 4.20

Illustrating the principle of commutation in a primitive d.c. machine

4.6.1 Principles – The Primitive D.C. Machine

Figure 4.20a shows a rotor consisting of a rectangular turn of wire, fed with a current I_a, in the airgap of an electromagnet. The rotor is usually mounted on a steel core, but this has been omitted for clarity.

If the rotor and field winding are energised by I_a and I_f respectively, the flux distributions so produced will be as shown in figure 4.20b. If the rotor is free to revolve it will turn anticlockwise until the rotor flux Φ_a is almost aligned with the field flux as shown in figure 4.20c. If the rotor current is now reversed in direction, mechanical inertia will carry the rotor beyond the position of figure 4.20c until the configuration of figure 4.20d is attained which clearly shows that the field directions are now such as to promote rotation in the same direction until the current must again be reversed when the conductors reach the position shown in figure 4.20e. The essential process of rotor-current reversal is known as *commutation* and provides current reversals at intervals of 180° of mechanical rotation in this example. A d.c. machine rotor and its commutator are together referred to as an *armature* and this term will be used hereafter. The simplest form of commutator (two segments) is shown in figure 4.21a in which it will be seen that, although the polarity of the supply and therefore of the brushes remains unchanged, the rotor winding current is reversed every 180° as alternative segments pass beneath each brush.

Practical D.C. Machine

A view of a practical commutator is shown in figure 4.21b where the copper commutator bars are insulated from the shaft by a fibre bush and from each other by intervening mica segments. There may be up to several hundred segments, depending on the machine size and use. Maintenance is confined to renewal of the brushes and possibly the brush springs and, provided this is done conscientiously, little wear of the segment bars is experienced. If the brushes receive inadequate attention the copper bars may become scored, when

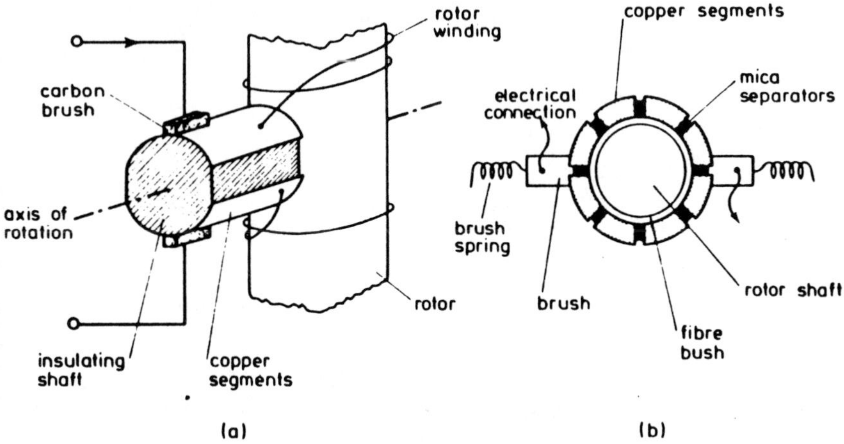

FIGURE 4.21

(a) A primitive and (b) a practical commutator

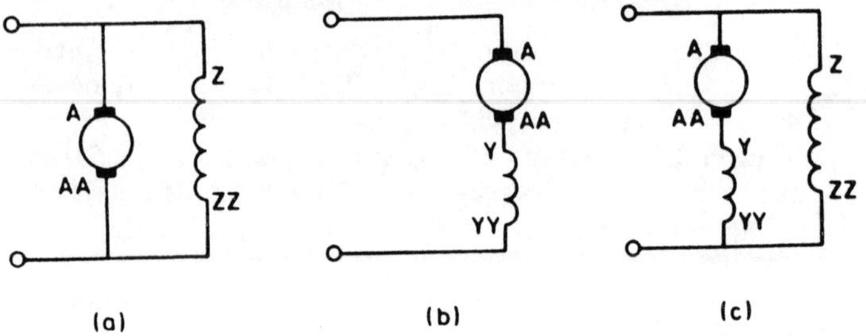

FIGURE 4.22

The basic interconnections between the field coil(s) and armature for (a) a shunt, (b) a series and (c) a compound-wound machine.

the armature must be removed for resurfacing in a lathe. The mica separators must then be slightly undercut to allow the bars to stand a little proud of the mica.

The design of armature windings is too complex to be included here and so general relationships for torque and speed will be developed applicable to each of the main three motor connections known as 'shunt', 'series' and 'compound' connection (figure 4.22).

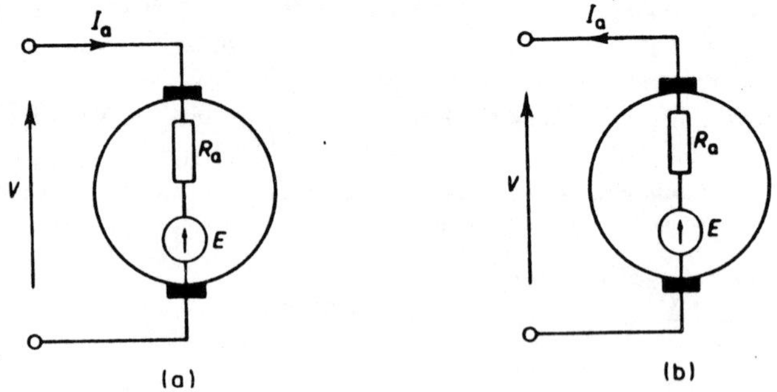

FIGURE 4.23

The armature's equivalent circuit when rotating. (a) Motor mode, (b) generator mode, R_a is the winding's self-resistance and E its back e.m.f.

4.6.2 Relationship Between Armature Voltage, Current and Speed

Consider an armature within a constant magnetic field, fed with a supply voltage V. When used as a motor (figure 4.23a), after transient conditions during starting, the armature would eventually attain a steady speed (n rev/s) and the electrical quantities would be related by the expression

$$V - E = I_a R_a$$

E is the back e.m.f. (which by Lenz's law must oppose the applied voltage V)

induced by the rotating armature conductors cutting the field flux, I_a is the armature current and R_a the armature resistance. When operating as a d.c. generator (dynamo), figure 4.23b shows that the above equation becomes

$$V = E - I_a R_a$$

The magnitude of E can be deduced from figure 4.20a. For each revolution of the coil the total flux cut by each conductor (xx or yy) will be $2p\Phi$ where Φ is the flux per pole and p the number of *pole pairs*. The average rate of flux cutting will therefore be $2p\Phi n$ since $1/n$ is the time in seconds for one revolution.

The total armature back e.m.f. will therefore be

$$E = 2p\Phi n Z_s \tag{4.1}$$

where Z_s is the effective number of armature conductors in series between the brushes, or

$$E = kn\Phi$$

where $k = 2pZ_s$ which will be a constant for any given machine. Substitution for E in the original armature equation gives

$$V - kn\Phi = I_a R_a$$

or

$$n = \frac{V - I_a R_a}{k\Phi}$$

The value of R_a is designed to be very low to minimise heating losses and the magnitude of $I_a R_a$ is typically only 0.05 of V.

An approximate expression for the speed may be obtained by ignoring the second term in the numerator, hence

$$n \approx \frac{V}{k\Phi}$$

4.6.3 D.C. Machine Torque Equation

Using the fundamental equation of the last section and multiplying through by I_a yields

$$V = E + I_a R_a$$

$$VI_a = EI_a + I_a^2 R_a$$

Two terms of this second equation may be readily identified by inspection

VI_a = total d.c. power supplied to the armature

$I_a^2 R_a$ = power dissipated in the armature as heat

The third term EI_a is thus revealed as the total mechanical power available. Losses will occur subsequently from friction at the bearings, windage (air friction) and from eddy-current and hysteresis losses in the armature. An approximate expression for the output torque T (in Nm) can be obtained by ignoring these losses in which case

$$EI_a = 2\pi nT$$

Substituting for E from equation 4.1 gives

$$2p\Phi nZ_s I_a = 2\pi nT$$

$$T = \frac{1}{\pi} Z_s p\Phi I_a$$

The important feature from the above expression is that for all d.c. machines the torque is proportional to the product of the armature current and the field flux. In the simple shunt machine (figure 4.22a) the flux is maintained constant for a particular speed-setting hence the torque is proportional to the armature current. In the *series* machine, however, it can be seen that the armature current also produces the field and hence the field is itself proportional to the current. Thus the torque $T \propto I_a \times I_a \propto I_a^2$. The characteristic parabolic form of this equation can be compared with the shunt characteristic in figure 4.24. The parabolic form can only extend until the magnetic core is saturated, in which region the flux is effectively constant and the curve reverts to the straight-line shunt form.

Compound-wound machines have their fields produced by two windings (figure 4.22c). The torque - armature - current characteristic is, as would be expected, intermediate between those of the singly-wound machines.

Example 4.5

A six-pole d.c. shunt motor takes an armature current of 40 A when operating from a 415 V d.c. supply. It has an effective flux per pole at this voltage of 0.025 Wb and the armature has 400 conductors effectively in series between

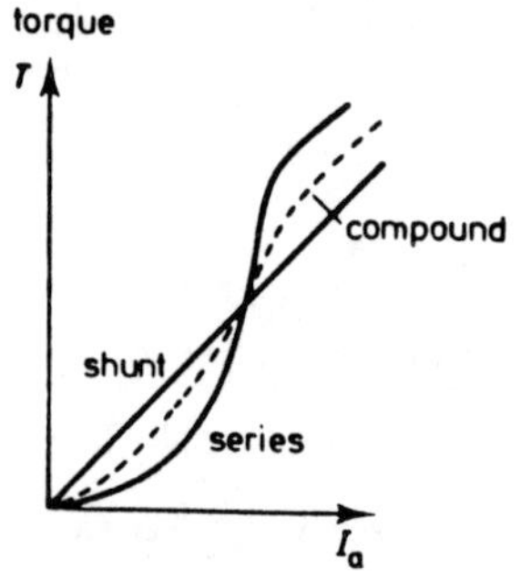

FIGURE 4.24

The torque - armature-current characteristics of d.c. motors the brushes. The total armature resistance is 0.25 Ω. Calculate (i) the speed and torque when running from 415 V and (ii) the approximate speed when connected to a 240 V supply (assuming the flux per pole to have fallen by 40 per cent).

(i) Supply voltage – Back e.m.f. = Armature drop

$$V - E = I_a R_a$$

$$\text{Back e.m.f.} = V - I_a R_a = 415 - (40 \times 0.25)$$

$$= 415 - 10 = 405 \text{ V}$$

Also

$$\text{back e.m.f.} = 2p\Phi_1 n_1 Z_s$$

Thus

$$n_1 = \frac{E}{2p\Phi_1 Z_s} = \frac{405}{2 \times 3 \times 0.025 \times 400}$$

$$= 6.75 \text{ rev/s}$$

or

$$N_1 = 6.75 \times 60$$

$$= 405 \text{ rev/min}$$

$$\text{Mechanical power at armature} = EI_a = 405 \times 40 = 16\,200 \text{ W}$$

Also

$$\text{armature power} = 2\pi n_1 T$$

Therefore

$$T = \frac{16\,200}{2\pi \times 6.75} = 382 \text{ Nm}$$

(ii) on 240 V, $n_2 = V/k\Phi_2$. Previously $k\Phi_1 = 415/6.75$, therefore

$$k\Phi_2 = k\Phi_1 \times \frac{\Phi_2}{\Phi_1} = \frac{0.015}{0.025} \times \frac{415}{6.75}$$

therefore

$$n_2 = \frac{240}{0.6 \times 415} \times 6.75 = 6.5 \text{ rev/s}$$

$$N_2 = 60n_2 = 390 \text{ rev/min}$$

4.6.4 Speed Characteristics of D.C. Machines

In section 4.6.2 it was shown that the speed of a d.c. machine is approximately proportional to the inverse of the flux and therefore in a similar relationship to the field current below saturation flux-density. The resultant speed – armature-current characteristics are shown for series and shunt machines in figure 4.25a.

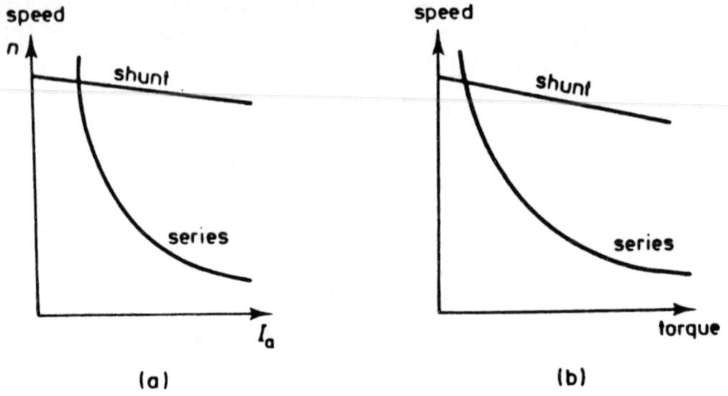

FIGURE 4.25

(a) The speed - armature-current characteristics of d.c. motors.
(b) The mechanical characteristics of d.c. motors

In the latter case the flux remains nearly constant and the decrease in speed is due solely to the increased armature voltage drop.

Perhaps the most important characteristic of all for a prime mover is the 'mechanical characteristic' or *speed - torque curve*. This can be inferred by eye from the torque - I_a and speed - I_a curves, by plotting together values of speed and torque for given armature currents giving figure 4.25b. It will be seen that the mechanical characteristic is similar to the speed - I_a curve for the shunt machine since the torque is approximately proportional to current. The series motor speed - torque is also similar to the speed - I_a curve after the magnetic field saturation effects described above have been reached.

A glance at figure 4.25 will immediately reveal that a series motor will run up to dangerous speeds on light loads and it is therefore essential that it cannot be disconnected accidentally from the load. Typical applications in which this could occur are the breaking of belt-drives, or in centrifugal-pump motors in which the interruption of fluid flow could allow the pump to become air-filled causing the motor to 'race'.

4.6.5 The Starting of D.C. Machines

On switching the supply directly on to a d.c. machine with stationary armature, there will be no back e.m.f. generated since flux cutting and hence rotation are needed for this. The back e.m.f. (normally about 0.95 of V) limits the armature current to a safe value once rotation occurs but the high initial armature-current if the motor were to be connected direct-on to the full supply voltage would damage brushgear, windings and possibly the supply arrangements. Consequently a series resistor is included to limit the current until a back e.m.f. has been established. This resistor is usually selected by a multi-position stud-switch allowing the resistance to be progressively removed (figure 4.26).

Once the starting procedure has been completed a safety arrangement is required to cater for temporary interruption of the supply. Such an event would

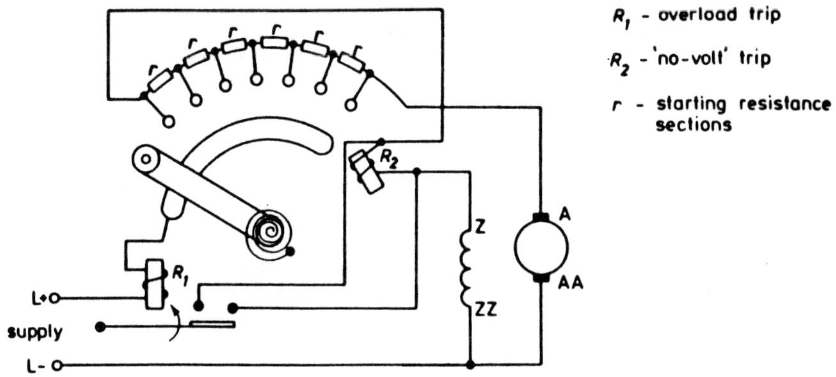

FIGURE 4.26

A d.c. motor starter with protections against overload and supply failure

bring the machine to standstill and the reinstatement of the supply would be an effective direct-on start on a stationary machine. Accordingly, an electromagnet trip is provided to hold the starter in the running position and to cause it to fail-safe back to the start position should the supply be switched off or accidentally fail. When the supply returns the complete starting procedure must be repeated before the full supply voltage can be applied to the armature.

The starting-resistor material will only be rated for short-duration running, so that, although the stud switch should only be advanced slowly, allowing adequate time for the speed to build up, it must move to the run position without too much hesitation which might cause the starter coils to overheat. Care must be taken that the *full field-current* is applied to shunt and compound machines at the instant of starting to give as high a starting torque and as slow a running speed as possible. In practice, the field coils are often fed from the starter to ensure this. A relay is incorporated in the starter to return it to the 'off' position if excessive armature current flows as a result of mechanical overload.

4.6.6 Speed Control of D.C. Machines

Section 4.6.2 showed that the speed of a d.c. machine is inversely proportional to the field strength and directly proportional to the applied armature voltage

$$n \approx \frac{V}{k\Phi}$$

Variation of either Φ or V can be used to control a d.c. motor's speed. Useful increases of speed above the rated value can be obtained on shunt and compound machines by field weakening with a variable resistor in series with the field coil if the accompanying fall in torque can be tolerated. The energy loss in such resistors is not great because field currents in these machines are typically only one-tenth of the armature current. Care must be taken that the field connections are not accidentally open-circuited by a dirty resistance-wiper contact or other cause; the flux density would fall to the residual value for the field system and the speed would rise destructively.

Where a decrease in speed below the rated value is required, control can be achieved by reduction of the armature voltage using a series resistor in the armature supply. This method suffers from two disadvantages: first, the control resistor has to carry the full armature current and therefore will dissipate much energy as heat, causing poor overall efficiency; second, the armature voltage for any given resistor setting will depend on the load supplied to the motor, since armature current and therefore the resistor voltage drop are proportional to load. This second feature destroys the constancy of speed under varying load which is such a desirable feature of shunt motors.

An efficient method of speed control by variation of the armature voltage is to feed the armature from a variable d.c. supply. Such a supply can most conveniently be a d.c. generator driven at constant speed by a prime mover. The magnitude and polarity of the generator output are controlled over very wide ranges by variation of the excitation (field current). Such a system forms a basic Ward - Leonard system (figure 4.27a).

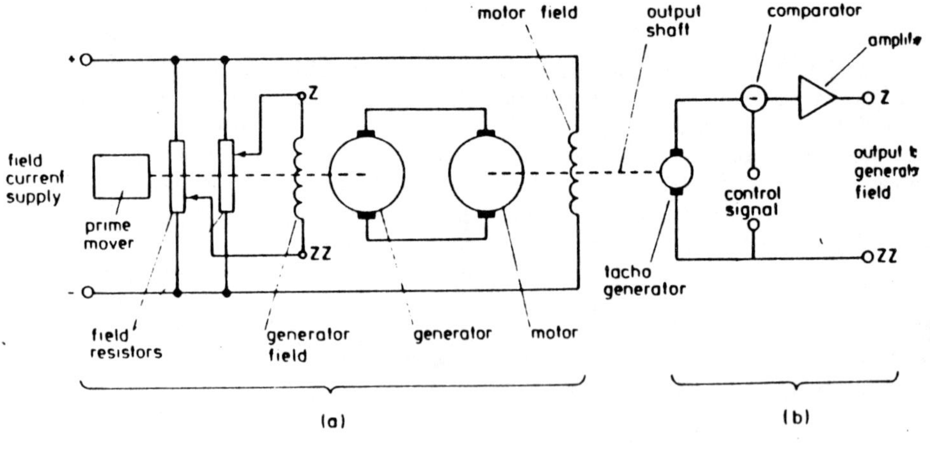

FIGURE 4.27

(a) The basic open-loop Ward - Leonard speed controller and
(b) the additional components required for automatic control in closed-loop operation

For more accurate control, the speed of the output motor may be monitored by a tachogenerator whose output is compared with the desired control-voltage. Any difference between these two voltages is amplified electronically and used to control the field current of the generator (figure 4.27b). Such a system, which automatically monitors any change in the output caused by load fluctuations or other external influences and then corrects the system automatically to the desired condition, is known as a *closed-loop control system*.

The analysis of the response of such a system to external variations is complex and is to be found in texts on control engineering.[11, 12]

Speed control of the series motor is more difficult since the speed varies widely with load but some measure of control is possible by variation of the motor field-strength. Since the field and armature windings are in series, some of the armature current has to be bypassed from the field winding by means of a parallel diverter resistor (figure 4.28a). Where a very high-power motor is used, as in some forms of electric traction, the design of a variable resistor for such high current-levels and the inefficiency thereby introduced limit the designer to discontinuous speed control methods. An example is the stud-switch arrangement of figure 4.28b in which the effective number of turns of the field coil may be varied.

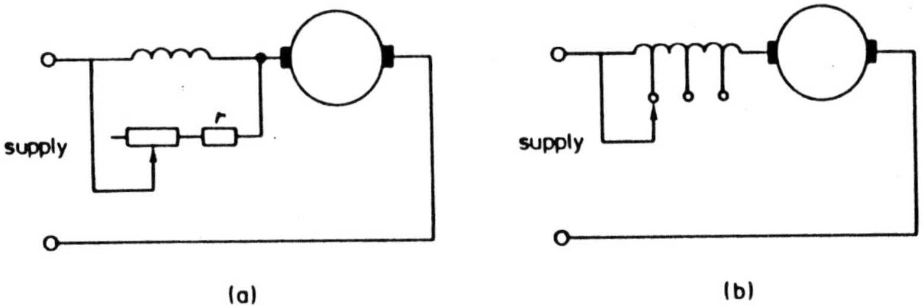

FIGURE 4.28

(a) The diverter resistor and (b) the tapped-coil methods of series-motor field-control.
Note: The safety resistor r is included to avoid ever short-circuiting the field completely

Example 4.6

The motor quoted in example 4.5 (p. 110) is running steadily on a 415 V supply. If the flux could be instantaneously reduced from 0.025 Wb by 0.1 per unit, calculate the momentary rise in armature current and the corresponding torque. If the load torque remains constant after acceleration calculate the final speed.

$$E_1 = \text{Original back e.m.f.}$$

$$= V - I_a R_a = 415 - (40 \times 0.25)$$

$$= 405 \text{ V}$$

Since $E_1 = kn\Phi_1$, when the flux decrease occurs the armature is momentarily at the same speed therefore the new back e.m.f. is given by

$$E_2 = E_1 \times \frac{\Phi_2}{\Phi_1}$$

$$= 405 \times 0.9 = 364 \text{ V}$$

Therefore the new armature current is given by $I_{a_2} R_a = V - E_2$

$$I_{a_2} = \frac{415 - 364}{0.25}$$

$$= 204 \text{ A}$$

Original torque is 382 Nm (from previous example). Since

$$\text{torque} = \text{armature current} \times \text{flux}$$

$$\frac{\text{new torque}}{\text{old torque}} = \frac{\text{new armature current} \times \text{new flux}}{\text{old armature current} \times \text{old flux}}$$

Thus

$$\text{new torque} = 382 \times \frac{204}{40} \times 0.9 = 1750 \text{ Nm}$$

This torque increase accelerates the armature until the motor output falls to its original value at a new speed n_3.

$$E_3 = V - I_{a_3} R_a$$

also

$$E_3 = k n_3 \Phi_3$$

thus

$$n_3 = (V - I_{a_3} R_a)/k\Phi_3 \tag{4.2}$$

therefore we must find I_{a_3} and k. For the same load torque

$$\text{original current} \times \text{original flux} = \text{final current} \times \text{final flux}$$

Thus

$$\text{final current } I_{a_3} = \frac{40}{0.9} = 44.5 \text{ A}$$

Originally $E_1 = kn_1 \Phi_1$, therefore

$$k = \frac{405}{6.75 \times 0.025}$$

and is a constant for the machine at any speed. Substituting in equation 4.2 gives

$$n_3 = \frac{[415 - (44.5 \times 0.25)]\ 6.75 \times 0.025}{405 \times 0.0225}$$

$$= 7.47 \text{ rev/s}$$

or

$$N_3 = 449 \text{ rev/min}$$

4.6.7 Thyristor Control of D.C. Machines

Until recently the methods of speed control of d.c. motors nearly all used resistors in the field circuit and in the armature circuits as discussed above. Some power loss in such resistors is unavoidable.

The recent advent of the thyristor and its use as a controlled rectifier (see section 5.3) has made d.c. motor control more efficient. Such devices are used to vary the potential applied either to the field or to the armature circuit and usually consist of bridge configurations as shown in figure 4.29. Whereas single-phase full-wave rectification is appropriate up to approximately 15 kW, bridge circuits supplied from a three-phase source are more usual for the higher-power machines.

Provision may also be made to monitor the motor speed via a tachogenerator and to incorporate a feedback loop in order that the system may be self-correcting maintaining constant speed under varying conditions of load and supply.

A phase-shifted supply to the gate electrodes of the thyristors varies the current through the thyristors and hence through the motor winding. Because the forward voltage-drop across power thyristors is only approximately 1 V, little power is lost in the speed-control circuits even when used in conjunction with high-power series motors in traction applications.

The inherent defects in stud-switch starting (contact fouling, wear and sparking) may be eliminated, since a fully thyristor-controlled starter can be designed without any moving parts.[13] This relies on the successive firing of several thyristors as the back e.m.f. of the machine rises with speed. A completely automatic starting sequence may be provided by thyristors progressively short-circuiting sections of the armature series-resistor by static electrical, as opposed to mechanical means.

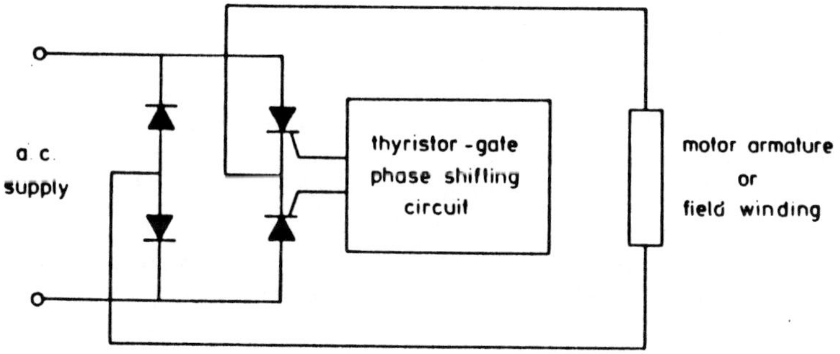

FIGURE 4.29

The basic half-controlled bridge supply for a variable-speed d.c. machine

4.7 Guide to the Selection of Motors[14]

Every industrial application of electrical motors will be influenced by the electrical supplies available, temperature and safety considerations, together with cost and maintenance charges. To attempt a definitive statement of motor choice on mechanical (speed – torque) characteristics is therefore to oversimplify the problem, as each situation may be nearly unique. It is however possible, with the above limitations firmly in mind, to lay some general guidelines along which selection may proceed.

The cheapest motor, because of its mechanical simplicity and volume production, is undoubtedly the squirrel-cage induction motor. Its applications vary from gramophone turntables (single-phase) to pump and machine-tool drives rated up to tens of kilowatts (three-phase). Such machines may be started direct-on for low powers or with a star – delta starter arrangement for powers over 1 kW. In applications where a high starting-torque is required the wound-rotor (or slip-ring) induction motor is preferred. The characteristics of all forms of induction motor limit the operation to very near the synchronous speed although some measure of control is possible via the external rotor resistor in the slip-ring machines. The squirrel-cage machine, having no brushgear, is especially suitable for explosive-hazard environments such as mines and chemical plants because of the absence of brush sparking. Major speed changes are usually made in machine-tool drives either mechanically through gearboxes and pulleys, or electrically by switching to vary the *effective* number of poles used and hence the synchronous speed of the field. The synchronous motor in its larger sizes is generally limited to power-factor correction (see section 4.5.1) because of its cost.

The great advantage of d.c. motors in spite of their relative complexity is the degree of speed control available. The shunt motor has uniform speed under varying load conditions and if the speed is controlled by the field current, has good efficiency. It is also greatly used in speed- and position-control systems where the relatively small field-currents may conveniently be controlled by thyristors or transistors.

The characteristics of the d.c. series motor are especially valuable for applications where its high torque at low speeds is essential. Such situations as hoists and electrical traction are foremost while an everyday example is that of an automobile starter-motor where high torque at low speeds is essential for cold-weather starting and high currents are available for a short time to provide this. The over-riding consideration in the specification of series motors is that the load must never be removed since destructively high speeds ensue.

Where the finest control of speed over wide ranges is required under conditions of widely fluctuating loads the Ward – Leonard system in either its open- or closed-loop form is employed. Such an example would be a metal rolling mill in which the load is applied suddenly as the metal billet passes between the rolls and where frequent rapid reversal of the drive is required.

Motors are available in a wide variety of enclosures for differing environments. Examples vary from the simplest screen enclosure to protect personnel from contact with the rotor or brushgear, through 'drip-proof' where cowls are arranged to prevent the ingress of rain in out-of-doors applications, to pipe-ventilated and totally enclosed types for use in dusty and corrosive atmospheres respectively.

The British Standards which specify the performance of electric motors between 1 and 2500 h.p. are set out in BS 2612 and BS 170.

The performance is specified in terms of

(i) The standard voltage of the machine.
(ii) The 'rating' or mechanical output obtainable under either short-term or continuous use.
(iii) The maximum permissible temperature-rise for various components of the machine under varying ambient-temperature conditions.
(iv) The tolerance allowed on the quoted efficiency and power-factor figures.
(v) The maximum permissible overloads which the machine is capable of withstanding.
(vi) The types of enclosures available.

For any given type of machine the rating for a given size of frame may be raised at the design stage by increasing the speed since the fanning and cooling arrangements for the armature are improved. Also for a given type of machine the cost is approximately proportional to its weight, and the weight varies inversely with the speed for a given rating. Hence a 40 kW machine may weigh 1150 kg for a speed of 500 rev/min whereas it would weigh 430 kg for the same power at 1500 rev/min. Typical costs at the time of writing would be £1000 and £300 respectively. Therefore the cost is almost inversely proportional to the speed for a given power, so that it is always economical to use the highest-speed motor which may be adapted for a particular application.

4.8 Problems

4.1 When measuring large currents a current transformer is often interposed between the ammeter and the circuit to be investigated. This is to step down the current to manageable values, enabling an ammeter of convenient range to be employed. The secondary terminals of such transformers are always short-circuited carefully by the operator except when the ammeter is actually being observed. Use your knowledge of transformer action to suggest reasons for this procedure.

4.2 An alternating current of the form

$$i = 10 + 50 \sin 100\pi t \text{ A}$$

is fed to the primary terminals of a voltage step-up transformer of turns ratio 1:10. Assuming the transformer to be ideal and feeding a 100 Ω resistive load, write down the expression for the secondary voltage with respect to time, justifying the answer by reference to transformer action and suggest a possible transformer application based on this particular property.

4.3 Calculate the speeds of flux rotation in revolutions per minute in a multipole a.c. machine operating from a 60 Hz supply. The stator winding may be arranged to give any combination of pole pairs to a maximum of six. Suggest ways in which the speed of a synchronous motor may be varied (a) in steps (b) continuously.

4.4 A primitive induction motor has its rotating field produced by a four-pole stator from a 50 Hz supply. The rotor consists of a single coil whose ends are terminated in slip-rings and whose resistance and reactance at 50 Hz are 0.3 Ω and 2.4 Ω respectively. The open-circuit voltage appearing across the slip-rings at standstill is 50 V. Calculate the motor torque when operating at 1440 rev/min with the slip-rings shortcircuited.

4.5 Calculate the starting torque of the motor in example 4.4 above if an external resistor of value 2 Ω is connected to the slip-rings.

4.6 A 440 V d.c. motor takes an armature current of 20 A when running at 500 rev/min, its armature resistance being 0.6 Ω. If the magnetic flux is reduced to 0.7 of its original value the torque increases to 1.4 times its original value. What are the new values of armature current and speed?

4.7 A 200 V series motor runs at 6 rev/s when taking a current of 40 A. The armature resistance is 0.02 Ω and the resistance of the field winding is 0.03 Ω. Find the speed when the field is shunted by a 0.02 Ω resistor in parallel, the load torque remaining constant throughout. Assume that the flux is proportional to the field current.

4.8 Explain the reasons for the choice of electric motor to be used for the following applications.

(a) A 20 kW induction motor driving a large lathe.
(b) A small a.c motor driving a polishing machine in a laboratory.
(c) A d.c. shunt motor driving a variable-speed stirrer.
(d). A d.c. series motor driving a battery-operated truck.

5 Power Supplies and their Applications

5.1 Rectification

Chapter 4 has shown that although electrical energy is commonly distributed in alternating current (a.c.) form, local supplies of direct current (d.c.) are often required for motors. Other common applications for d.c. are electroplating and battery charging.

Rectification is the process of converting an alternating current into a unidirectional one. The final result may not necessarily be direct current which would imply constancy of magnitude as well as direction. The distinction between these terms is shown in figures 5.1a, b, and c which depict typical variations in the instantaneous value *i* of alternating, direct and unidirectional currents. Figure 5.2 shows that many unidirectional currents may be resolved into an alternating and a direct component.

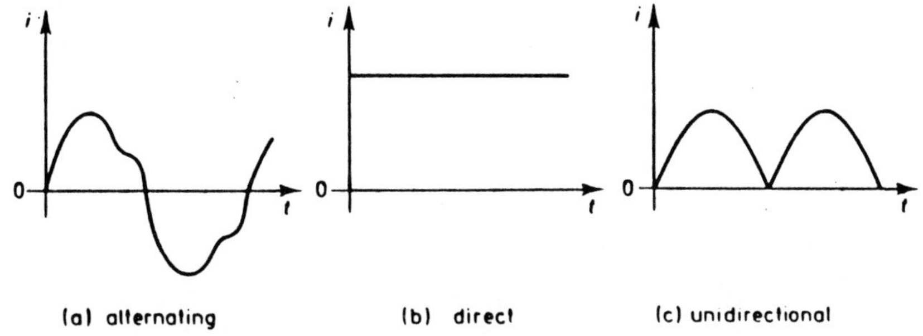

FIGURE 5.1

Typical waveforms of electrical current

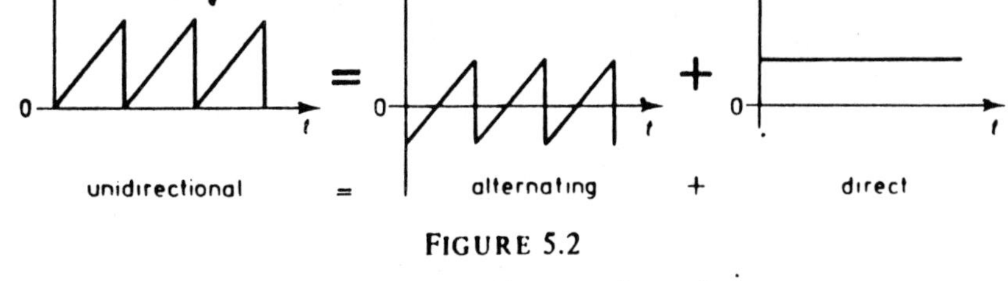

FIGURE 5.2

Resolution of a unidirectional waveform

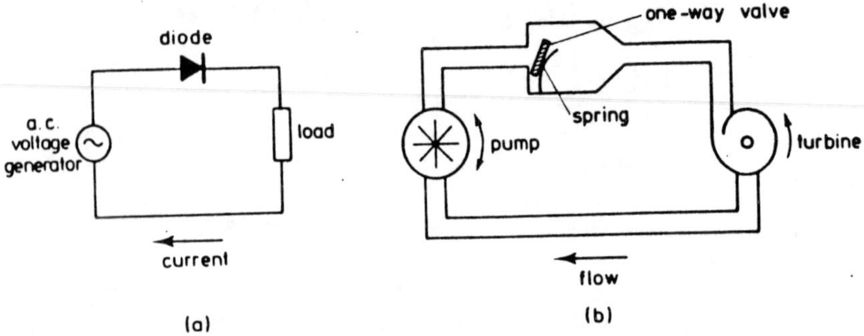

FIGURE 5.3

A diode and a clack valve used as unidirectional switches in electrical and fluid circuits respectively

Rectification requires some form of unidirectional switch between the supply and the load. This is analogous in its operation to the type of one-way valve often used in hydraulic and pneumatic circuits (figure 5.3 b). The pump is driven alternately in opposite directions, but flow only occurs when the pressure is in the correct sense to open the valve. The resultant turbine speed will be unsteady but always unidirectional; in other words the turbine speed will have an alternating component superimposed upon a constant or d.c. component.

An electrical component which only permits current flow in one direction is known as a diode. In the past, thermionic devices have been used for low currents and gas-filled diodes, mercury-arc rectifiers and ignitrons for successively higher currents. The advent of cheap, sturdy, reliable and efficient semiconductor diodes has all but eclipsed the use of these earlier devices in all fields except welding circuits where the ignitron (section 5.4.3) still holds some advantage. The basic operation of the semiconductor diode has been explained in section 3.4.3 but figure 5.4 repeats the diode's electrical characteristic.

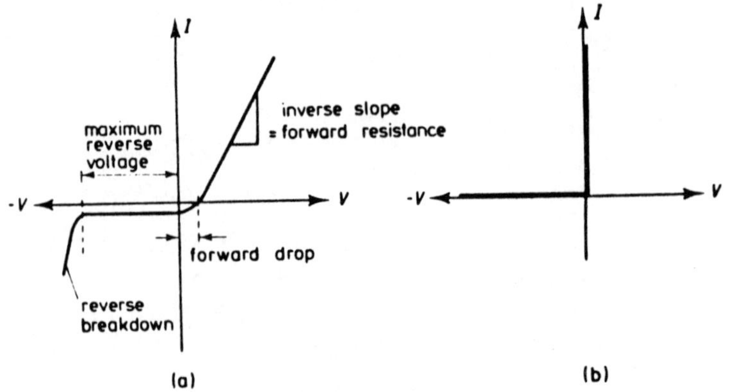

FIGURE 5.4

(a) Actual and (b) idealised electrical characteristics of a semiconductor diode

A typical value of the forward voltage-drop in a diode is from 0.3 to 0.8 V. Voltage drops of this magnitude may have to be taken into account in circuits where small voltages are being rectified, but in the majority of power-frequency applications the supply voltage may be two orders of magnitude greater than this (above 100 V) rendering the forward drop negligible. The forward resistance or inverse slope of the characteristic will give rise to heating effects within the diode (typically 1.2 W per ampere of forward rectified current). While these losses are small and may often be neglected in power-rectifier calculations, they do give rise to heating effects within the diode and care has to be taken in the internal construction and mounting of the diode so that these losses are adequately dissipated to the surroundings. In order to ensure that the semiconductor junction temperature never exceeds the critical value (150 °C for silicon) at which thermal runaway and destruction occur the body temperature of the device is limited to between 100 and 120 °C.

Figure 5.5 illustrates the internal construction and mounting of a power diode. The lower side of the semiconductor slice (often only about 1 cm diameter) is soldered directly to the copper base which extends downwards into a threaded mounting-stud. The base is annealed during manufacture so that care must be taken to prevent accidental knocks or overtightening of the mounting nut from distorting the copper base. This would prevent the intimate contact required between the undersurface of the base and the heat sink on which it must be mounted for efficient heat conduction from the device. Typical heat sinks are aluminium plates, either plain or finned (often with a blackened surface) which radiate the waste heat to the surrounding air. The temperature rise of the cooling air (up to 0.2 °C per watt dissipated) together with the small size of high-current silicon devices means that the exhaust air may be extremely hot in large installations and adequate ventilation arrangements must be provided.

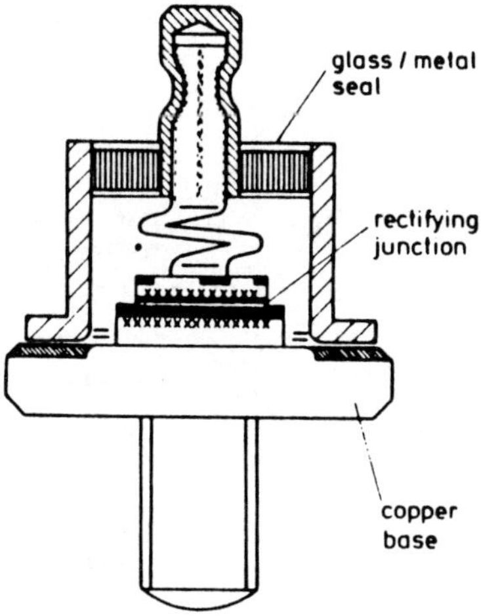

FIGURE 5.5

The internal construction of a power rectifier diode

In installations where the rectified current is above 250 A forced-draught cooling is often employed; even a 30 cm diameter fan being adequate for installations of 500 kW rectified power. Care must also be taken in the design of low-power electronic equipment to keep these diodes with their relatively high body temperatures out of the vicinity of more temperature-sensitive components.

Provided that adequate cooling provisions are made we may adopt figure 5.4b as a convenient idealised characteristic for calculations based on a power-rectification diode. The diode's internal losses are neglected and the forward drop is assumed to be negligible compared to the supply voltage.

5.1.1 The Half-wave Rectifier

A half-wave rectifier uses a single diode (figure 5.6) to ensure that current only flows through the load on the positive half-cycle. With resistive loads this means that $i = 0$ whenever the upper end of the transformer secondary winding is more negative than the lower end and that $i = (V_m \sin \omega t)/R$ on positive half-cycles. The mean value of the rectified output current for a half-cycle is given by

$$I_{mean}' = \frac{1}{\pi}\int_0^\pi \left[\frac{V_m \sin \omega t \; d(\omega t)}{R}\right]$$

This has been shown in section 2.1.1 to equal $2\,V_m/\pi R$, thus for an integral number of cycles the over-all mean rectified current will be half the above, namely, $I_{mean} = V_m/\pi R$. It may be clearly seen that although the load-current waveform is unidirectional it contains a high alternating component (ripple) which though acceptable for some purposes (battery charging or electroplating) renders it totally unsuitable for most applications.

$$\text{Ripple factor} = \frac{\text{r.m.s. value of all the a.c. components}}{\text{mean d.c. current}}$$

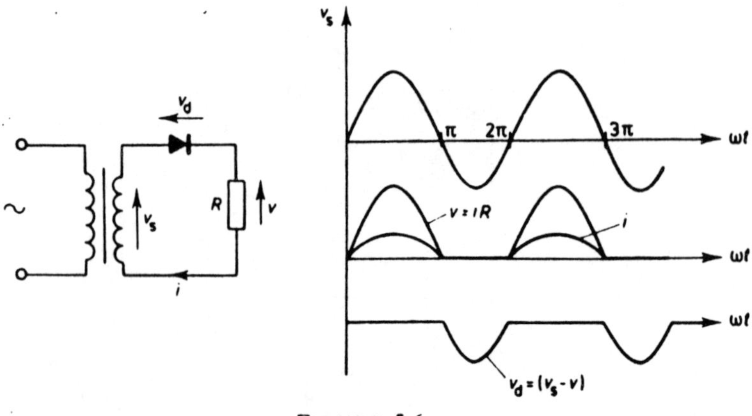

FIGURE 5.6

Basic half-wave rectifier circuit and waveforms

While this may be reduced by the smoothing circuits of section 5.2, the low efficiency, poor voltage-regulation and the unidirectional value of transformer-core flux cause the half-wave rectifier to be unsuitable for power rectification. Since the transformer currents are unidirectional, the transformer flux will be unidirectional and unless the core area is made adequate, difficulties may be experienced because of magnetic saturation within the core material. This last disadvantage may largely be overcome by employing two diodes acting on alternate half-cycles to pass a unidirectional current through the load.

5.1.2 The Bi-phase Rectifier

If two half-wave circuits are used back-to-back, as in figure 5.7a, D_1 will conduct when terminal A of the transformer winding becomes positive with respect to the centre-tap B, while D_2 will conduct when terminal C becomes more positive than B. The forward currents of D_1 and D_2 will therefore pass on alternate half-cycles through R in the same direction giving a 'full-wave' output. The effective equivalent circuits during each half-cycle are shown in figure 5.7b. We can intuitively see that the mean rectified current is now twice that in the half-wave case and that the ripple factor has decreased appreciably. Lastly, the transformer secondary current flows in successive halves of the winding in opposite directions ensuring that the core flux is cycled about zero and allowing important economies to be made in core size. It will be seen that when D_1 conducts points A and E are at the same potential ($+V_m$ at t_1). Since point C is at a potential $-V_m$ at that time, the total reverse voltage of $2V_m$ is between C and E. This requires diodes with a higher reverse rating (peak inverse voltage) than either the half-wave, or the bridge rectifier to be considered in section 5.1.3.

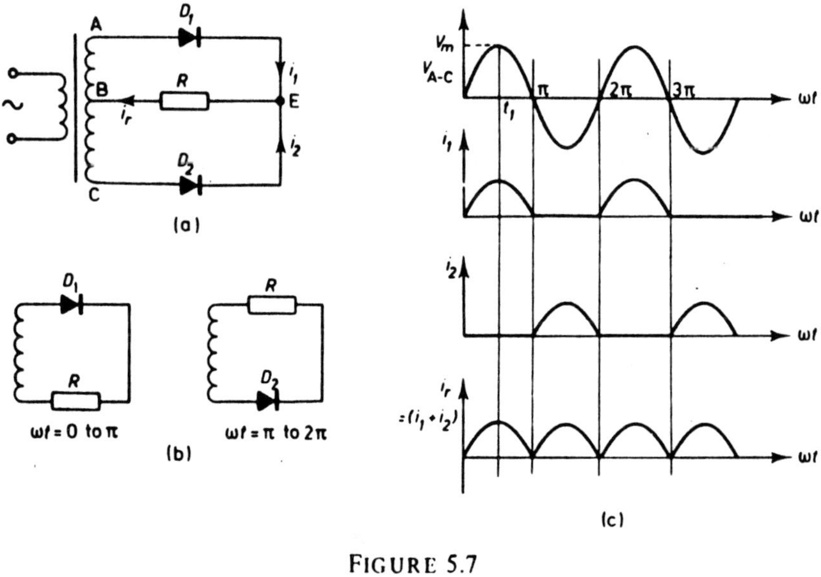

FIGURE 5.7

(a) The bi-phase rectifier circuit, (b) its equivalent circuits and (c) its waveforms

Example 5.1

Assuming the use of an ideal diode, calculate the mean rectified load-current through R in the half-wave circuit of figure 5.6. The transformer secondary voltage is 240 V and the load resistance 250 Ω. What would be the power lost in the rectifier if the diode had a finite forward resistance r of 1 Ω.

On conducting half-cycles the instantaneous load voltage is $V_m \sin \omega t$ or $240\sqrt{2} \sin \omega t$. By Ohm's law the instantaneous load current will be $(V_m \sin \omega t)/R$ or $(240\sqrt{2} \sin \omega t)/250$. The mean value of the expression over the forward half-cycle is given by

$$I_{mean}' = \frac{1}{\pi} \int_0^{\pi} \frac{240\sqrt{2}}{250} \sin \omega t \; d(\omega t)$$

$$= \frac{-240}{\pi} \times \frac{\sqrt{2}}{250} \left[\cos \omega t\right]_0^{\pi}$$

$$= 0.433\,[-2]$$

$$= 0.866 \text{ A}$$

Since $i = 0$ over the whole reverse cycle $I_{mean} = I_{mean}'/2$

$$I_{mean} = 0.433 \text{ A}$$

$$P_{mean}' = I_{r.m.s.}^2\, r$$

where $I_{r.m.s.} = I_m/\sqrt{2}$ and I_m is the maximum forward current (see section 2.1.2).

$$I_m = \frac{240\sqrt{2}}{251} = 1.355 \text{ A}$$

therefore

$$I_{r.m.s.} = \frac{1.355}{\sqrt{2}} = 0.958 \text{ A}$$

Thus

$$P_{mean}' = (0.958)^2 \times 1 = 92 \text{ mW}$$

so that

$$P_{mean} = \frac{P_{mean}'}{2} = 46 \text{ mW}$$

5.1.3 The Bridge Rectifier

Figure 5.8 shows the arrangement of four diodes to form perhaps the most common rectifier circuit – the full-wave bridge, which requires twice the number of diodes compared to the simple bi-phase circuit but retains all the

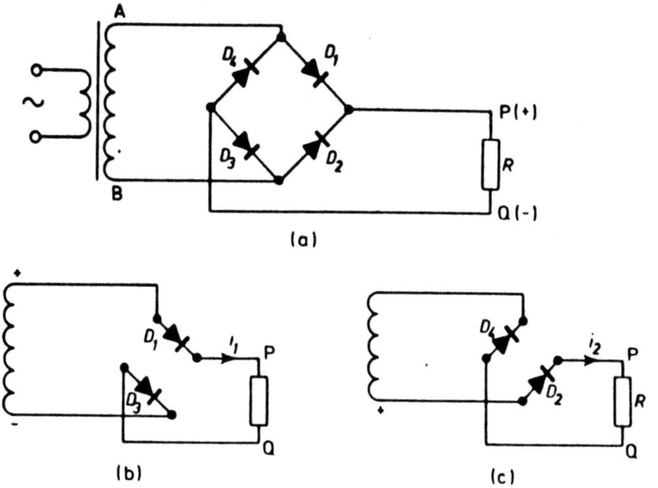

FIGURE 5.8

(a) The complete bridge-rectifier circuit and (b) and (c) its equivalent circuits during alternate half-cycles

good features of the latter without the need for a centre-tapped transformer. Its operation is as follows.

(i) *First half-cycle* (A positive with respect to B). Diodes 1 and 3 are forward-biased and therefore conduct, the current flowing through the load R from P to Q. Since the potentials across D_1 and D_3 are negligible the reverse voltages across D_2 and D_4 are equal to the maximum value of the supply V_m. The equivalent circuit is shown in figure 5.8b.

(ii) *Second half-cycle* (B positive with respect to A). Diodes 2 and 4 conduct and current again flows through R from P to Q. The reverse voltage across D_1 and D_3 is that of the maximum supply voltage V_m alone. The output waveform is the same as that of the bi-phase rectifier in figure 5.7 yielding the same mean rectified current.

Each of the diodes must be capable of withstanding the peak output voltage in the reverse direction but each carries current for only half the cycle.

An important advantage of both bi-phase and bridge arrangements that yield a 'full-wave' output can be seen by reference forward for a moment to figure 6.10 (p. 180). While a half-wave output by analysis can be shown to contain a significant sinusoidal content at the fundamental supply frequency, a 'full-wave' output contains no alternating term at a frequency less than twice the supply frequency. It will be demonstrated in section 5.2 that these higher ripple frequencies are more easily eliminated than terms at the fundamental frequency. An arrangement which minimises even further the ripple content and raises its minimum frequency to three or even six times the supply frequency is provided in systems of *polyphase rectification.*

Comparison of Single-phase Rectifier Circuits

From the foregoing sections it will be seen that for the same rectified current I_{mean} supplied to a constant load voltage equal to the peak value V_m of the a.c. supply voltage, the following requirements for the transformer and diode(s) must be satisfied.

	Half-wave	Bi-phase	Bridge
Transformer supply (r.m.s.) voltage	$0.707\ V_m$	$0.707V_m + 0.707V_m$	$0.707V_m$
Maximum diode current	πI_{mean}	$\pi I_{mean}/2$	$\pi I_{mean}/2$
Peak reverse voltage per diode	$2V_m$	$2V_m$	V_m

5.1.4 Polyphase rectification

If a three-phase supply is available, more efficient and more easily smoothed rectification is possible. The circuit waveforms and diagrams in figures 5.9a and b show the construction of three-phase half-wave and three-phase bridge rectifiers respectively. The secondary winding of the three-phase transformer must be star-wired for half-wave operation in order to make the star point N_1 available for connection of the load. The output waveform is best understood by initially considering the circuit to act as three separate half-wave circuits feeding their positive half-waves into a common load R. Because of the $2\pi/3$ phase-shift between each winding, the resultant load currents would be those of the feint lines in figure 5.9a. However forward-biasing and consequent conduction of D_1 will only occur when v_1 is greater than both v_2 and v_3; that is between 0 and $2\pi/3$ radians.

At point B, v_2 exceeds v_1 and conduction through the load now occurs via D_2. The load current throughout the full cycle (and therefore the load voltage for a resistive load) thus follows the heavy line connecting points A - B - C - A. By inference it can be seen that the lowest ripple-frequency will be three times that of a single-phase half-wave rectifier, that is three times the fundamental frequency. The load current progresses during the cycle through the phases in the order 1 - 2 - 3 - 1. This process is known as *commutation*.

A three-phase full-wave rectifier may be constructed using six diodes and three centre-tapped phase windings (not illustrated). Such a circuit is known as a 'six-phase diametral star' connection and the minimum ripple-frequency is, as may be supposed, six times the fundamental but the circuit has the disadvantage of requiring a three-phase centre-tapped transformer.

A simpler arrangement employing six diodes and a simple three-phase supply in which even the star point need not be available is the three-phase bridge rectifier (figure 5.9b). Current flows to the load via T_1 from whichever transformer terminal is most positive and returns via T_2 to the most negative terminal. At time t', when terminal A is at its maximum instantaneous value, terminals B and C are at equal negative potentials of half their maximum values. The load voltage is thus v_1, the conduction being shared between D_1 at maximum current and

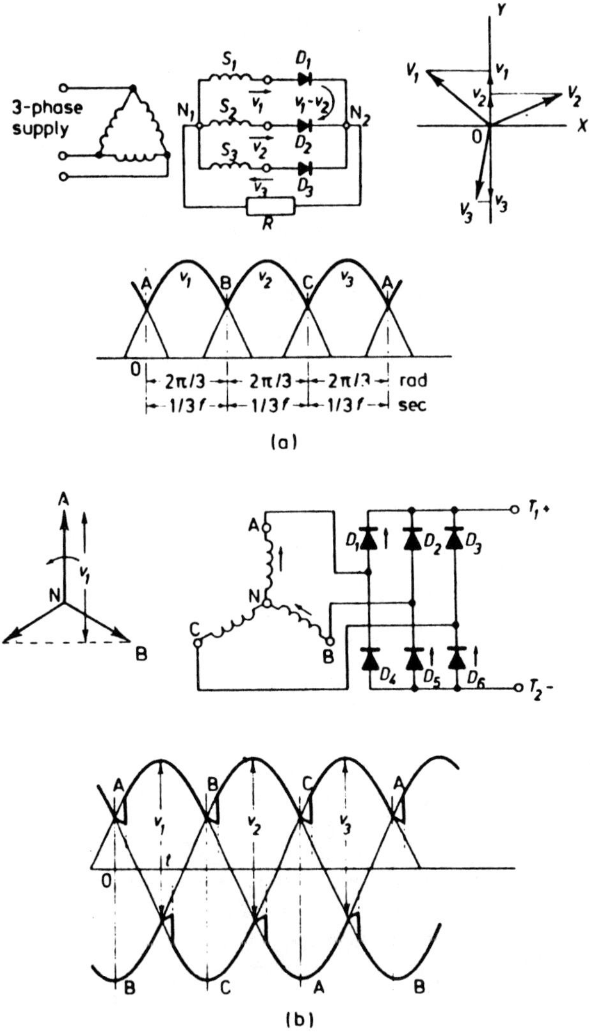

FIGURE 5.9

(a) Three-phase rectification and (b) voltage diagram for a three-phase bridge-connected rectifier

diodes D_5 and D_6 each carrying half the load current. Thus unlike the half- and full-wave polyphase rectifiers where the load voltage is the distance between the envelope and zero, the polyphase bridge load-voltage is the distance between the positive and negative envelopes and may be shown to contain an a.c. component of minimum frequency $6f$. The reason for the load current and voltage not precisely following the a.c. envelope at the start of each conduction region (points A, B and C) may be explained as follows.

The simple explanation employed previously would require that the diodes should suddenly start conduction at points A, B and C. Because of the finite inductive reactance of the transformer windings and of the load (if not purely resistive) the phase current is unable to rise instantaneously from zero to v_1/R at point A. There is thus a time lag, known as the 'period of overlap' or 'overlap angle', during which conduction is gradually transferred from D_3 to D_1 resulting in a momentary reduction of output voltage in this period.

It will be realised that the mean current through each of the diodes is one-third of the mean load current I_{mean} and it can be shown that the r.m.s. current carried by each diode is 0.58 I_{mean}. The peak inverse voltage which the diodes must withstand is simply the supply voltage maximum as in the single-phase bridge.

The relative costs of the most popular rectifier arrangements are single-phase full-wave (bi-phase) 1.54, single-phase bridge 1.47 and three-phase bridge 1.00. These prices refer to the cost of rectifier diodes and heat sinks alone for the same d.c. output voltage and current. Prices vary approximately linearly with the current required in power applications.

5.2 Smoothing or Filtering Circuits

The previous section has shown that polyphase rectifiers, particularly in the bridge configuration produce an output whose waveform is low in ripple. 'D.C.' in this form is perfectly satisfactory for many applications especially battery charging and electroplating. The use of unsmoothed d.c. for motor supplies may also be permissible, leading only to some increase in acoustic noise. If the rectifier circuit is intended as a power supply for electronic or other electrical measuring equipment, the output waveform must be smoothed (have its a.c. content removed) to provide a supply similar to that obtained from batteries. This is achieved by using either inductance, capacitance or both.

In figure 5.10a if the current is not smooth, its rate of change induces an e.m.f. in L so that the varying (a.c.) component of voltage is developed across the inductor rather than across the resistive load. In figure 5.10b however, the alternating current passes through C more easily than through the load.

In both cases the current flowing through the load R is made more constant, that is, it is smoothed or filtered.

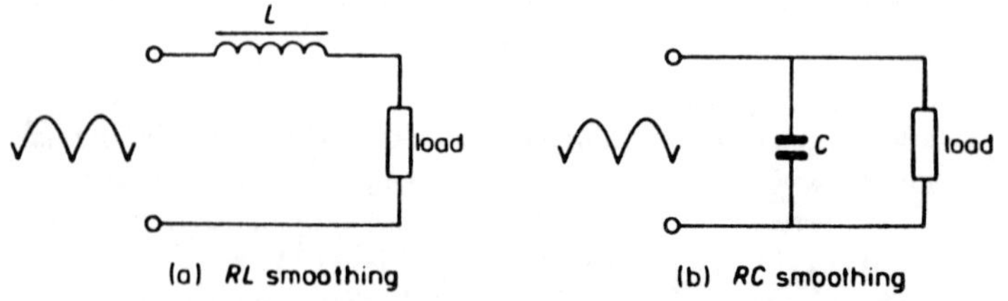

FIGURE 5.10

Simple single-reactance smoothing circuits

5.2.1 *RC* Smoothing

This method is the more widely used of the simpler smoothing circuits (figure 5.11) because of the relatively low weight and price of capacitors compared with inductors.

Between times t' and t'' the capacitor charges up from the supply for a time t_1 and during the period t_2 discharges exponentially into the load R which will be assumed to be purely resistive for simplicity of analysis. If the supply unit has a low internal impedance, the charging current into the capacitor during t_1 could momentarily exceed the supply unit's maximum rated current. To avoid this a series resistor R_s is included of a sufficiently small value to ensure that the charging time (5 × time constant = $5R_sC$) is completed within the available period t_1.

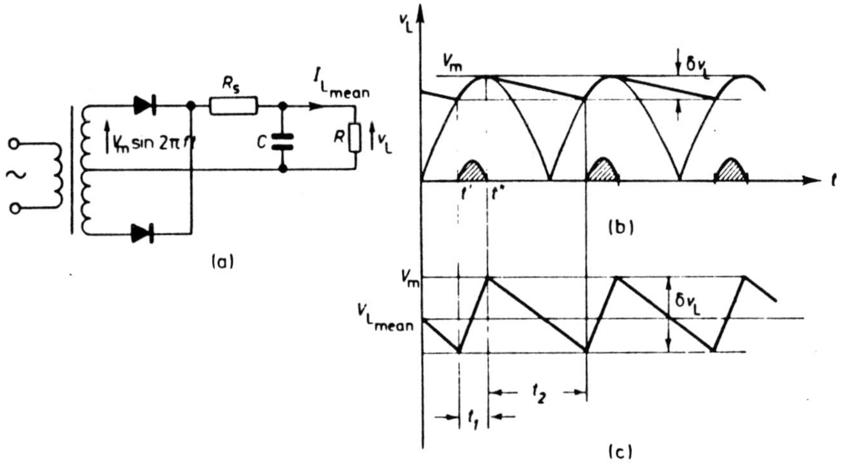

Figure 5.11

RC smoothing circuit on a full-wave rectifier; (a) the circuit, (b) the output waveform and (c) the approximate output waveform

For equilibrium the charge taken from the smoothing capacitor by the load R is replaced by the current pulse from the supply during t_1. Assumptions are made that R_s is much less than R and C is large so that $1/\omega C$ is much less than R. Neglecting t_1 compared to t_2 then t_2 is approximately $1/2f$ where f is the a.c. supply frequency.

$$\text{Charge taken from capacitor during } t_2 = \text{current} \times \text{time}$$

$$= I_{L\,\text{mean}} \times \frac{1}{2f}$$

$$= \frac{V_{L\,\text{mean}}}{R} \times \frac{1}{2f} \text{ coulombs}$$

where $V_{L\,\text{mean}}$ and $I_{L\,\text{mean}}$ are the mean load-voltage and current respectively

The charge supplied is given by

$$C\,\delta v_L = \delta Q$$

$$= \frac{V_{L\,\text{mean}}}{R} \times \frac{1}{2f}$$

thus

$$\delta v_L = \frac{V_{L\,\text{mean}}}{2fRC}$$

But mean load-voltage

$$V_{L\,\text{mean}} = V_\text{m} - \frac{\delta v_L}{2}$$

$$= V_\text{m} - \frac{V_{L\,\text{mean}}}{4fRC}$$

$$= \frac{V_\text{m}}{1 + \dfrac{1}{4fRC}}$$

Thus by binomial expansion

$$V_{L\,\text{mean}} \approx V_\text{m}\left(1 - \frac{1}{4fRC}\right) \qquad (5.1)$$

Example 5.2

Design a full-wave smoothing filter to deliver 250 V at an average current of 50 mA to a resistive load. The amplitude of the ripple voltage must not exceed 4 per cent of the maximum supply voltage. Design an additional LC section to reduce the ripple voltage by a further factor of 50. The supply frequency to the rectifier is 50 Hz.

$$\frac{\text{Ripple amplitude}}{\text{Maximum supply voltage}} = \frac{V_\text{m} - V_{L\,\text{mean}}}{V_\text{m}} = 0.04$$

$$\frac{V_{L\,\text{mean}}}{V_\text{m}} = 1 - \frac{1}{4fRC} = 1 - 0.04 = 0.96$$

The load resistance R takes 250 V at 50 mA and thus has a value of

$$\frac{250}{0.05} = 5\ \text{k}\Omega$$

Hence

$$\frac{1}{4fRC} = 0.04$$

or

$$C = \frac{1}{4 \times 50 \times 5 \times 10^3 \times 4 \times 10^{-2}}$$

$$= 25\ \mu\text{F}$$

The ripple voltage may further be decreased by the addition of an LC filter section as shown in figure 5.12. For the required attenuation the reactances of C_1 and L must be in the ratio of 1:50. Hence $2\pi fL = 50/2\pi fC_1$ or $LC_1 = 50/4\pi^2 f^2$. If L is chosen to have some convenient value, say 10 H, then $C = 12.5\ \mu\text{F}$ if the ripple voltage is assumed to approximate to a second harmonic (100 Hz) sinusoid.

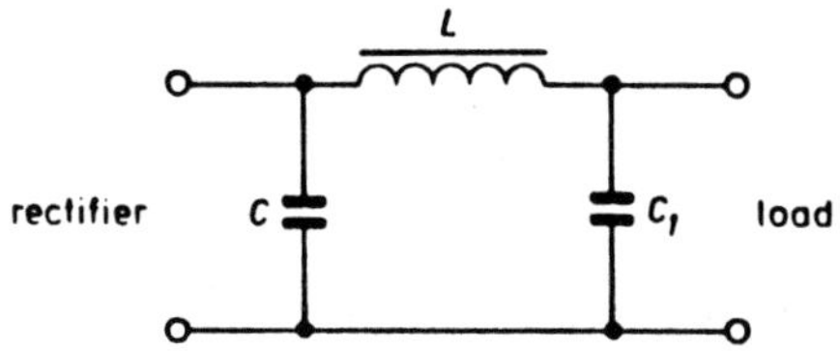

FIGURE 5.12

Addition of an LC section to a simple capacitance filter

5.2.2 Ripple

As stated in the introduction to section 5.2, the aim of a smoothing circuit is to reduce the a.c. components (ripple) in the power-supply output to a minimum. A perfect smoothing arrangement would require infinitely large capacitors and inductors, but a measure of the performance of any particular practical smoothing arrangement is stated in terms of the ripple factor, defined thus

$$\text{Ripple factor} = \frac{\text{r.m.s. value of all the a.c. components present}}{\text{d.c. component}}$$

Let us for example examine the ripple content of the simple RC smoothing filter of figure 5.11. An enlarged diagram of the output waveform appears in figure 5.13, where it may be seen from equation 5.1 that the a.c. component approximates to a triangular waveform, symmetrical about the mean d.c. output level $V_{L\ \text{mean}}$ and of a peak value $V_\text{m}/(4fRC)$.

It is left to the student to show, using the methods of section 2.1.2, that the

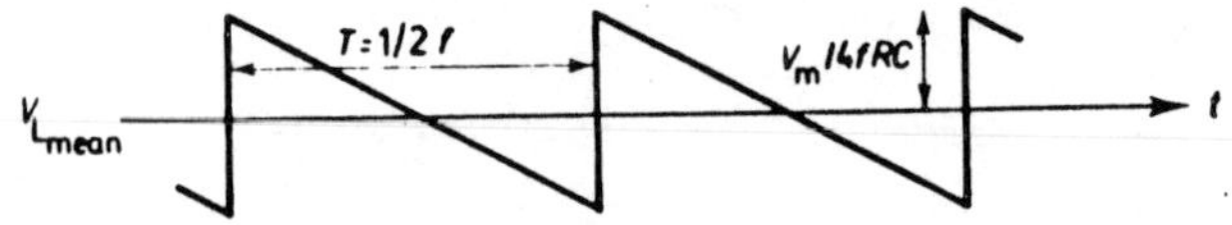

FIGURE 5.13

Ripple waveform of an RC smoothed full-wave rectifier

r.m.s. value of such a wave is $(1/\sqrt{3})\,(V_m/4fRC)$. Hence

$$\text{ripple factor} = \frac{1}{\sqrt{3}} \times \frac{V_m}{4fRC} \times \frac{1}{V_m\,(1 - 1/4fRC)}$$

since d.c. component $V_{L\,\text{mean}} = V_m\,(1 - 1/4fRC)$, therefore,

$$\text{ripple factor} = \frac{1}{\sqrt{3}(4fRC - 1)}$$

It is readily seen from this equation that the higher the frequency of the ripple, the smaller the value of capacitor C required for a given ripple content. This accounts for the greater popularity of full-wave and bridge configurations compared to half-wave rectifiers since the lowest a.c. frequency present is twice the supply frequency. It can also be seen that for high-power rectifiers, where the value of the load resistance R is very low, large values of input capacitor C are required. This in turn demands excessive values for the series limiting resistor (R_s in figure 5.11), with consequent power loss. High-power installations therefore usually employ some variation of the choke-input filter in which an inductor (sometimes termed a choke) is the first component after the rectifier.

5.2.3 Choke-input filters

Figure 5.14 shows the simplest possible form of choke-input filter. The back e.m.f. induced in the choke by any attempt at current change is used to minimise such changes. The choke, together with the resistive load, may be considered as a potential divider across the rectifier output. The reactance of the choke to the ripple frequency ω under consideration is of course ωL ohms, and the resistor ripple voltage is thus $V_{rip}R/\sqrt{(R^2 + \omega^2 L^2)}$, that is, reduced by a factor $R/\sqrt{(R^2 + \omega^2 L^2)}$.

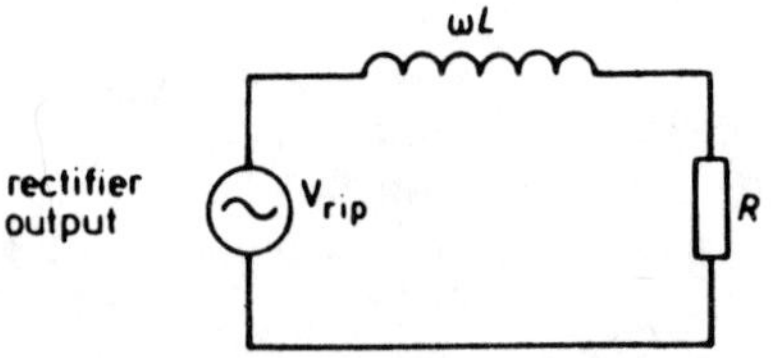

FIGURE 5.14

Simplest choke-input filter a.c. equivalent circuit

Inspection of the reduction factor reveals that it yields lowest output ripple for low values of R, corresponding to high output currents. Since the rectifier current is identical to the load current, no heavy capacitor charging-pulses occur and the need for a series limiting resistor with its associated losses is obviated.

5.3 Controlled Rectification

In many applications a d.c. supply produced by rectification must be subsequently controlled in magnitude, that is, the mean level of the d.c. voltage or current must be controllable. This could be performed by a series resistor; figure 5.15 shows the relatively small field current of a shunt d.c. motor being limited by a variable resistor to control the motor speed. If torque control was also required by means of a variable armature current, a resistor inserted in series with the armature (dashed) would carry the full armature current and dissipate much heat, since I_a is typically ten times I_f.

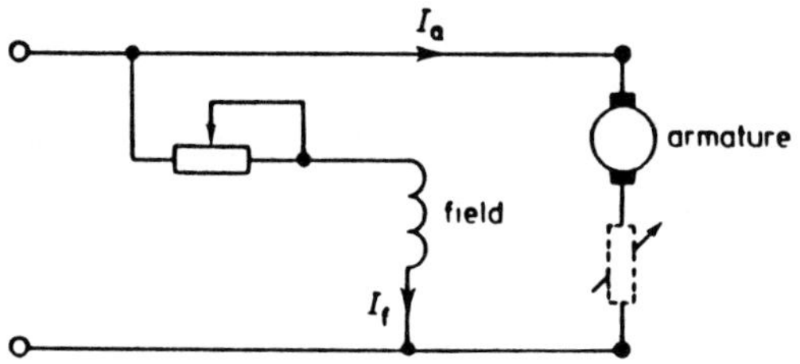

FIGURE 5.15

Motor speed and torque control by series resistors

If the rectifying device supplying the d.c. were capable of being controlled such that rectification only occurred during a portion of the a.c. cycle, the mean value of d.c. current could be varied at will. Figure 5.16a shows the waveform of an a.c. supply and directly beneath it the waveform of a normal full-wave rectifier. Conduction occurs during the whole of the forward half-cycle and the mean rectified current (dashed line) has been shown in section 2.1.1 to equal 0.638 I_m. If, however, the rectifier is constructed so as to delay conduction in each half-cycle by an angle α (the firing angle), conduction will only be permitted during the angle γ and it is left as an exercise to the student to show that the mean rectified current is now

$$I_{mean} = \frac{I_m (\cos \alpha + 1)}{\pi} \qquad (5.2)$$

Note, from figure 5.16c, that the resultant output waveform is no longer truly sinusoidal and therefore the r.m.s. value of the d.c. output current must be obtained from first principles. Because the form factor is no longer 1.11 a simple relationshop between mean and r.m.s. currents no longer exists.

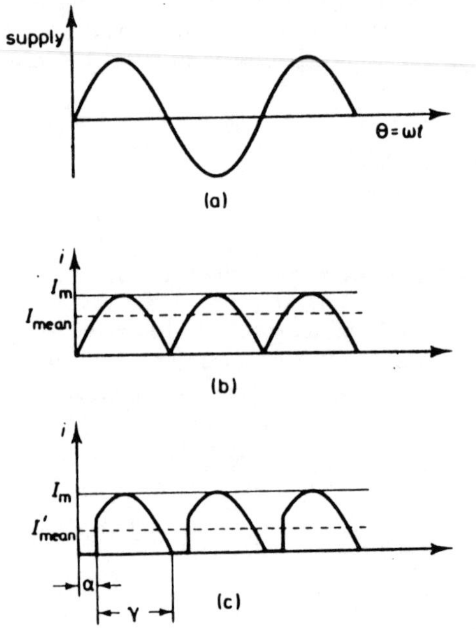

FIGURE 5.16

The effect of controlled rectification

5.3.1 The Thyristor as a Controlled Rectifier

A rectifying device capable of giving delayed forward conduction is the thyristor (see section 3.4.6). In the simplest circuit of figure 5.17 the thyristor directly replaces the diode in a half-wave rectifier and has its gate fed from a variable d.c. current source.

If R_g were adjusted so that I_g had a large value the thyristor would conduct over the whole of its forward half-cycle and the circuit behave exactly as the simple diode rectifier of figure 5.6. The rectified load-current waveform would be that of the solid line in figure 5.17 c. If I_g were now reduced, the thyristor conduction would be delayed until its forward breakdown-voltage was reached v_{b_2}. No current would flow until the firing angle (α_2 in figure 5.17c) had been reached. After firing, the voltage across a thyristor falls to a constant value v_c of approximately 1 V until the load current falls to the thyristor holding-current; at this time the thyristor switches off. During the next half-cycle (π to 2π) the thyristor blocks the reverse supply voltage, behaving exactly as a normal diode. Since the forward breakdown-voltage depends on the value of gate current, variation of I_g may be used to vary the firing angle as shown by α_1, α_2 and α_3 in figure 5.17c. The portion of each cycle for which conduction occurs is the conduction angle γ.

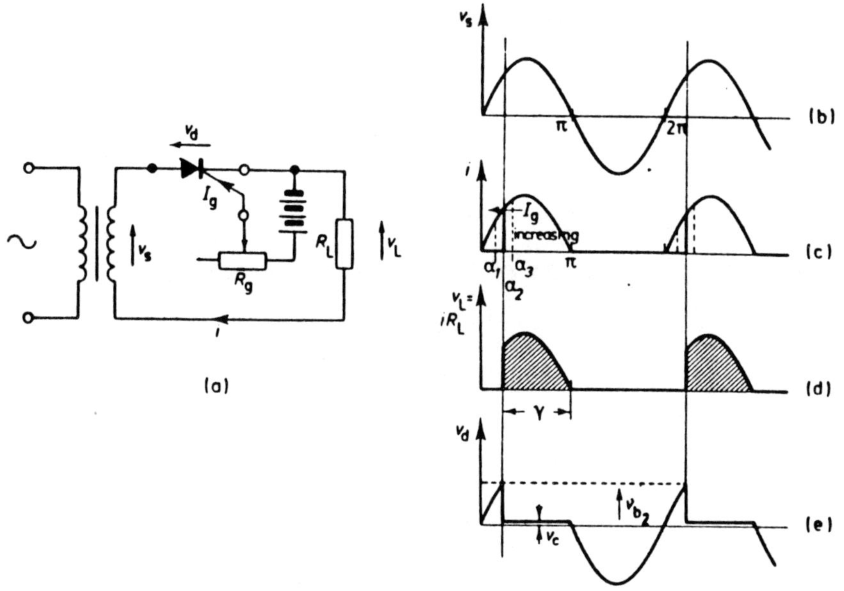

FIGURE 5.17

Basic half-wave controlled rectifier and its associated waveforms

A serious disadvantage of this method of firing is that, as I_g is reduced, the firing angle may be increased as shown but cannot be increased above $\pi/2$ radians. This is because any further reduction in I_g would result in the breakdown voltage required exceeding the maximum value of v_s (at an angle $\pi/2$ in figure 5.17b); consequently firing would never occur. Because the mean rectified current at a firing angle α of $\pi/2$ is one-half of the maximum mean rectified current which occurs at $\alpha = 0$, this simple firing circuit only allows the mean rectified current to be reduced to one-half of its maximum value. This circuit does however find application in simple ON–OFF controls as shown in figure 5.18 in its more usual full-wave (bi-phase) form. On closing S at time t_1 the gate current is allowed to reach a high value, limited only by R_g. At time t_1 the thyristors are put into a state in which they will conduct because either Thr_1 or Thr_2 will always be forward-biased. On opening S and removing the gate current no immediate switch-off occurs, since once a thyristor has attained the conducting state, the gate no longer has any control. Switch-off occurs as soon as the current through the conducting thyristor falls below the holding current. An application of this simple form of control is the use of the thyristors to switch on automatic fire-extinguisher pumps for aircraft engines. The output of a temperature-sensing element is used to provide the critical d.c. gate current for operation.

More usual thyristor firing circuits use either pulse or a.c. phase control to allow the rectified current to be fully controlled down to zero. Both these

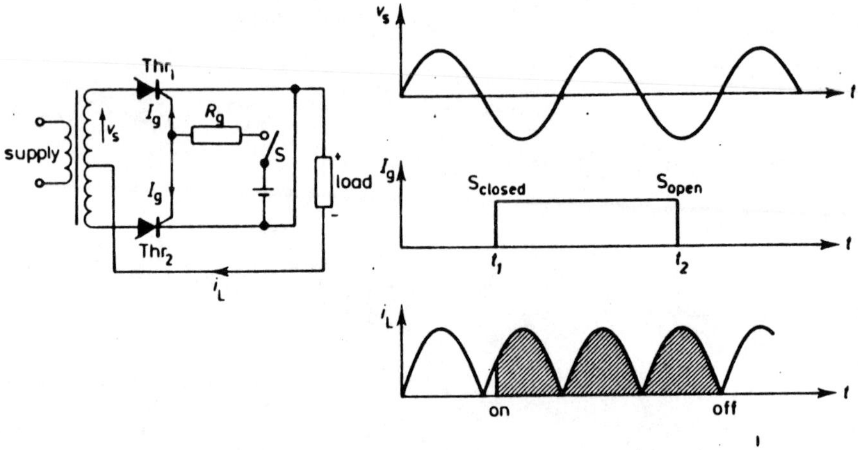

FIGURE 5.18

A full-wave ON-OFF controller using two d.c.-gated thyristors

methods employ a.c. gate currents allowing a transformer to be used to isolate the firing circuitry from the thyristor output. This is often necessary for operator safety since the rectified output voltages may be dangerously high.

Figure 5.19 shows the 0 to 180° phase-shift network (whose action has been explained in section 2.5.5) used to produce an alternating gate-current whose phase may be varied compared to that of the main supply winding. If the gate-cathode impedance of the thyristors is assumed purely resistive the gate current will be in phase with the gate-cathode voltage and may thus be varied from 0 to 180° by adjustment of R. Note that the magnitude of the a.c. voltage applied to the gate and cathode is constant, it is only its phase relationship to the supply voltage which is varied by R. If R is adjusted so that the gate current is in phase

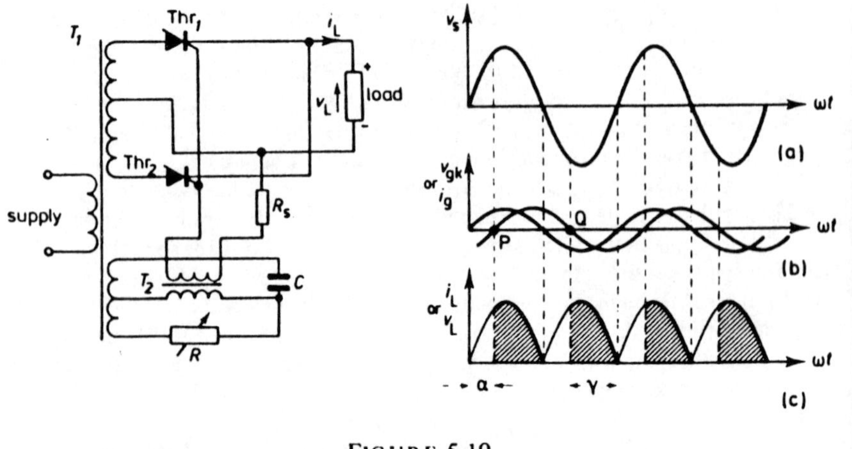

FIGURE 5.19

An a.c. phase-controlled rectifier and its waveforms

with the supply voltage, the gates are always of such a polarity with respect to their anode potentials that they conduct and the typical full-wave rectified output of figure 5.19c is obtained. If the values of C and R are such that the gate current lags the supply by α then the firing of each device will be delayed by this angle, the mean rectified current being proportional to the shaded area in figure 5.19c.

Example 5.3

A phase-controlled full-wave thyristor rectifier supplies current to a non-inductive furnace whose windings have a total resistance of 10 Ω. If the rectifier is fed from a 400 V - 0 - 400 V transformer winding, calculate the mean furnace currents at delay angles α of (i) 90° and (ii) 45°

(i) The circuit will be similar to figure 5.19a where the maximum value of V_s is

$$V_m = 400\sqrt{2} = 564 \text{ V}$$

neglecting a typical thyristor conduction voltage of 1 V compared to 400 V gives

$$I_m = \frac{V_m}{R} = \frac{564}{10} \text{ A}$$

Using equation 5.2 (p. 135)

$$I_{mean} = \frac{I_m(\cos\alpha + 1)}{\pi}$$

$$= \frac{56.4 \times 1}{\pi}$$

$$= 18 \text{ A}$$

(ii) At $\alpha = 45°$

$$I_{mean} = \frac{56.4 \times 1.707}{\pi}$$

$$= 30.7 \text{ A}$$

In practice a.c. phase control has the disadvantage that the precise value of i_g at which the thyristors will break down varies because of the production tolerances encountered in thyristor manufacture. A more reliable method of firing which retains the isolation between the control circuitry and the output voltage is pulse control.

At the desired point in the electrical cycle a pulse much larger than the minimum value required for firing is applied to the gate electrodes and it is only the timing of this pulse within the cycle which determines the firing angle. Figure 5.20 shows the pulses generated by an electronic clock (usually the relaxation oscillator of section 5.5.3) and synchronised with the supply voltage.

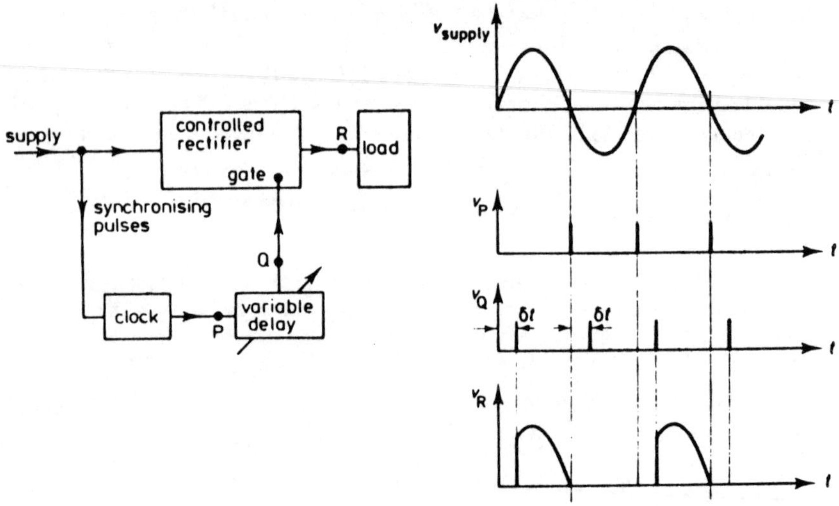

FIGURE 5.20

Block diagram and waveforms of a pulse controlled rectifier

These pulses are then delayed by a suitable electronic circuit so that they appear on the gate electrodes at the desired time. Altering the delay produced by the electronic circuits alters the firing angle as desired.

In precisely the same manner as section 5.1.3 the rectifying elements may be rearranged in the bridge form resulting in a cheaper transformer and halving of the required peak-inverse-voltage specification of the thyristors. Figure 5.21a shows a fully controlled single-phase bridge in which four thyristors are used.

Careful examination of figure 5.21 will show that in one half-cycle (A positive) the current flows from left to right through the load via Thr_1 and Thr_3 in series. There is thus no need for both these to be thyristors since if Thr_1 has not fired no current can flow in Thr_3. By a similar argument Thr_2 may be replaced by a diode D_2 yielding the final circuit 5.21b which is often employed for reasons of economy since diodes of a given voltage and current rating are

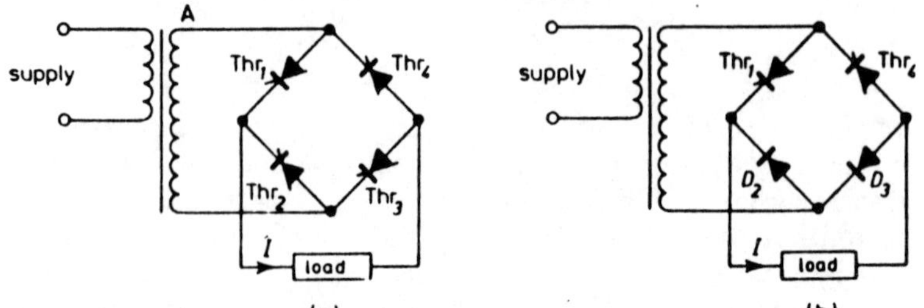

FIGURE 5.21

(a) Fully controlled and (b) half-controlled variants of a single-phase bridge rectifier. (Firing circuits omitted for clarity)

reasons of economy since diodes of a given voltage and current rating are considerably cheaper than thyristors of that rating. The fully controlled bridge does, however, retain some advantages which will become apparent in section 5.5.1.

For high-power applications it is likely that the d.c. will be obtained from a polyphase a.c. source and a three-phase fully controlled bridge suitable for this application is shown in figure 5.22a together with its associated waveform diagrams (b) at low and (c) at high conduction angles. As in figure 5.21 economy

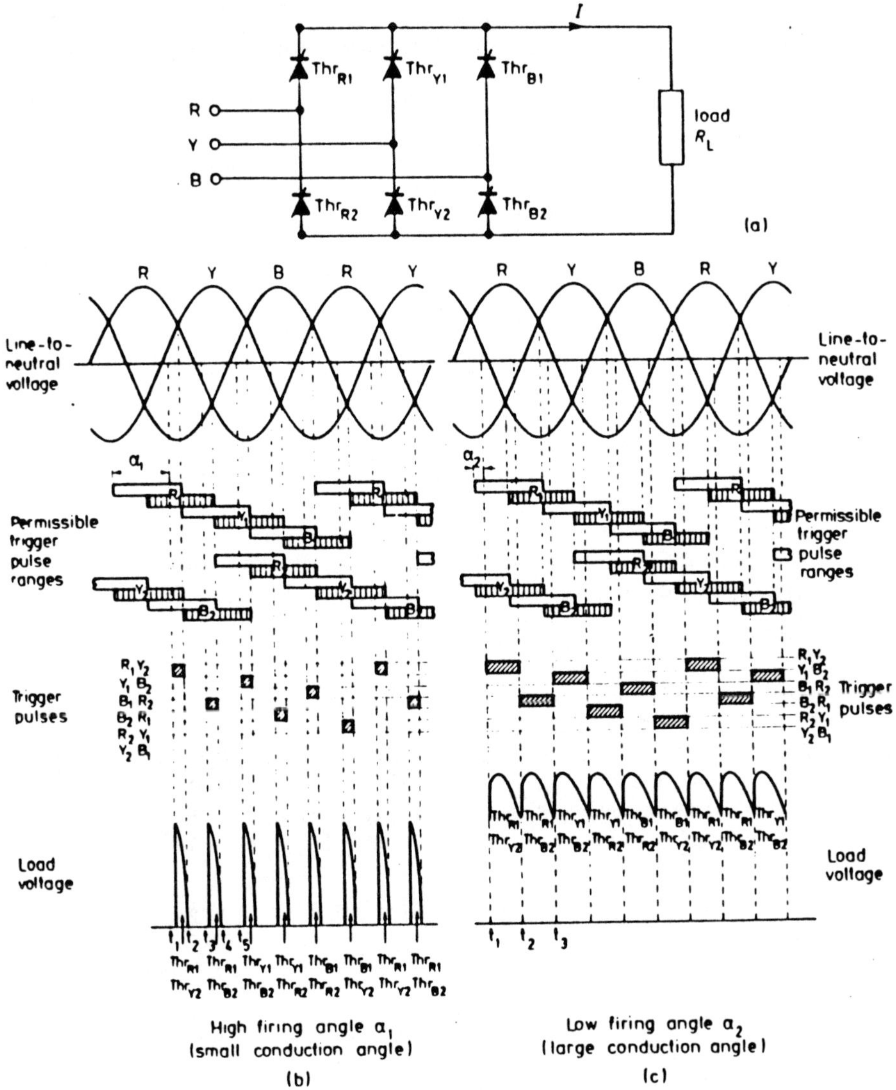

FIGURE 5.22

Three-phase fully controlled bridge rectifier (a) and its waveforms (b) and (c)

may be effected by the substitution of diodes for Thr_{R1}, Thr_{Y1} and Thr_{B1} if controlled rectification is all that is required.

Thyristors are currently being quantity-produced in ratings up to 1500 V and 700 A, giving a possibility of rectifying 2 MW of power with a three-phase bridge. This has caused the thyristor to replace other devices such as the mercury-arc rectifier in nearly all applications in modern equipment. Those encountering these older devices in existing equipment are referred to reference 4 for their principle of operation. The only shortcoming inherent in semiconductor devices when used as rectifiers is their inability to withstand even momentary overloads – for such applications as high voltage d.c. the later gas-filled devices are still preferred.

The waveform diagrams so far have illustrated the operation of controlled rectifiers only with purely resistive loads. In many applications, such as supplying energy to d.c. motor fields and armatures, the load contains an appreciable inductive element whose effect will be to oppose any sudden change of load current. This effect will cause the rising edge of the load-current pulses to become less abrupt and more rounded in form. Additionally the 'flywheel' action caused by energy storage by the inductance may cause the thyristors to conduct after the time at which 'switch-off' would occur with a resistive load. Modifications to the basic circuit may have to be made to provide a continuous path for inductive currents when the supply is suddenly interrupted otherwise high potentials may arise (section 1.2) which could exceed the maximum voltage rating of the thyristors.

A disadvantage encountered with any type of controlled rectifier is that, although the instantaneous values of current and voltage occurring during the cycle may be high the mean power absorbed by the load will be low at low conduction angles. Thus the effective power-factor which is Power (watts)/ Apparent Power (volt amperes) may be low. This may give cause for concern to power-supply authorities as discussed in section 2.3.3.

5.4 The Control of A.C. Current

The simplest method of limiting or varying the current flowing between an a.c. source and its load is the insertion of a series resistor as in figure 5.23a. This is impracticable in any but the lightest current applications because of the inefficiency due to energy dissipation as heat in the resistor. An alternative method is to use an iron-cored inductor (figure 5.23b) which may have its

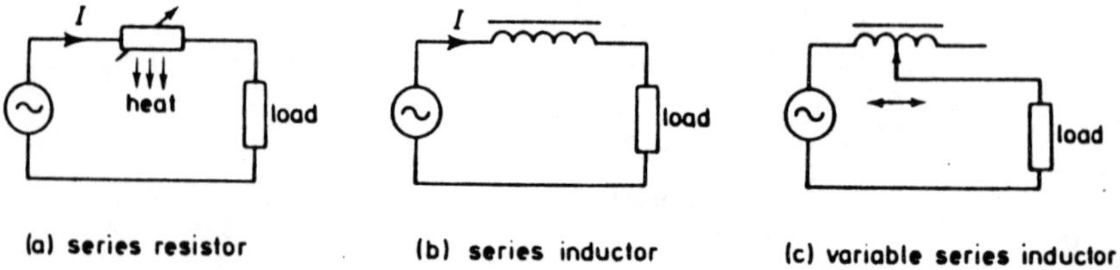

FIGURE 5.23

The simplest methods of limiting current I flowing from an a.c. source into a load inductance varied by altering the number of turns in operation by a tapping switch (figure 5.23c). Sliding iron cores have been used in inductors in attempts to provide continuously variable inductance-control as opposed to the discrete step control available with switch taps. The *saturable reactor* can provide load-current control by a comparatively small d.c. control current from a remote position.

5.4.1 The Saturable Reactor (or Transductor)

The impedance of an inductor is largely composed of its reactance because flux changes produce a back e.m.f. which opposes any current change. Figure 5.24a shows the area of operation on the *B*/*H* curve of a choke core as depicted in the circuit of figure 5.23c.

The operating area is centred around the origin and should be restricted in magnitude so that the core never saturates. If however by some external means the average core flux is biased away from zero (to Q in figure 5.24b), the same alternating m.m.f. now causes the core to saturate during part of the cycle. Once any further increase in flux is prevented by having entered the saturation region, the back e.m.f. (α d ϕ/dt) and thus the reactance collapse and the inductor impedance reduces to the (low) self-resistance of the winding.

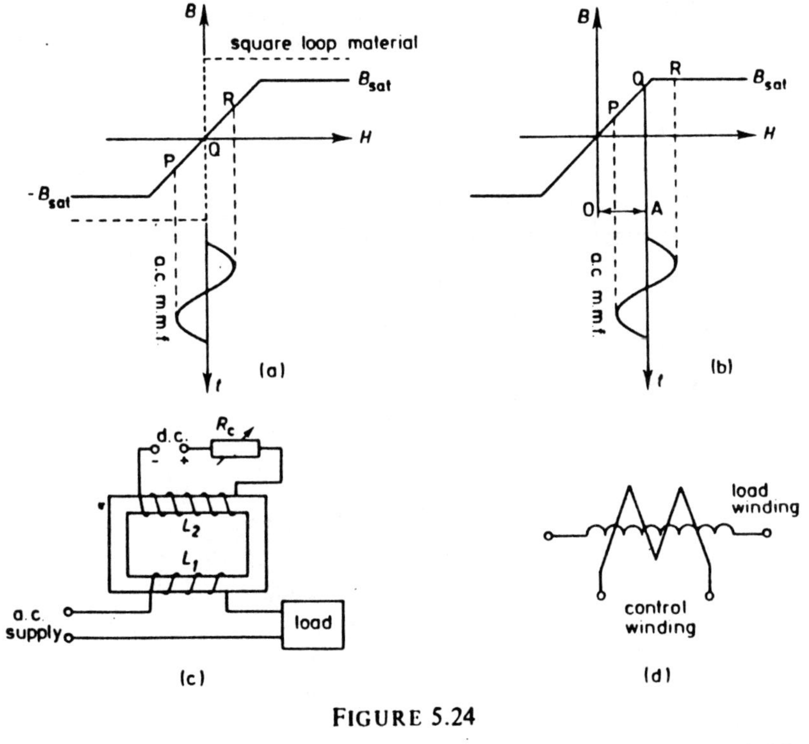

FIGURE 5.24

(a) *B*/*H* curves of a simple series choke and (b) the effect of d.c. magnetic bias. (c) Crude control circuit and (d) the symbol for a saturable reactor

In figure 5.24c the simple choke of figure 5.23b has been replaced by L_1 wound upon a two-limb core. The other limb bears a bias winding fed from a variable d.c. supply. Progressive reduction of R_c applies an increasing d.c. m.m.f. to the core allowing the reactor to saturate during the positive peaks of the a.c. m.m.f. cycle. This reduces the self-inductance of L_1 and hence the voltage drop across it. The reducing voltage drop thus increases the load voltage from a given supply. A d.c. m.m.f. from L_2 thus controls the load voltage and if L_2 has a large number of turns even small d.c. control currents may effect this change.

Three-limb Reactors

The two-limb saturable reactor previously discussed suffers from the major disadvantage that L_1 and L_2 together act as a transformer and the a.c. current changes in L_1 are reflected into the control circuit; this effect may be avoided by using three-limb reactors. The two a.c. load windings are connected in parallel (figure 5.25a).

The alternating fluxes produced by the two load windings L_1 and L_3 are arranged to flow in opposite directions through the centre control-limb L_2. Thus the net a.c. flux linking the control winding is zero. Such three-limb cores require great precision in construction to ensure exact flux cancellation, and more often two separate cores are used as in figure 5.25c.

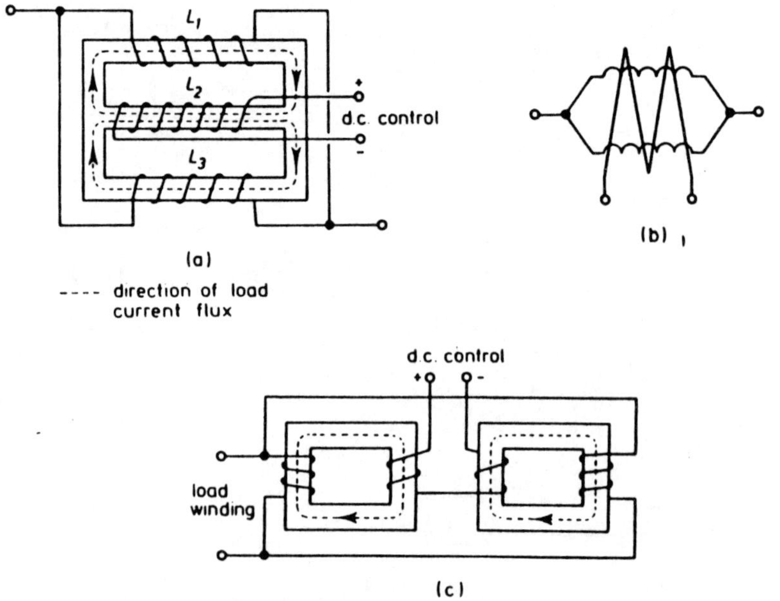

FIGURE 5.25

(a) Parallel-connected, three-limb saturable reactor and (b) its circuit symbol. (c) Two parallel separate cores provide the same effect

Waveform Diagrams

Consider the two-core parallel-winding circuit using square-loop core material having the B/H characteristics shown dashed in figure 5.24a and using a circuit as depicted in figure 5.26. If no saturation were to occur both flux-density waveforms would be sinusoidal (B_0). If a control current flows the flux density is limited at B_{sat} and $-B_{sat}$ for cores 1 and 2 respectively. The reason why the mean level of the two flux-densities moves in opposite directions is that the control windings are cross-connected to ensure this.

If no saturation occurred, the transductor voltage would be sinusoidal, leading the flux by $\pi/2$ rad or 90°. When the flux density reaches B_{sat} or $-B_{sat}$, at angles α and $(\pi + \alpha)$ the reactance falls to zero in a square-loop core material and the current is only limited by the load and the transductor self-resistance. This latter being negligible, V_{PQ} becomes zero, leaving the full supply-voltage across the load QE.

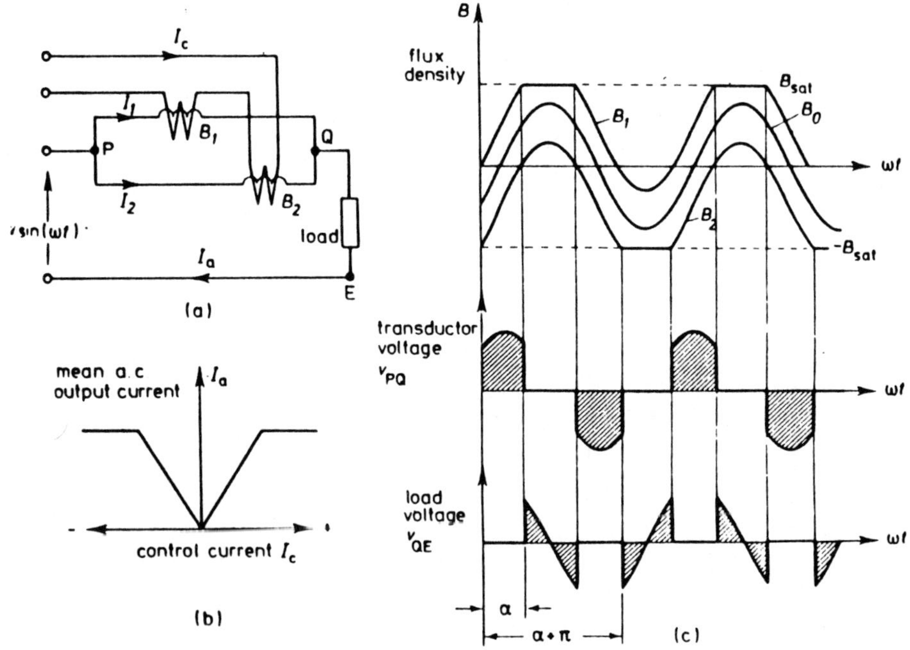

FIGURE 5.26

(a) Circuit, (b) transfer characteristic and (c) waveforms of a saturable reactor

Characteristic Curves

The mean a.c. output current I_a may be increased by increasing the control current I_c, thus saturating the reactor for an increasing proportion of the cycle.

If the control current is reversed, however (figure 5.26b) the mean value of I_a still increases even though the other winding is saturating. The flattening at the top of the characteristic shows that a maximum output current has been reached with one or other of the cores being saturated for the full cycle.

Additional windings may be added to the above simple arrangement to cause the transductor to act either as an amplifier or as a switch. The study of these is beyond the range of this text but reference to the bibliography [15] will provide information on these magnetic amplifiers which form a robust method of control. They were extensively used instead of thermionic valves in applications where vibration or mechanical shock was high, before the more rugged semi-conductor devices were developed.

5.4.2 The Triac and Quadrac ®

These are bidirectional semiconductor switches similar to the thyristor except that the rectification action is absent. The principal voltage – current characteristics and circuit symbol are shown in figure 5.27.

Operation in the forward direction (upper-right quadrant of the electrical characteristics in figure 5.27) is identical to that of the thyristor; the only difference being that the gate terminal may be made negative as well as positive with respect to MT1 to initiate breakdown. The construction is such that the main terminal MT2 is usually in electrical contact with the body and mounting stud.

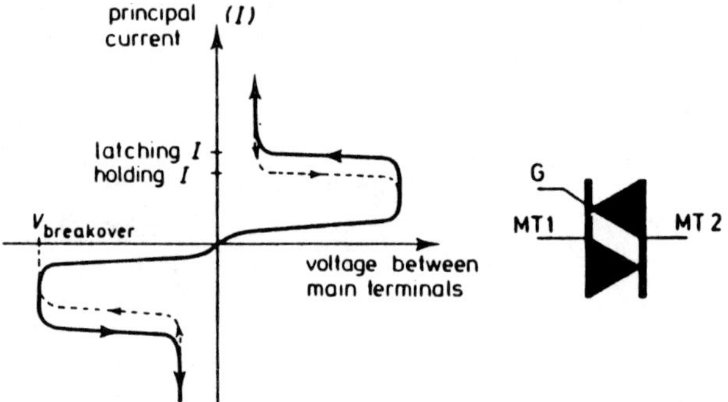

FIGURE 5.27

Characteristics and circuit symbol of a triac

Examination of the reverse characteristic shows however that (unlike the thyristor) reverse breakdown may also be initiated by the trigger electrode; in fact operation in four separate polarity-modes is possible as the curves in the first and third quadrants are symmetrical. These modes correspond to MT2 being either positive or negative with respect to MT1, and the voltage between the gate and MT1 being either positive or negative. One of these combinations is however usually avoided (third quadrant with gate positive) since triggering difficulties often occur in practice.

® Registered Trade Mark of the E.C.C. corporation

Triac Firing Circuits

There are several possible circuits for firing a triac very similar to those employed for thyristors. Regardless of which circuit is used, the gate firing current and the latching current I_L must be maintained above certain levels.

Figure 5.28a shows a simple solid-state contactor using a manually controlled gate switch to operate a triac used as a simple on - off switch. This circuit would be used in preference to a manual main switch because the mechanical moving parts are only carrying the relatively low gate-currents and will not suffer from burnt contacts caused by arcing.

Gate current i_g is positive whenever MT2 is positive and vice versa. This avoids the undesirable operating mode of i_g positive, negative MT2. The triac will turn off each time the principal current drops to zero (at the end of each cycle). A suitable value for R_G would be 400 Ω for a 240 V supply.

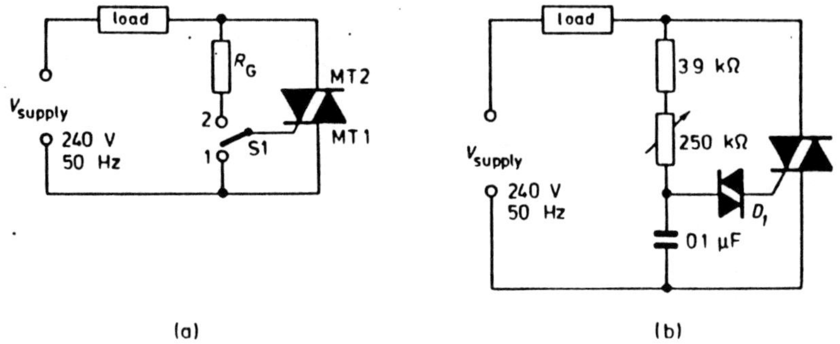

FIGURE 5.28

(a) ON - OFF power control circuit, (b) phase control using a bidirectional diode

A disadvantage found with both triacs and thyristors is that there is considerable variation between devices of the threshold gate-current required for firing with a given supply voltage. This may be overcome as in figure 5.28b by using a phase-shift control consisting of an *RC* network and feeding this into a bi-directional trigger diode D_1. As soon as the control voltage reaches the diode breakdown voltage D_1 conducts heavily applying a sharp trigger current to the triac gate and accurately determining the firing point.

So common has this last method of firing become that it is now possible to purchase triacs with a trigger diode ready connected to the gate electrode and mounted within the same case. The word Quadrac® is used to describe such a triac - trigger-diode combination.

Example 5.4

An a.c. lamp-dimming controller uses the circuit of figure 5.29. The supply voltage is 240 V and the load consists of filament lamps having a total resistance of 10 Ω when the firing angle is 90°. Calculate the power supplied to the lamps if the forward voltage-drop of the triac is negligible.

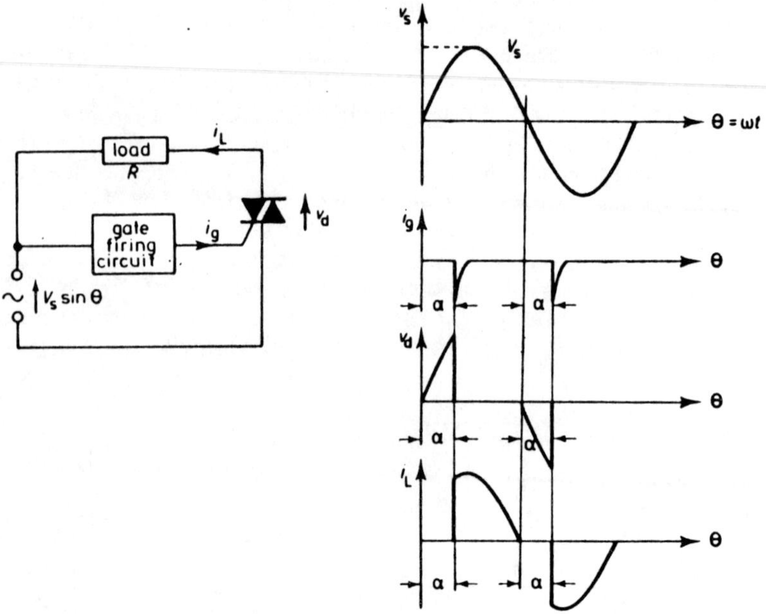

FIGURE 5.29

Basic a.c. control circuit using a triac and its principal waveforms when feeding a resistive load

The waveform of the load current will be that of i_L in figure 5.29. The maximum value of the instantaneous load current will occur at the moment of firing (at 90°), hence the mean power will be

$$P = \frac{1}{2\pi}\int_0^{2\pi}\frac{v^2}{R}\,d\theta$$

$$= \frac{1}{\pi}\int_{\pi/2}^{\pi}\frac{v^2}{R}\,d\theta = \frac{1}{\pi R}\int_{\pi/2}^{\pi} V_m{}^2 \sin^2\theta\,d\theta$$

since

$$V_m = 240\sqrt{2}$$

$$P = 3667\int_{\frac{\pi}{2}}^{\pi}(1 - \cos 2\theta)\,d\theta$$

$$= 3667\left[\theta - \frac{\sin 2\theta}{2}\right]_{\frac{\pi}{2}}^{\pi}$$

$$= 3667 \left[\pi - 0 - \frac{\pi}{2} + 0 \right]$$

$$= 5760 \text{ W}$$

5.4.3 Resistance-welding Control: the Ignitron

In applications where high momentary overloads may occur the vulnerability of semiconductor devices to such hazards may be avoided by using a specialised form of mercury-arc rectifier such as the ignitron described in section 3.3.4. In welding circuits the ignitron is used merely to control the duration of an alternating current, not for rectification.

The sequence of operation during resistance welding is as follows.

(i) The two items to be welded are placed in contact with each other between the electrodes of the welding machine (figure 5.30) and subjected to an accurately controlled mechanical pressure from the electrodes.

(ii) A high alternating current is passed through the junction of the two items via the electrodes for an accurately controlled time. This precise control of welding time is necessary because of the varying natures, surface conditions and thicknesses of workpieces encountered.

(iii) The passage of current having ceased, the electrodes must continue to exert their pressure until the workpiece material solidifies.

(iv) The electrode pressure is released and the electrodes part to allow the work to be removed.

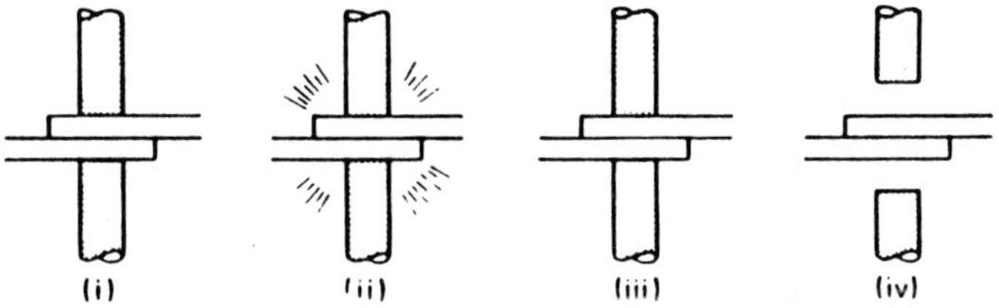

	Grip	Weld	Hold	Release
Pressure	ON	ON	ON	OFF
Current	OFF	ON	OFF	OFF

FIGURE 5.30

Simple resistance welding sequence

Each of these four steps of the sequence is accurately timed by electrical means but it is the control of the second (weld) phase which concerns us here. The heating effect is caused by the large current which passes through the low-resistance copper electrodes and traverses the comparatively high-resistance workpiece and surface layers (of total resistance R ohms). The heat liberated during a flow of weld current for t seconds will be

$$E = I^2R \times t \text{ joules}$$

Hence for workpieces of various materials and surface conditions control of both weld time and current will be required. The currents encountered vary from tens of amperes in light spot-welding to several thousand amperes for a butt-weld between 5 cm diameter bars. These heavy currents are obtained from the secondary of a voltage step-down transformer and current control is obtained by switching the primary circuit (figure 5.31a). Any form of mechanical switch is unsuitable for this application since the primary currents (up to 100 A) would cause pitting and corrosion of the contacts, nor could the required precision of timing be achieved.

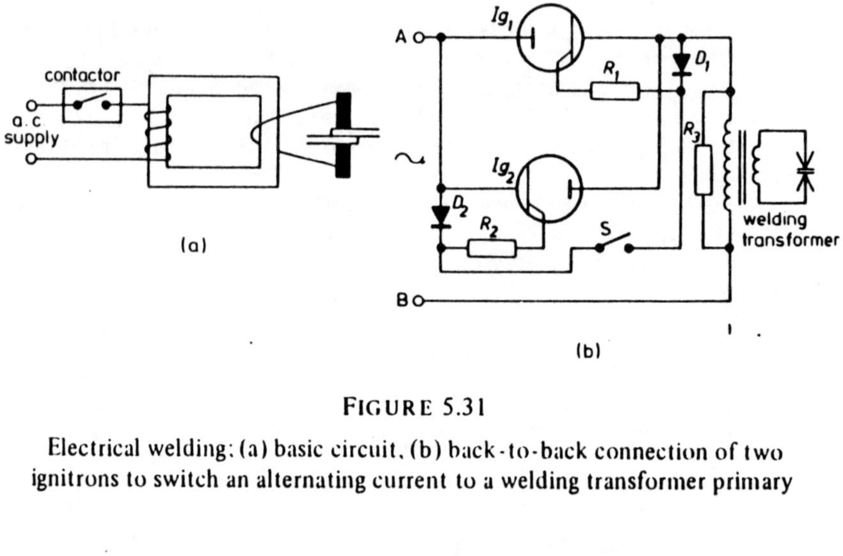

FIGURE 5.31

Electrical welding: (a) basic circuit, (b) back-to-back connection of two ignitrons to switch an alternating current to a welding transformer primary

Two ignitrons connected in inverse parallel (back to back) will not give any rectifying action since one anode will be positive with respect to its cathode in each half of the cycle. The circuit of figure 5.31b shows a self-contained ignitron contactor in which the closing of S will immediately cause the ignitron with a positive anode to fire, the transformer primary current then being carried by each ignitron in turn until S is reopened. Assume the supply terminal A to be positive. When S is closed, current may flow from A through D_2, S and R_1 to the igniter of I_{g1} which, having a positive anode and negative cathode, will fire. As long as S is closed I_{g2} will take over conduction at the commencement of the next half-cycle; its igniter current flowing from B, through the transformer primary, D_1, S and R_2.

Resistor R_3 allows a safety path for the high voltages induced in the primary when the current is interrupted. This occurs at the end of the half-cycle during which S is opened, since absence of igniter current will not in itself extinguish an ignitron; the current does not decrease gradually to zero since about 10 A is the minimum current an ignitron can carry. This sudden decrease from 10 A to zero can give rise to large e.m.f.s in the highly inductive primary which might cause ignitron malfunction. It is possible to replace S by an electronic switch synchronised to the supply frequency so that firing may be initiated at some precise point in the supply cycle and may continue for a definite number of half-cycles; this will determine the weld time with great accuracy.[16]

5.5 Inversion

Inversion is the process of converting electrical energy from a unidirectional (d.c.) form to an alternating form. It may thus be introduced as inverse rectification. *Conversion* is a term which is variously used to describe either rectification or inversion or merely a change in level, not in kind. An example of the last is a device to produce d.c. at say 250 V from an input of 12 V d.c. Because of this variation in usage the word converter and conversion will not be used here.

Figure 5.32 shows a rectifier/inverter interposed between an a.c. and a d.c. system. If the energy flow is from left to right rectification occurs, whilst the reverse direction of energy flow indicates the presence of inverter action. Inverters may be subdivided into natural-commutation and forced-commutation types; the latter generates its own frequency whereas the former feeds power from a d.c. supply into an existing a.c. system and is therefore forced to operate at the existing a.c. frequency.

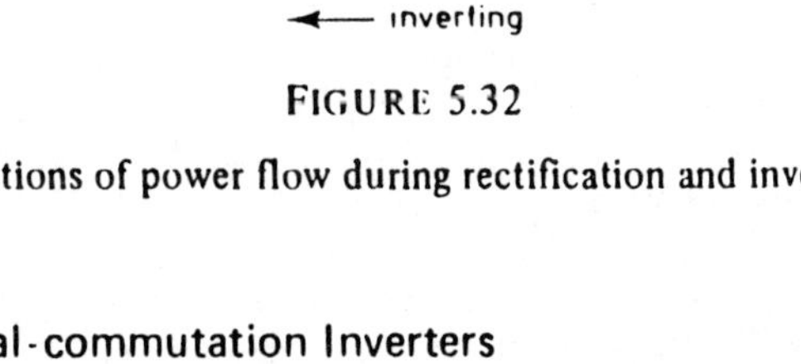

FIGURE 5.32

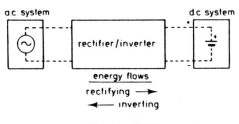

Directions of power flow during rectification and inversion

5.5.1 Natural-commutation Inverters

Any rectifying circuit, other than the single-phase half-wave, in which all the diodes are controlled rectifiers may be operated as an inverter. The most commonly used circuits are the single-phase and three-phase bridges (figures 5.21a and 5.22a). The direction of current flow on the d.c. side of the rectifier is fixed by the direction of the diode connections. To reverse the direction of energy flow in order to obtain inverter action therefore, the polarity of the d.c. system must be reversed. This is usually accomplished by inserting a reversing contactor or switch between the rectifier/inverter and the d.c. system. Figure 5.33 shows a single-phase fully controlled bridge rectifier connected (a) as a rectifier and (b) as an inverter.

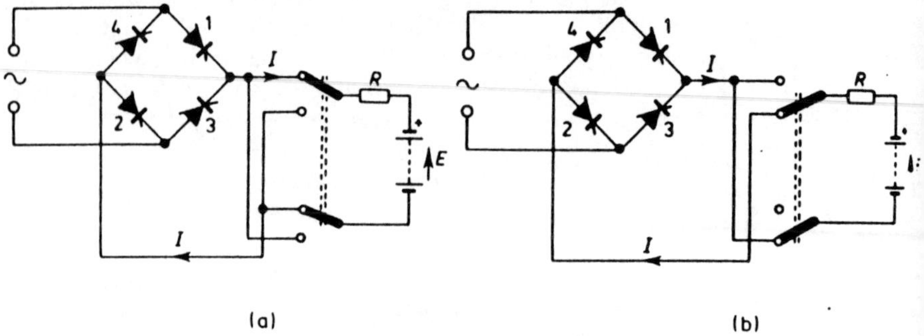

FIGURE 5.33

A full controlled single-phase bridge and a reversing contactor used as (a) a rectifier and (b) an inverter. (*R* limits the current to an acceptable value.)

The output-voltage and thyristor-current waveforms are shown in figure 5.34a for successive half-cycles with gradually increasing firing angles. It will be noted that unlike the simple resistive-loaded rectifier waveform of figure 5.19c conduction occurs on the negative half-cycles, gradually increasing as the firing angle increases until, when $\alpha = 90^\circ$, the currents in the forward and reverse directions are equal and no net power transfer occurs. If the d.c. system has its polarity reversed and the firing angle is set to more than 90° inverter action occurs; increasing in magnitude until maximum net power is transferred from the d.c. to the a.c. system when $\alpha = 150^\circ$. For technical reasons concerned with the firing of the thyristors α is not allowed to approach any closer to its theoretical maximum of 180°.

A graph of the theoretical d.c. output voltage against firing angle is given in figure 5.34b (ii) and may be compared with a similar curve (figure 5.34b (i)) for the half-controlled bridge when two of the thyristors have been replaced by diodes. Some clarification of the fully controlled inverter curve is required since the reader might assume from a casual glance that the bridge was capable of delivering output voltages and therefore output currents of either polarity dependent upon whether α is greater or less than 90°. It must be emphasised that operation at α greater than 90° can only be obtained when feeding a d.c. supply which is capable of reversal thus causing the output current to flow in the only possible direction (set by the thyristor connections.)

A typical application of such an arrangement would be an electric hoist in which the bridge could be operated without a reversing contactor. A d.c. hoist motor could have its speed and torque regulated by feeding its armature with the output of the fully-controlled bridge; the field receiving constant excitation. This would allow loads to be raised under control, the motor speed falling to zero when $\alpha = 90^\circ$. As the loaded hoist descends, the winding-motor will rotate in the reverse direction, acting as a generator with reverse polarity. If the firing angle during lowering is varied between 90° and 150°, the speed of descent may be controlled electrodynamically as the fall in potential energy of the load is returned via inverter action to the a.c. supply (regenerative braking). This

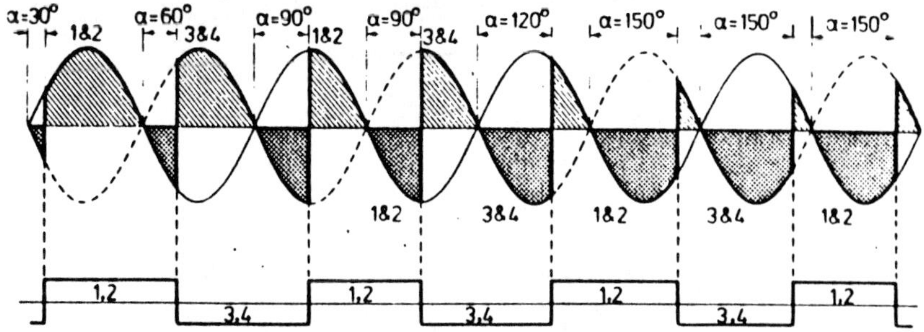

(a)

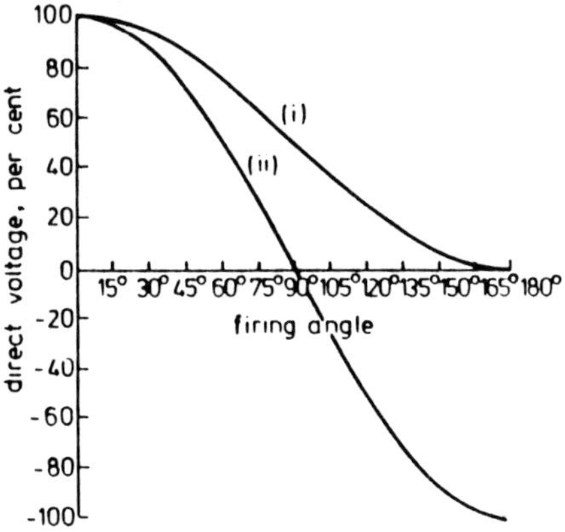

(b)

FIGURE 5.34

(a) Operation of a single-phase bridge circuit with different angles of delay α. Numbers 1 to 4 refer to the thyristors in figure 5.33
(b) Control characteristics for (i) a half-controlled rectifier and (ii) a fully controlled rectifier/inverter

electrodynamic braking may be experienced on a bicycle – it is slightly harder to pedal the machine at a given speed when the dynamo (or alternator) is loaded by the lighting circuit; this being irrespective of the fact that the dynamo is still being rotated when the lights are extinguished.

In an application such as an electric locomotive where traction and regenerative braking are required in both directions a reversing contactor has to be used. Where the use of a contactor is precluded because of the required speed of operation, two bridges having outputs of reverse sign are operated in back-to-back (inverse parallel) connection (figure 5.35). Such an application might be the position control of the massive workpiece in a large automatic machine-tool.

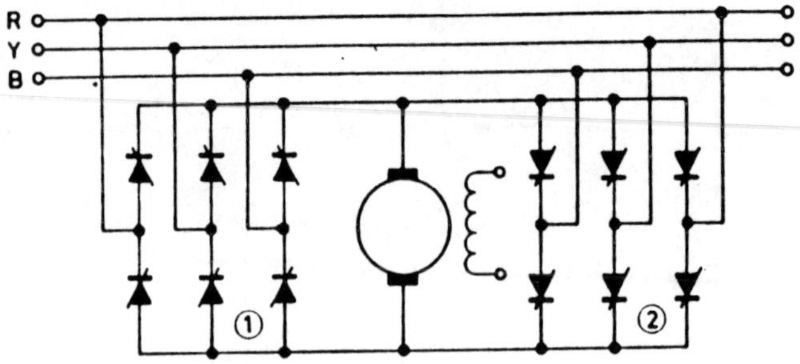

FIGURE 5.35

Controlled rectifier/inverters connected in inverse-parallel to provide fast control of a d.c. motor's speed

Assume that initially the worktable was travelling forward at its highest speed and that this movement had to be reversed as quickly as possible. Assume that bridge 1 drives in the forward direction. Its firing angle would have previously been very small to give this high forward speed. On initiation of braking the firing angle of bridge 1 would instantaneously be set to 90° (removing the drive) while bridge 2 would be set to fire near 180°. This provides regenerative braking with bridge 2 acting as an inverter, returning the inertia energy of the moving table to the a.c. supply. As soon as the worktable comes to rest the firing angle of bridge 2 is switched to a small value and the motor accelerates in the opposite direction.

Rectifier/inverter drive offers the following advantages to the Ward - Leonard system (section 4.6.6) for the drive of reversible loads

(1) lighter and therefore easier to install
(2) more efficient
(3) faster response
(4) easier to control at low speeds
(5) absence of moving parts in the control equipment.

Comparative disadvantages include
(1) the non-sinusoidal switching current can cause mains interference
(2) it presents a low power-factor to the a.c. supply at low speeds
(3) it lacks the energy storage capability of the Ilgner version of the Ward - Leonard system (use of a generator flywheel).

5.5.2 Forced-commutation Inverters

In many applications an inverter is required to provide a.c. power from a d.c. source without the presence of an existing a.c. system into which to feed. Perhaps the most common example of this is the high voltage a.c. requirement for fluorescent lighting in public service vehicles where the main electrical system operates at low voltage d.c. from accumulators. Forced-commutation inverters usually employ an *oscillator* to generate an a.c. voltage or current which is then sometimes amplified (increased in magnitude) before utilisation.

5.5.3 Oscillators

An oscillator is a device for producing alternating voltages or currents from a direct current power source. The output waveform may have one of many shapes, for example sinusoidal, square or triangular waveforms. The practical circuits for electronic oscillators are discussed in section 8.5 because a prior knowledge of *amplification* is needed.

A useful introduction to the subject can be obtained, however, by considering the hydraulic and the electromechanical oscillators shown in figures 5.36a and 5.38a. The former depicts a syphon whose flow rate is much greater than that of the return pump and restriction. The water level in the upper tank will fluctuate between the levels A and A′ as shown in figure 5.36b; the frequency of these oscillations will be determined by the diameter of the restriction and the difference in capacities of the upper tank between levels A and A′. The energy to drive this 'oscillator' is obtained from the constant (or d.c.) hydraulic power supplied by the pump.

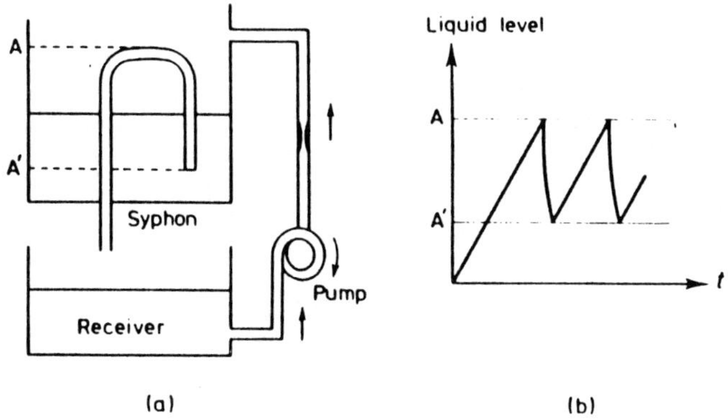

FIGURE 5.36

A liquid relaxation oscillator; (a) the pump and syphon and (b) the 'output waveform'

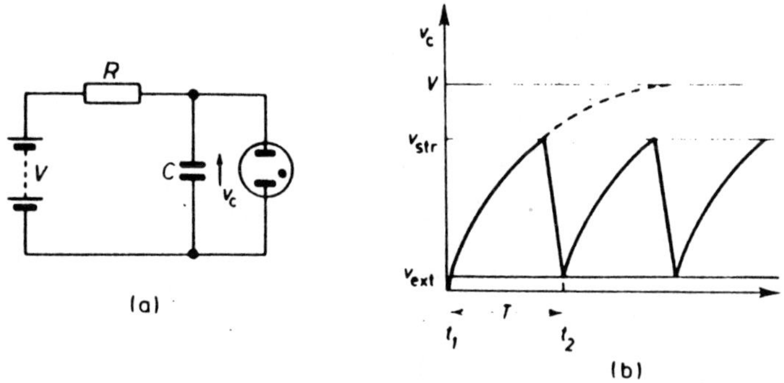

FIGURE 5.37

An elementary electrical relaxation oscillator and its output waveforms

This is analogous to the primitive electrical *relaxation oscillator* of figure 5.37a in which capacitor C charges slowly from a d.c. source V via a large resistor R. When the capacitor voltage v_c reaches the gas-tube's striking voltage V_{str}, C is discharged almost instantaneously until v_c falls to the tube's extinction voltage V_{ext}. The capacitor voltage thus varies with time as shown in figure 5.37b. The frequency of the oscillations are determined by the values of V_{str} and V_{ext} together with the time constant RC (see section 1.3).

Example 5.5

Calculate the frequency of oscillation of the neon-tube oscillator shown in figure 5.37a if the supply voltage is 250 V, the striking and extinction voltages are 180 V, and 80 V respectively and if the values of R and C are 1 MΩ and 1 μF. Neglect the tube's resistance when conducting.

$$T = t_2 - t_1$$

from section 1.3. At t_2

$$V_{str} = V\left[1 - \exp\frac{-t_2}{RC}\right]$$

at t_1

$$V_{ext} = V\left[1 - \exp\frac{-t_1}{RC}\right]$$

so

$$t_2 = \frac{1}{RC}\ \ln\frac{V}{V - V_{str}}$$

and

$$t_1 = \frac{1}{RC}\ \ln\frac{V}{V - V_{ext}}$$

therefore

$$T = t_2 - t_1 = \frac{1}{RC}\ \ln\frac{V - V_{ext}}{V - V_{str}}$$

$$T = \frac{1}{10^6 \times 10^{-6}}\ \ln\frac{250 - 80}{250 - 180}$$

$$= \ln 2.43 = 0.888$$

therefore

$$\text{frequency} = \frac{1}{T} = 1.13\ \text{Hz}$$

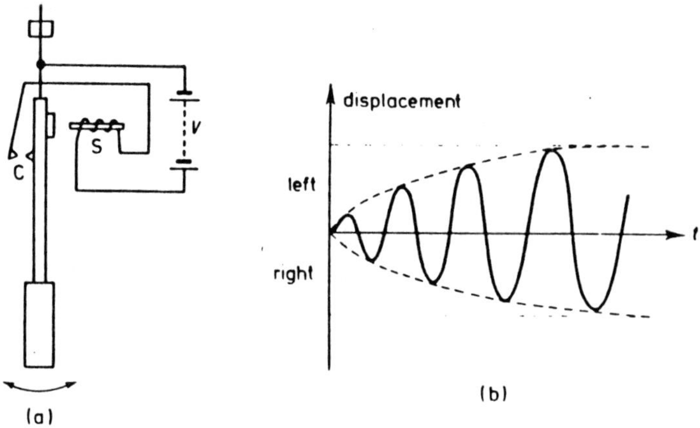

FIGURE 5.38

(a) An electromechanical sinusoidal oscillator and (b) its 'output waveshape'

The electromechanical 'oscillator' of figure 5.38a has a pendulum whose swings are maintained by electromagnetic attraction to the solenoid S whenever left-hand excursions cause the contacts C to close. The energy to counteract air friction and pivot losses is derived from the (d.c.) battery and the frequency is determined almost entirely by the pendulum's dimensions and by the acceleration due to gravity. The graph of its movement might resemble figure 5.28b.

The electrical equivalent would be an *LC* tuned-circuit oscillator whose block diagram is shown in figure 5.39a. The output of an amplifier (section 8.2) is fed, via the current-limiting safety resistor R_s, to a coil L'. The mutual inductance of the latter causes an e.m.f. in L reinforcing the original input voltage. The correct phase relationship for this reinforcement can only occur in a narrow band of frequencies around the *resonant frequency* of *LC* (see section 2.4.2).

In both the electromechanical and electronic sinusoidal oscillators the final value of the amplitude (steady-state condition) is limited by the losses and the potential V of the power supplies.

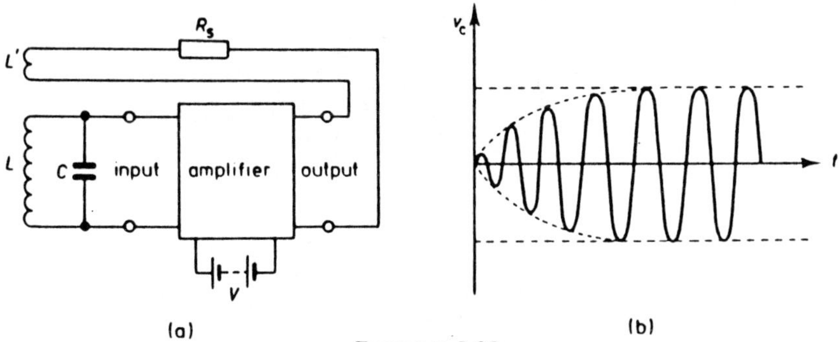

FIGURE 5.39

(a) An electronic *LC* oscillator and (b) its initial output waveform

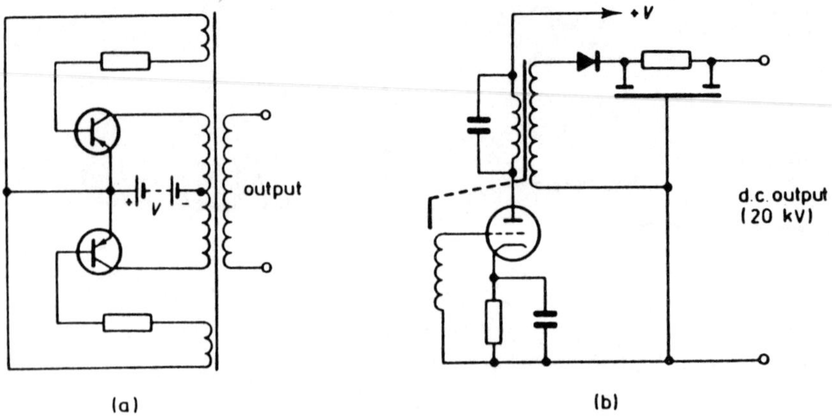

FIGURE 5.40

Circuit diagrams of two practical electronic oscillators

5.5.4 Oscillators for Forced Commutation

More sophisticated types of electrical oscillator than the above are used for commutating inverters. The oscillator output is often at a frequency much higher than 50 Hz so that the values of smoothing components (section 5.2) to remove the ripple can be minimised. Frequencies of 10 kHz are often encountered.[17] Figure 5.40a is a magnetically coupled multivibrator oscillator suitable for power applications up to 20 W. The alternating output for subsequent rectification and smoothing is a square wave of frequency 1 kHz.

Figure 5.40b however is a high-frequency (20 kHz) sinusoidal *LC* oscillator of the type once used to produce the extra-high-tension supplies of 20 kV for television cathode-ray tubes.

A simply built practical circuit for supplying a few hundred volts from a car battery is shown in figure 5.41.

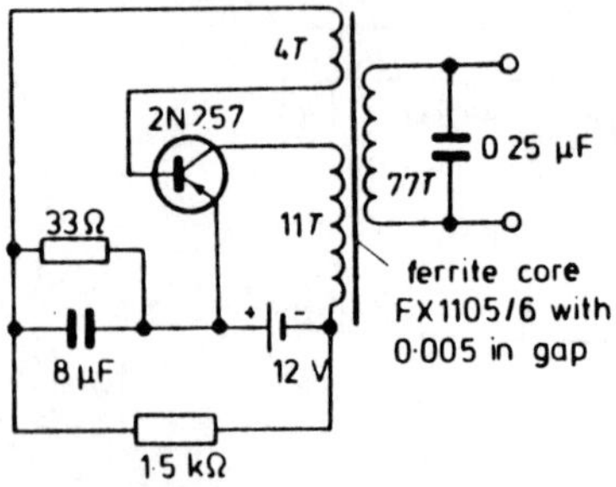

FIGURE 5.41

A practical circuit for supplying mains voltage from a car battery

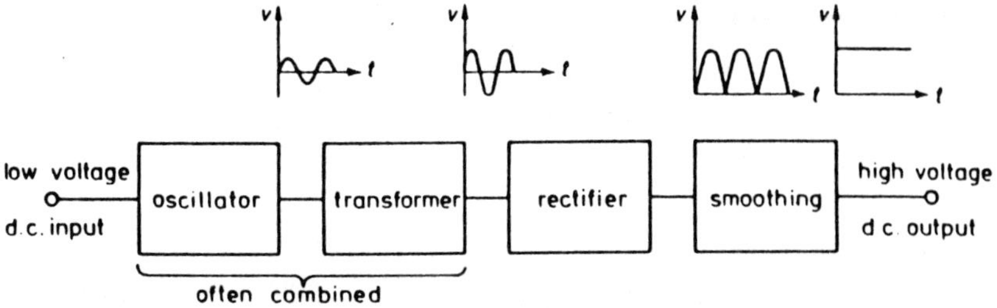

FIGURE 5.42

The basic components of a converter which uses a forced-commutation inverter

The alternating output of a forced-commutation inverter may easily be transformed upwards in potential for subsequent rectification and smoothing to provide a high-voltage d.c. output from a low-voltage d.c. input. The main components of such a *converter* circuit are shown in figure 5.42; either the full-wave or bridge rectifiers of sections 5.1.2 and 5.1.3 may be adapted for this purpose and complete circuits are readily available.[18]

5.6 Dielectric Heating

This technique is often encountered in such applications as the pre-heating of moulding powders or the welding of plastic sheet (usually PVC). It relies for its action upon the fact that many insulating materials have internal losses, that is they may be represented by their dielectric capacitance C together with a parallel loss-conductance or resistance R (figure 5.43b).

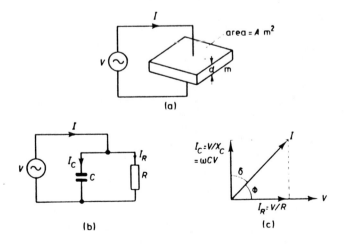

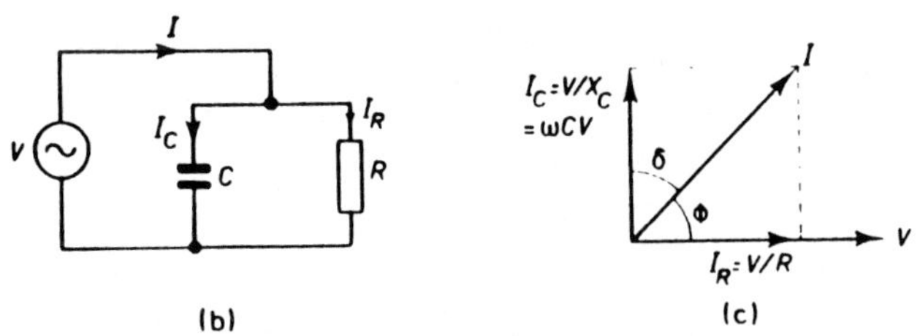

FIGURE 5.43

Dielectric heating

Assume that the block of dielectric material in figure 5.43a to be welded or melted down is subjected to a sinusoidally alternating electric field of angular frequency ω radians per second. If ϵ_0 and ϵ_r represent dielectric constants of vacuum and material, the block's capacitance will be

$$C = \frac{\epsilon_0 \epsilon_r A}{d}$$

Heating occurs as a result of current flowing in the parallel loss-resistance R. From the phasor diagram $I_r = I_c \tan\delta$ where δ is the loss angle (defined as 90° minus the phase angle ϕ).

$$I_r = I_c \tan\delta = \omega CV \tan\delta$$

The power lost in R is

$$W = I_r^2 R = \frac{I_m^2 R}{2}$$

where I_r is the r.m.s. and I_m the peak value of current. Therefore

$$W = (\omega CV_m \tan\delta)^2 \frac{R}{2}$$

but from figure 5.43c $\tan\delta = 1/\omega CR$, thus

$$W = \frac{1}{2} \omega CV_m^2 \tan\delta$$

where V_m is the maximum or peak value of the applied voltage. If the loss angle is small, δ is approximately equal to $\tan\delta$, therefore

$$W \approx \frac{1}{2} \omega CV_m^2 \delta$$

Substituting for C

$$W \approx \omega V_m^2 \delta \times \frac{\epsilon_0 \epsilon_r A}{2d}$$

Division by Ad (the volume) will give the power loss per cubic metre as

$$= \omega \left(\frac{V_m}{d}\right)^2 \frac{\delta \epsilon_0 \epsilon_r}{2} \text{ W/m}^3$$

$$= \frac{\omega E_m^2 \delta \epsilon_0 \epsilon_r}{2} \text{ W/m}^3$$

where E_m is the maximum potential gradient (V_m/d). Since

$$\epsilon_0 = \frac{1}{36\pi} \times 10^{-9} \text{ and } \omega = 2\pi f$$

$$\text{Loss per cubic metre} = \frac{f E_m^2\, \delta \epsilon_r}{18} \times 10^{-9} \text{ W/m}^3$$

For substantial power dissipation therefore either high voltage or high frequency, or both are needed.

Suitable frequencies range from 2 MHz upwards and potential gradients of 100 to 200 kV/m are not uncommon. Care must be taken to avoid an airgap between the material and the electrodes because it is a simple matter to show that the potential gradient in the airgap would be ϵ_r times that in the dielectric.

Example 5.6

A piece of material having a relative permittivity of 5 is subjected to a sinusoidal electric field of maximum value 150 kV/m at a frequency of 20 MHz. If the loss angle δ is 1° calculate the power loss in 1 cm³ of the material.

$$\text{Power loss/m}^3 = 20 \times 10^6 \times 150^2 \times 10^6 \times \frac{2\pi}{360} \times \frac{5}{18} \times 10^{-9}$$

$$= 2.18 \times 10^6 \text{ W/m}^3$$

$$\text{Power loss/cm}^3 = 2.18 \times \frac{10^6}{10^6}$$

$$= 2.18 \text{ W}$$

The material to be heated or welded is passed between two electrodes fed from a high-voltage, high-frequency oscillator. Temperatures are usually less than 200 °C so that radiation losses are minimal. Conduction losses are reduced by heating the electrodes to the required final dielectric temperature by steam or electricity to reduce the temperature gradient within them.

5.7 Induction Heating[19, 20]

Section 2.8 has shown that when magnetic materials are subjected to an alternating field heating occurs from hysteresis losses. Additionally if the material is a good electrical conductor eddy-current heating will also take place. The first of these phenomena is proportional to the frequency f of the field whereas the eddy-current effect is proportional to f^2.

Both these effects are used industrially to melt quantities from a few grams up to fifty tonnes of both ferrous and non-ferrous metals.

5.7.1 The Channel-type Induction Furnace

This operates at between 25 Hz and 60 Hz and is essentially a voltage step-down transformer. The primary winding is fed from the mains supply and the material to be melted is in the shape of a loop or ring which is effectively a short-circuited secondary turn. Figure 5.44a shows the constructional features of a

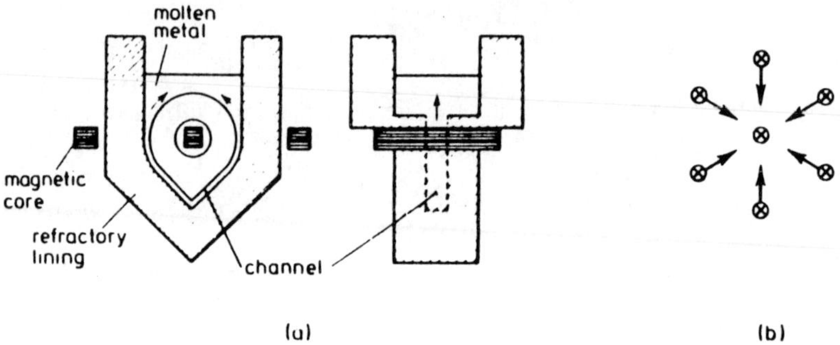

FIGURE 5.44

(a) Two views of a Wyatt type of induction furnace; (b) attractive forces between conductors carrying current in the same direction

Wyatt type of furnace used for melting 300 kg batches of brass. The bulk of the brass is contained in a refractory-lined ladle mounted on trunnions to permit tilting for pouring. A V-shaped channel is formed in the floor of the ladle through the centre of which passes the centre leg of the magnetic circuit of a single-phase or two-phase Scott-connected[4] transformer. The primary is wound upon an asbestos former and is air-cooled. The furnace is fed from a tapped auto-transformer to provide five or six alternative input powers to the furnace.

Two interesting electrodynamic effects cause a liquid-metal circulation or self-stirring action. Each limb of the channel of molten metal may be considered as an infinite number of parallel conductors carrying current in the same direction (figure 5.44b). The conductors attract each other with the result that mechanical forces arise tending to concentrate the liquid in the centre of the channel – this is the *pinch effect* which squirts the metal out of the end of the channel. In addition, for complex magnetic reasons, electrical currents tend to flow in opposite directions in each of the limbs of the V. The metal in one leg thus repels that in the other at the place where the two legs meet (the apex of the V) and metal is forced in the direction of the arrows to the reservoir above. So violent is this motion that the whole contents of a 300 kg furnace pass through a 10 kg capacity channel up to 3.5 times per minute.

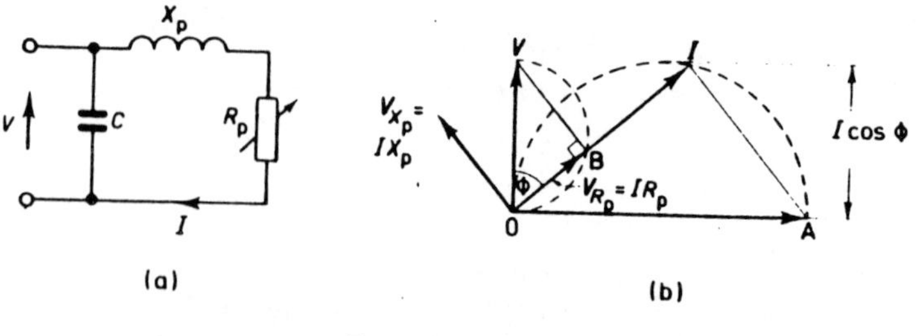

FIGURE 5.45

Equivalent circuit and phasor diagram of a Wyatt furnace

Figure 5.45a shows the electrical arrangement of the circuit. The secondary single turn of molten metal behaves like an impedance of constant reactance and variable resistance depending on the volume of metal charge. This series reactance and resistances are reflected into the input circuit where they appear as R_p and X_p after having been multiplied by the square of the transformer turns-ratio (section 4.2). Figure 5.45b is a phasor diagram drawn for such an arrangement when fed from a constant voltage V. As R_p varies point B will travel in a semicircle as shown by the dashed line. This is because IR_p is always perpendicular to IX_p. When $R_p = 0$; $I = V/X_p$ and lies at right-angles to V (OA). At some non-zero value of R_p the current phasor will be OI as shown. Triangles OIA and OBV are similar, therefore angle OIA = angle OBV = 90°. Thus the locus of OI is also a semicircle.

For a given supply voltage the furnace power $VI \cos \phi$ will thus be a maximum when $I \cos \phi$ is a maximum, that is when $\phi = 45°$ or the power factor ($\cos \phi$) = 0.707. In many cases the furnace operator will not be able to attain this theoretical optimum figure but perhaps only 0.4 lagging. Power-factor-correction capacitors C are thus always included to minimise supply costs.

5.7.2 The Coreless Induction Furnace

These furnaces of up to 40 tonnes capacity have the primary coil wound around the melting vessel as shown in figure 5.46. There is no magnetic core as such and early furnaces ran at frequencies of from 20 kHz to 1 MHz to maximise the eddy and hysteresis effects. These were powered by valve oscillators but later models used motor-generators to provide a supply in the range 500 Hz to 10 kHz: though this method of supply has now been superseded by static inverters (section 5.5.2). Recent years have seen an increase in the number of coreless furnaces designed to operate at mains frequencies.

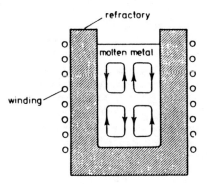

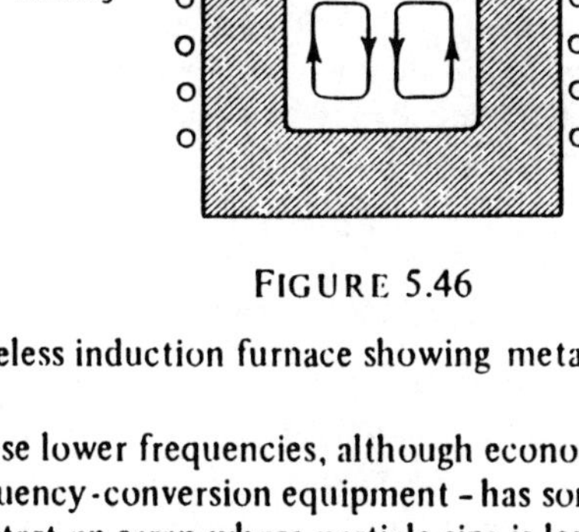

FIGURE 5.46

Coreless induction furnace showing metal movement

The use of these lower frequencies, although economical avoiding as it does the need for frequency-conversion equipment - has some disadvantages. The furnace will not start on scrap whose particle size is less than 200 mm because the eddy currents cannot find sufficient length of path. One-third of the vessel capacity is always left filled to provide a solid mass for starting. Most mains-frequency coreless furnaces have a single winding which presents too large a load for single-phase supply. They are fed from a three-phase supply via a phase-balancing unit as shown in figure 5.47 which divides and balances the load between the supply phases.

Whatever frequency is used the pinch effect and the mutual repulsion between the primary coil and the secondary metal force the melt towards the vessel centre and upwards. This produces a vigorous stirring effect as shown by the arrows in figure 5.46.

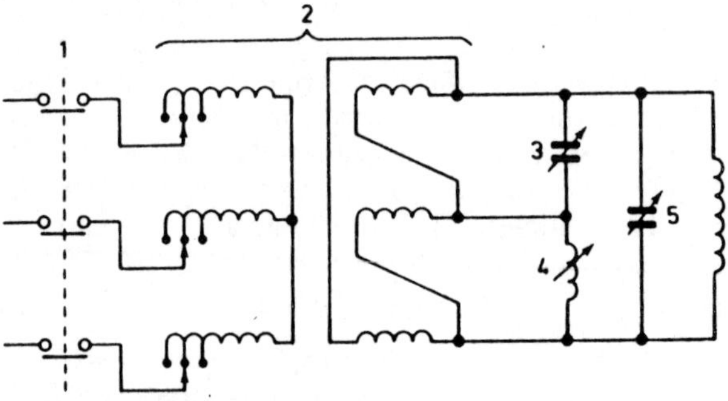

FIGURE 5.47

Circuit diagram for a coreless furnace. 1 – Main supply switch; 2 – regulating transformer; 3, 4 – phase-balancing components; 5 – power-factor correction capacitor; 6 – furnace winding

5.7.3 Induction Hardening

Localised heating for the surface hardening of components may be achieved with a coil of a few turns wound round the area to be treated. If the coil is supplied with energy at a frequency of between 50 Hz and 800 kHz eddy currents flow in the surface layers of the component only. This is because of the *skin effect.*

Figure 5.48 shows a conductor carrying a constant d.c. current I normal to the plane of the page. This current I causes a magnetic flux Φ which (by the screw rule) flows clockwise round the wire and within it. As long as I is constant, Φ is constant so there are no induced e.m.f.s. If however I is alternating, the flux Φ (including the internal flux) will rise and fall inducing e.m.f.s within the material of the conductor. An element of the conductor's

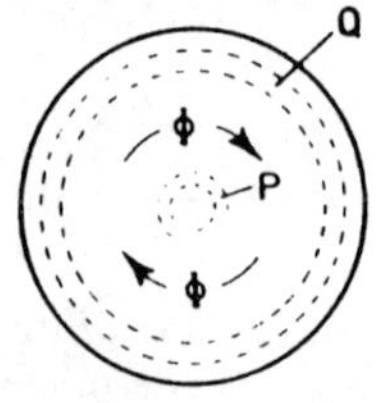

FIGURE 5.48

Cross-section of a current-carrying conductor illustrating the skin effect

cross-section at the centre P is surrounded and therefore cut by more alternating flux than one at the edge Q and thus the back e.m.f.s will be greater towards the centre. The reactance (which depends upon the back e.m.f.) of the conductor material is thus higher at the centre and the alternating current *I* therefore takes the path of least impedance (at the surface). This effect is noticeable at frequencies of 60 Hz upwards and is so pronounced at radio frequencies above 1 MHz that important economies can be made by forming conductors of hollow tubing. This is the electrical analogy of the stress distribution in a torque-bearing rod which allows hollow tubes to be used for drive shafts.

Some examples of surface penetration by varying frequencies are shown in table 5.1 for steel.

TABLE 5.1
Penetration of surface hardening at various frequencies

Frequency (Hz)	Region of most current flow (depth below surface) (mm)
50	7.60
1×10^3	1.65
10×10^3	0.508
440×10^3	0.08
4×10^6	0.025

Coils can be made to enclose selectively only those areas to be treated (for example the teeth of a gear-ring) and a supply frequency can be chosen to give just the hardness depth required. A coil arrangement for the surface hardening of a round shaft is shown in figure 5.49; when the desired temperature has been attained the supply is interrupted and the workpiece quenched by liquid spray.

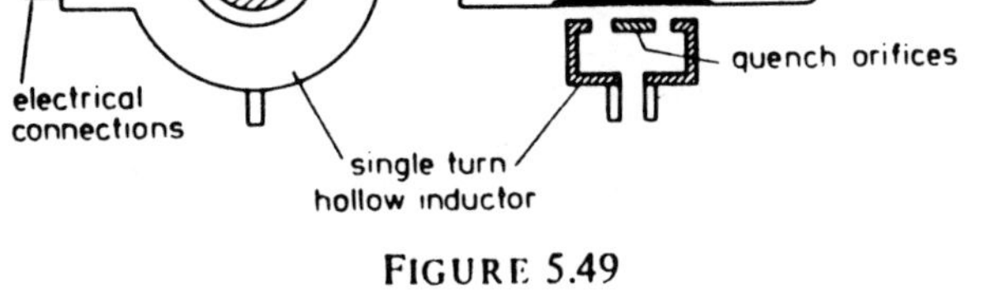

FIGURE 5.49

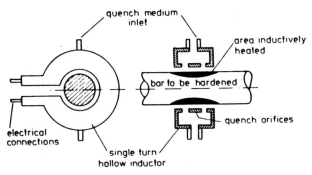

Two views of an induction-hardening furnace

5.8 The Arc Furnace

Furnaces of up to 200 tonnes capacity have been constructed using the intense heat produced in an electric arc to fuse the metal. Figure 5.50 shows the principal constructional features of a medium-sized furnace. All but the smallest laboratory-scale models use three-phase power fed to graphite electrodes which protrude through the furnace roof. The electrode clamps are supported on arms that may be raised or lowered to control the power of the arc. For a given supply voltage, lengthening the arc increases its resistance, decreases the current and hence the furnace power. Most production units have their electrode height automatically controlled by an electrical winch or by hydraulic rams.

In the electrical heating-methods of sections 5.6 and 5.7 the energy taken from the supply is substantially constant. In the arc furnace however, especially at start-up, current surges occur which would cause damage unless limited to safe values. Consequently much reactance is included in the electrode supply leads to minimise current fluctuations.

Arc furnaces are supplied via a special transformer from a high-voltage supply of between 3 kV and 33 kV. Oil-blast or in high-power cases, air-blast circuit breakers are employed which must be robust to withstand frequent operation by current surges. Variable reactors are often included within the transformer housing which can be employed when first striking the arc to limit the effects of the effective short-circuit as the electrodes first come into contact with the metal.

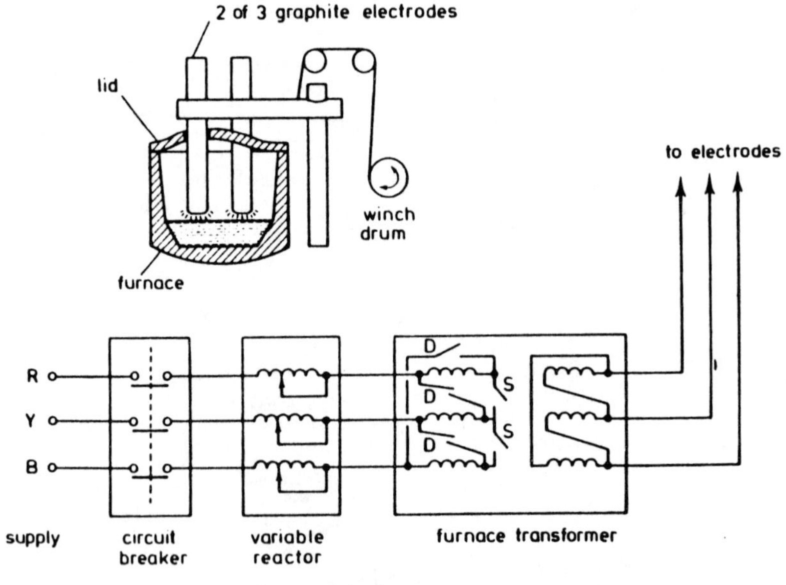

FIGURE 5.50

Mechanical and electrical arrangement for an arc furnace.
Switches S closed for starting (star connection),
D closed for later running (delta connection)

Power control by both electrode position and transformer tap-changing is essential since the melting sequence can be divided into three distinct periods.

(i) The initial power switch-on, where electrodes are pushed down through the charge of metal and scrap until they strike an arc on molten metal on the furnace floor. In this period the voltage is low to avoid long arcs with consequent power wastage.

(ii) The main melting phase, when the whole length of the electrodes is covered with scrap and the charge is melting from the bottom upwards. Full power can be applied with the highest-voltage tap.

(iii) The last period when the power must be reduced to protect the refractory lining of the roof and walls from direct heat from the unshielded arcs. All scrap has fallen away and the charge is completely molten.

Maximum transformer-secondary voltages are usually less than 300 V with six or seven primary tapping positions to reduce this to say 190 V. A star-delta switch is incorporated in the primary circuit which in the delta position gives these higher output-voltages. If the primary is star-connected all the electrode voltages will be decreased by a factor of $1/\sqrt{3}$ (see section 2.6) giving a range of 170 to 110 V.

5.9 Problems

5.1 An ideal diode is used to connect a 10 V, 50 Hz a.c. supply to a d.c. load. The load consists of a 10 kΩ resistor in parallel with a 10 μF smoothing capacitor. Estimate the mean d.c. load voltage if the capacitor discharge can be assumed linear for the first 20 ms.

5.2 Repeat example 5.1 using two ideal diodes and a 10 V-0-10 V centre-tapped transformer in a bi-phase circuit.

5.3 Estimate the ripple factor of the circuit used in example 5.2.

5.4 A full-wave pulse-controlled rectifier supplies a resistive load of 10 Ω from a 10 V-0-10 V, 50 Hz centre-tapped transformer. The rectifiers are ideal, and a firing pulse is applied to them 3 ms and 13 ms after the commencement of the electrical cycle. Calculate the power dissipated in the load.

5.5 A high-frequency heater is used to weld 0.1 mm thick polythene sheet at a frequency of 150 MHz. The applied voltage is sinusoidal having an r.m.s. value of 200 V. Polythene has the following properties

power factor	0.0002
relative permittivity	2.3
specific heat	2.22 kJ/kg K
density	920 kg/m^3

If a 0.2 p.u. of the heat generated escapes by radiation and conduction, calculate the time required to raise the material from an ambient temperature of 18 °C to its softening temperature of 120 °C.

6 The Measurement System

The measurement of any variable in a physical system inevitably disturbs the system, so that we are never able to become aware of the state of the system before the measurement was taken. This statement may appear rather sweeping but on close examination it will be seen that even the most sophisticated measurement technique involves an interchange of energy either from, or to the measured system. The standard determination of length by means of optical interferometric methods compares the length of the standard sample to the wavelength of light emitted from a particular atomic transition in the Krypton atom. Even at this high degree of sophistication the length of the standard sample has to be observed for comparison and the mere act of illuminating the standard causes a disturbance to it by radiation pressure. In this particular standard measurement, the technique has been so designed that this disturbance error is negligible, but the fact that the measurement procedure does affect the measured variable should be paramount in the reader's mind since quite significant errors may be incurred inadvertently. This phenomenon will henceforth be referred to as *parameter loading*.

This chapter will confine itself to measurements in which electrical techniques of monitoring the measured variables are employed and this fact itself leads to another pitfall for the unwary. The degree of refinement available in electrical and electronic techniques may so intrigue the engineer that he quite unconsciously employs electrical techniques where a simple dial-gauge or other mechanical technique would be both more robust and more economical.

6.1 Application of Electrical Measurement Techniques[21]

With the restrictions of the foregoing in mind, the field in which electrical methods have significant advantages become

- (i) measurements to be made on moving members (for example the instantaneous value of a strain in a rotating crankshaft)
- (ii) measurements in systems made inaccessible by either distance or intervening objects (for example the lunar environment or medical measurements of the body's interior by temperature and pressure 'pills')
- (iii) parameters that are varying too quickly for high-inertia mechanical instruments to follow (for example the vibration of the tool or the bed in machine tools)
- (iv) variations that have subsequently to be analysed or fed to a computer for data processing.

6.2 Static Characteristics of Measuring Systems

Some general terms that apply to any measuring system will be defined here because they are sometimes used incorrectly in manufacturers' specifications.

Calibration – the periodic checking of the absolute accuracy of an instrument by comparison with either (i) a Primary Standard, for example the wavelength of light or the melting temperature of pure materials, or (ii) a Substandard, such as slip gauges or a standard cell.

Range – if the highest value of the variable that can be measured is *b* and the lowest value either zero or *a*, then the range is quoted as *b* in the former or from *a* to *b* in the latter case.

Span – the difference between the highest and lowest possible readings, in the above example *b* minus *a*.

Accuracy – the degree to which the indicated reading of the instrument approaches the true value of the measured variable. As the accuracy may vary over the span of the instrument, a complete statement of accuracy is best presented in the form of a *calibration curve* in which the indicated value of the variable is plotted against a standard or substandard in the directions of both increase and decrease of the measured variable. This directional precaution is necessary since there may be *backlash* or *hysteresis* (see below) in the instrument. Accuracy is often quoted as a proportion of the true value, or more usually as a proportion of the span or full-scale deflection,

Error (Static) – the difference between the true value and the indicated value of the measured variable.

Reproducibility or Precision – the degree to which a given value of the variable may be repeatedly measured.

Drift – the shift in calibration over a period of time. This may occur in two ways (see figure 6.1)

Zero drift – the whole calibration shifts an equal amount.
Span drift – the error increases progressively from zero.

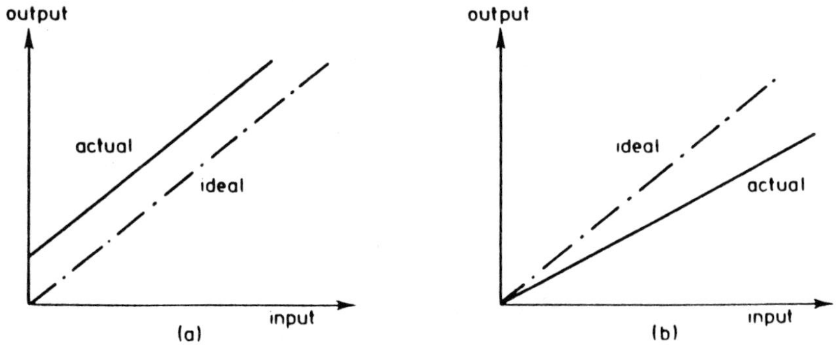

FIGURE 6.1

(a) Zero drift – the whole calibration shifts by a constant amount; (b) span drift – the error increases progressively from zero

Sensitivity – the ratio of the change in output of a system or sub-system to the (usually unit) change in the input variable. For example, the output of a transducer may be quoted in volts/mm movement.

Resolution – (formerly called sensitivity) the smallest change in input to which the measuring system will respond.

Dead Zone – the largest change in input to which the system will just not respond.

Hysteresis (literally 'lagging-behind') – the difference between the two possible values of output for a given input depending upon whether the input values are approached from above or below (usually expressed as a percentage of the full scale or span; see figure 6.2).

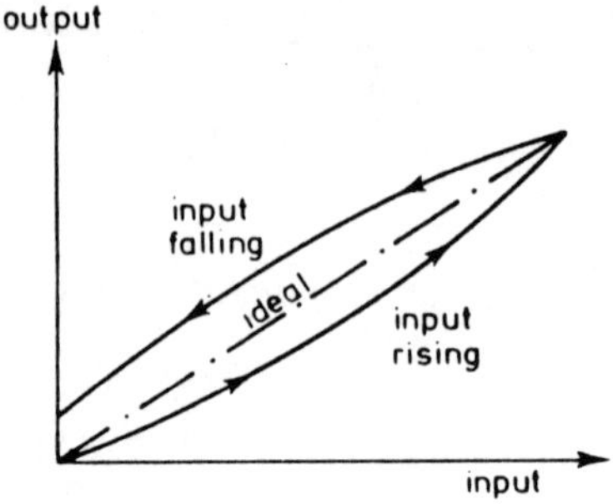

FIGURE 6.2

Hysteresis

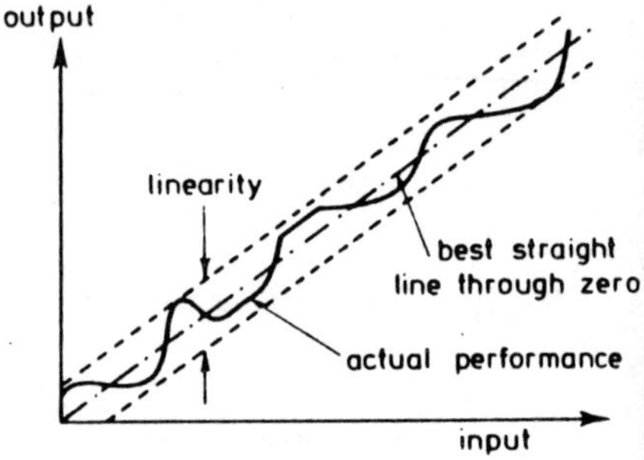

FIGURE 6.3

Zero-based linearity

Linearity – the proximity of the output - input graph to a straight line, as a percentage of full scale or span (figure 6.3).

Dynamic Errors – a system may have negligible static errors but yet exhibit large errors if the measured variable is varying quickly. These are called dynamic errors, their study is complex.[21]

6.3 Measurement-system Components

Any measurement system may be expressed as an interconnection of the components shown in figure 6.4, the direction of signal flow being shown by the solid arrowhead. It may well be that in some less sophisticated systems such as a thermometer consisting of a thermocouple and a moving-coil meter all the separate modules are not clearly defined and some indeed may be absent. Nevertheless, however complex the system, figure 6.4 embraces all its salient features.

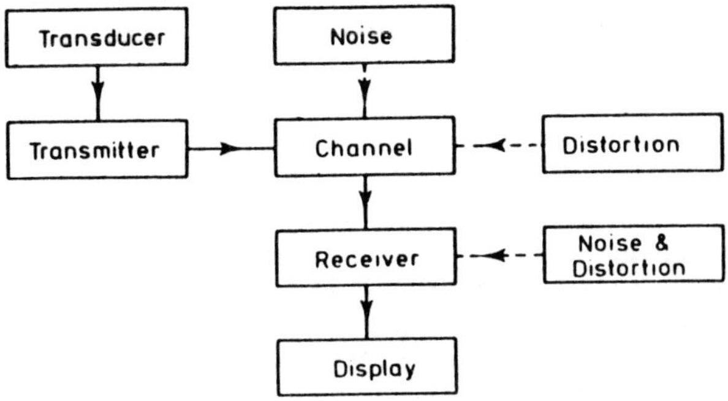

FIGURE 6.4

Block diagram of a generalised measuring system

The transducer is merely a device for converting the measurement energy into an electrical *signal*. Since the measured variable may take so many forms, the transducer merits a separate chapter (7) where it is considered together with the transmitter of which it often forms an integral part.

Noise and distortion may enter the system at any stage but, in practice, their effect is most critical at low signal levels. The introduction of the word 'signal' leads us to a discussion of the *information* passing through the measurement system.

6.4 Channel and Signal Information[22]

The measured variable must vary in some unpredictable way, since if we already possessed complete information about its future behaviour the measurement would not be necessary.

The amount of information concerning the variable which it is desired to collect in a given time, and the path length and environment through which the signal must travel from transmitter to receiver largely dictate the nature of the channel. This may vary from a simple pair of wires carrying Morse code, to a highly sophisticated microwave link with an orbiting satellite that is capable of simultaneously relaying many hundreds of spoken messages and several television signals between continents. To appreciate the differences in channel requirements that are dictated by the signal it is convenient to illustrate some simple concepts of information theory by means of a well-known channel, namely spoken and written English.

Consider the case of a telephone conversation between two individuals; it is commonly observed that, even though the received signal is not an exact replica of the sound waves impinging on the transmitting microphone, intelligibility is good. The signal is commonly distorted from its full frequency range of 25 Hz to 6 kHz and compressed into the frequency spectrum extending from 250 Hz to 3 kHz to simplify the telephone transmission equipment. The range of frequencies over which a particular signal is transmitted is referred to as its *bandwidth*. The fact that the listener can reconstruct the message from the noisy signal is because his memory has conditioned him to hearing only the combinations of sound frequencies allowed by the English tongue and the structure of the English sentence. Thus, part of the information if transmitted to him perfectly would be *redundant* since it would merely be conveying information concerning the structure of the language. This redundant information could well be of great help to a non-native speaker whose memory concerning the grammar and syntax of the language was not so well developed.

The fact that *information rate* is of great importance in any communication system is easily seen. It is easier to construct the original message in the presence of noise or other distraction (foreign language) if the sender presents the information (speaks) more slowly. A numerical measure of *information quantity* can be obtained from the number of possible values a signal may have in a basic interval. This number is usually defined as the number of on - off pulses or *binary digits* required to transmit the information.

The *quantity of information* in a given interval is

$$I_o = \log_2 L$$

where the unit of I_o is the *bit* (*bi*nary digi*t*) and L is the number of levels distinguishable by the receiver. The human ear can distinguish between changes in received acoustic power of approximately 2:1 or 3 dB (see section 2.3.5) and the dynamic range of the human voice (the ratio between its maximum and minimum powers) is approximately 1000 to 1 or 30 dB. Thus there are 30/3 = 10 such discernible levels. The information content of speech is thus

$$I_o = \log_2 10 = 3.32 \text{ bits}$$

Considering another communication mechanism, the five-hole paper tape used for telegraphy and formerly for computer processing. The quantity of information per line is given by the number of permutations possible in five punched holes, that is $2^5 = 32$ states. This allows 26 letters plus 6 punctuation, space and capital commands. Thus the information contained in the unit interval is

$$I_o = \log_2 32 = 5 \text{ bits}$$

The *information rate*, defined as the quantity of information transmitted in unit time will thus be measured in bits per second. With any variable of maximum frequency f it will be shown in section 6.9.2 that the variable must be sampled at a frequency of at least $2f$. Thus with a restricted audio-spectrum of 250 Hz to 3000 Hz a sampling rate of 2 x 3000 Hz must be used.

$$\text{Information per sample} = \log_2 10 \tag{6.1}$$

$$\begin{aligned} \text{Information rate } I &= \text{Sampling rate } n \times \text{Information per sample } I_o \qquad (6.2)\\ &= 2 \times 3000 \times \log_2 10 \\ &= 6000 \times 3.32 \\ &= 19\,920 \text{ bits per second} \end{aligned}$$

Using paper tape each five-hole group is used to characterise one written alphabetical letter. If the transmission speed used is one 5-letter word per second and one extra basic sampling interval is used for a group indicating a space between words, the information rate is

$$\text{sampling rate} \times \text{information per sample}$$
$$(5 + 1) \text{ per second} \times \log_2 32 = 30 \text{ bits per second}$$

We shall take an extreme case of high information rate for our worked example.

Example 6.1

A television system has a picture composed of 625 horizontal lines. The complete picture area (the frame) is covered by the cathode-ray spot 25 times per second. Assuming that 10 intensity graduations of the spot are to be displayed and that there are effectively 500 dots per line, calculate the sampling rate required. If this picture is to be faithfully transmitted over a radio link calculate the information rate required.

$$\begin{aligned} \text{Sampling rate } n &= \text{number of dots per second} \\ &= \text{frames/s} \times \text{lines/frame} \times \text{dots/line} \\ &= 25 \times 625 \times 500 \\ &= 7.8 \times 10^6 \text{ Hz} \end{aligned}$$

Information per sample $I_o = \log_2 10 = 3.32$ bits

$$\begin{aligned} \text{Information rate} &= \text{information per sample} \times \text{sample rate} \\ &= I_o n \\ &= 3.32 \times 7.8 \times 10^6 \\ &= 25.9 \text{ Mbits per second} \end{aligned}$$

The reasons for the different channel-media chosen for the above three systems are now apparent: it is the information rate together with the accompanying necessary *channel capacity* (bits per second) which is often crucial in deciding the bandwidth and therefore the type of channel chosen.

Even if the required information rate is low the use of a sophisticated type of channel may be necessary because of the *environment* or the *distance* over which the information must travel (see section 6.8).

6.5 Noise[23, 24]

The function of the transmitter is to convert the electrical signal from the transducer into a form suitable for transmission over the channel chosen and to amplify it before transmission so that, on arrival at the receiver, the signal energy is still distinguishable from any *noise energy* acquired en route.

If there is a noise power of N watts entering a channel of given bandwidth which carries a signal power of S watts, then the received power will be $S + N$ watts. Because the power is proportional to the square of the voltage the total received voltage will be $\sqrt{(S + N)}$ containing $\sqrt{N}$ noise volts. Using conventional signal-processing techniques a signal voltage less than the noise voltage will not be distinguishable, thus the number of discrete signal levels which may be distinguished will be

$$L = \frac{\sqrt{(S + N)}}{\sqrt{N}} = \sqrt{(1 + S/N)}$$

The quantity of information transmissible in a basic interval is thus

$$I_0 = \log_2 \sqrt{(1 + S/N)} = 0.5 \log_2 (1 + S/N)$$

Hence with a sampling rate twice that of the bandwidth as before, the total information content that can be transmitted over a *noisy channel* of *signal-to-noise ratio* S/N is given by

$$I_t = I \times T$$

from equation 6.2

$$I = n \times I_0$$

and therefore

$$I_t = BT \log_2 (1 + S/N) \text{ bits} \qquad (6.3)$$

where B is the *bandwidth* and T the total *time for transmission*. This is known as the Hartley - Shannon equation. Thus the total information content may be increased by either (i) increasing the channel bandwidth or (ii) increasing the time available for transmission. Indeed the Hartley - Shannon equation shows a possibility of information retrieval even if the signal-to-noise ratio is less than unity but the signal-compression techniques required for this are beyond the scope of this text and we shall henceforth assume that the signal-to-noise ratio at the receiver must be greater than unity, typically 2:1 or + 3dB.

Example 6.2

Determine the bandwidth required for a television broadcast channel using the data given in example 6.1. The signal-to-noise power ratio as measured at the receiver is to be 10^4:1 (40 dB)

From the calculation of example 6.1, information rate = 25.9 Mbits/s, therefore in 1 second since

$$I_t = I \times T$$

$$I_t = B\,T \log_2 (1 + S/N)$$

$$25.9 \times 10^6 \times 1 = B \times 1 \times \log_2 (1 + 10\,000)$$

$$B = \frac{25.9 \times 10^6}{13.39} = 1.95 \text{ MHz}$$

Since the maintenance of a good signal-to-noise ratio is so important for the efficiency of a data-transmission system it is important to minimise the entry of noise where the signal is at its lowest level – in the channel and the input stages of the receiver. Noise may be divided into two categories, that arising externally, and that produced within the measuring system itself.

External Noise is the more readily understood since its sources are readily defined. Examples are electromagnetic fields surrounding the channel, induced for instance by the flux variation in transformers and machines; electrostatic fields set up by lightning dischatges or the sparking which occurs at the commutators of some electrical machines. These can usually be minimised by screening the sensitive system-component (for example signal transformers and microphones in magnetic fields) with a low-permeability cover; or by an earthed electrostatic metal screen around amplifiers and cables in electrostatic fields. Screened cables often have conductors which are coaxially arranged, the outer metal screening braid being the lower potential return conductor (figure 6.5a). In multiway cables all the conductors may be enclosed in an outer metal screening braid (figure 6.5b). Care must be taken to connect the screen to earth at one end only otherwise a closed 'earth-loop' may be formed around which currents may circulate, induced by magnetic fields (see figure 6.5c).

External noise can arise from the mains power-supply of the system containing 'interference' waveforms caused by motor brushgear or by thyristor-controlled rectifiers. This may be cured by including a suppressor or filter network at the point where the mains supply enters the instrument. In strain-gauge bridges noise from the mains supply can often be so acute that a separate battery supply has to be used. Where this is not possible the noise must be traced to its source and suppressed there by screening or filters; an excellent example of this was legislation to ensure the suppression at source of noise from vehicle ignition systems which interfered with television reception.

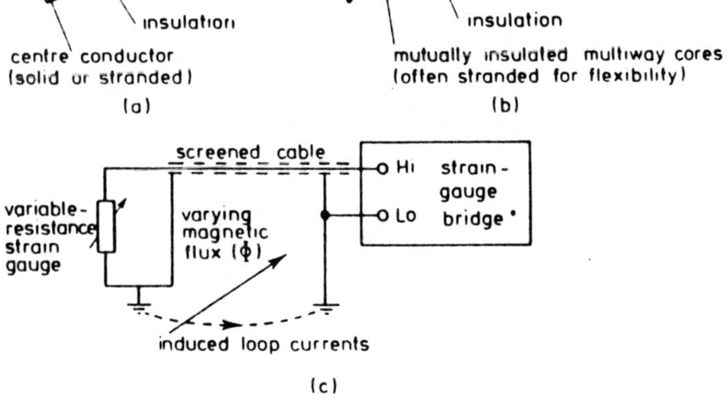

FIGURE 6.5

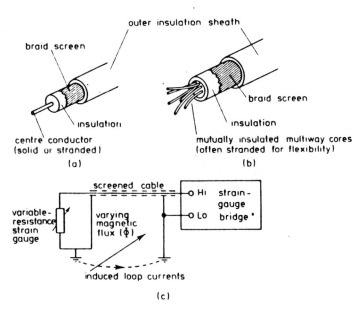

(a) Coaxial and (b) multiway screened cables; (c) induced earth-loop currents arising from multi-point earthing

Internal Noise. Even if the careful application of the above techniques reduces external noise to an acceptable level, it is often not appreciated that the measurement system itself generates noise. This arises from two main sources, (i) the transducer, and (ii) any semiconductor (or thermionic) device in the receiver. The former source is Johnson (or thermal) noise and arises from the random motion of electrons, caused by thermal agitation, within the self-impedance of the transducer. This noise voltage is random and therefore distributed evenly across the frequency spectrum (*white noise*). Its average value over a period of time will be zero but at any instant it will have a magnitude $\bar{v}_n$ and we may speak of the mean square noise-voltage $\bar{v}_n^2$. This can be shown from thermodynamic considerations to equal $4kTRB_n$, where k is Boltzmann's constant = 1.38×10^{-23} J/K, T is the source temperature (K), R is the real or resistive component of the source impedance and B_n is the equivalent noise bandwidth. The latter can be shown to be approximately equal to 1.57 times the half-power bandwidth of the subsequent measuring equipment.[24]

The second type of noise occurs in two forms, *shot noise* and *1/f noise*. The former is caused by the random generation of carriers (electrons and holes) within semiconductor devices or *saturated* vacuum devices (diodes, photocells). It is frequency-independent (white) and its magnitude is equal to

$$\bar{i}_n^{\,2} = 2eIB_n$$

where $\bar{i}_n^{\,2}$ is the mean square noise-current, e = carrier charge (1.6×10^{-19} C) and I is the d.c. device current. The cause of $1/f$ (flicker, excess or semiconductor) noise is still the subject of speculation, but it appears to be associated with surface leakage effects at surface contacts and junctions in semiconductors[24] and has negligible magnitude above 1 kHz. It is only encountered in very-high-gain amplifiers whose response curve (see section 8.2.3) extends down to zero frequency. These are called *direct coupled* or d.c. amplifiers.

Example 6.6

An optical transducer consists of a vacuum photoelectric cell in series with a 1 kΩ load resistor as shown in figure 6.6. The dark current of the cell is 10 μA and the half-power bandwidth and input impedance of the following amplifier are 1 kHz and 1 MΩ respectively. If the circuit is maintained at room temperature and the amplifier has negligible internal noise calculate the minimum observable signal voltage at the amplifier input terminals.

Thermal resistive noise is generated in the load resistor

$$\bar{v}_n^2 = 4kTRB_n = 4 \times 1.38 \times 10^{-23} \times 291 \times 10^3 \times 1.57 \times 10^3$$
$$= 2.52 \times 10^{-14}\ \text{V}^2$$

For the photocell

$$\bar{i}_n^{\,2} = 2eIB_n = 2 \times 1.6 \times 10^{-19} \times 10^{-5} \times 1.57 \times 10^3$$
$$= 5.03 \times 10^{-21}\ \text{A}^2$$

This shot-noise current passes through the load resistor in parallel with the amplifier's input impedance. Since the latter is so much greater the impedance

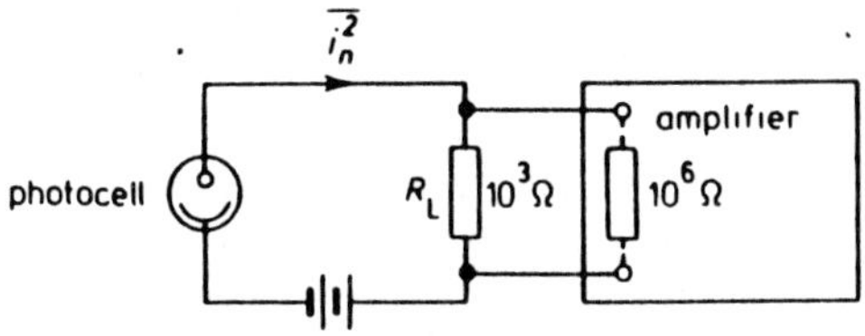

FIGURE 6.6

The circuit of example 6.6

of this parallel combination is approximately 10^3 ohms. Applying Ohm's law the shot-noise voltage produced by this current in the load resistor is

$$\overline{v_{ns}^2} = \overline{i_n^2}\, R_L^2$$
$$= 5.03 \times 10^{-21} \times 10^6$$
$$= 5.03 \times 10^{-15}\ V^2$$

The total noise voltage may be obtained by adding the squares of the separate noise voltages

$$\overline{v_{tot}^2} = \overline{v_n^2} + \overline{v_{ns}^2} = 2.52 \times 10^{-14} + 5.03 \times 10^{-15}$$
$$= 3.02 \times 10^{-14}\ V^2$$

therefore

$$v_{tot} = 0.174\ \mu V$$

The minimum observable signal would be approximately twice this, that is, 0.35 μV.

The Minimisation of Internal Noise. Since both thermal and shot-noise powers are proportional to bandwidth, they may be reduced by deliberately reducing the bandwidth of the receiver, using filters, to that value just required to include the highest frequency components of the signal. Shot noise may further be reduced by limiting the device current in the early stages of the receiver amplifier. In the case of high-gain wide-band amplifiers the minimum detectable signal will be limited by the thermal source noise and elaborate attempts have been made to minimise this in some amplifiers by lowering the temperature with liquid helium.

6.6 The Transmitter and Receiver

As previously stated one of the functions of the transmitter is to convert the signal energy from the transducer into a form capable of being transmitted through the channel to the receiver which returns the signal to its original form. These processes are known as *modulation* and *demodulation* respectively and the signal may undergo amplification either before or after demodulation, or both, within the receiver before being passed to the display system.

The type of transmitter and receiver will obviously be influenced by the channel chosen. Various forms of channel with their appropriate transmission and reception circuits will be considered in section 6.8.

6.7 The Signal Nature[25]

In practice, signals are never either continuous sine waves or d.c. terms since their behaviour would then be perfectly predictable and they could convey no information. They are always either discontinuous or of more complex waveform. Fortunately, by means of the Fourier series, we are able to express mathematically any *repetitive waveform* as the sum of two infinite series containing sine and cosine terms at multiples, known as *harmonics*, of the waveform's *fundamental frequency* together with a constant or d.c. term.[4]

The time required for one complete cycle or repetition of the waveform whatever its shape will henceforth be referred to as its *period*, or the reciprocal of the *fundamental frequency*. In order to avoid rigorous analysis let us take the easily demonstrable case of the square wave shown in figure 6.7.

It will clearly be seen that the addition of successive terms leads to a closer approximation of the ideal square-waveshape, although for perfect synthesis an infinite number of harmonics (each decreasing in amplitude) is required. A useful engineering approximation is that the signal bandwidth required for rectangular waveforms is approximately 10 times the fundamental frequency.

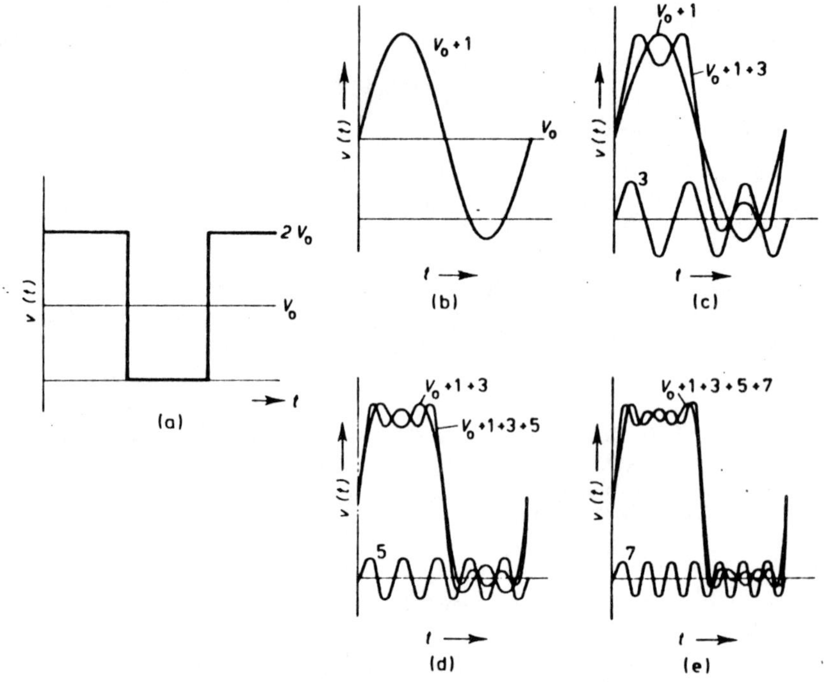

FIGURE 6.7

The graphical construction of the square wave at (a) from its various frequency components; (b) d.c. term plus fundamental;
(c) d.c. term, fundamental and third harmonic:
(e) d.c. term, fundamental, third and fifth harmonics

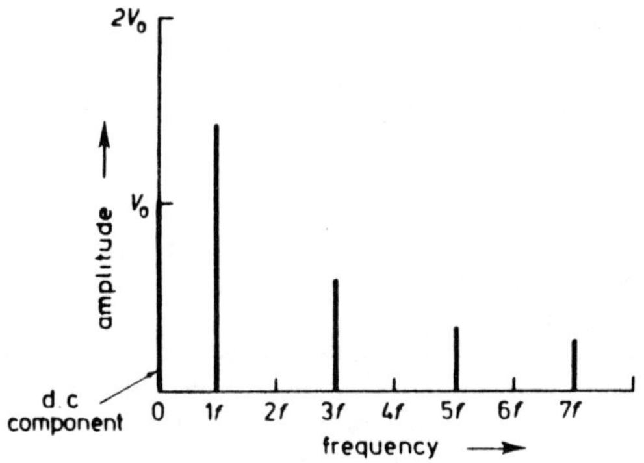

FIGURE 6.8

The frequency spectrum of the continuous square wave of figure 6.7a

In addition, if any repetitive signal is symmetrical about zero amplitude, its Fourier expansion will contain no d.c. component and hence the lower frequency limit of its bandwidth will be that of the fundamental (see figure 6.9a). So far, waveform diagrams showing the variations of the signal in the *time domain* have been considered, but a more simple representation of the amplitudes of the frequency components of figure 6.7 may be represented by an equivalent specfral diagram in the *frequency domain.* Figure 6.8 shows the frequency

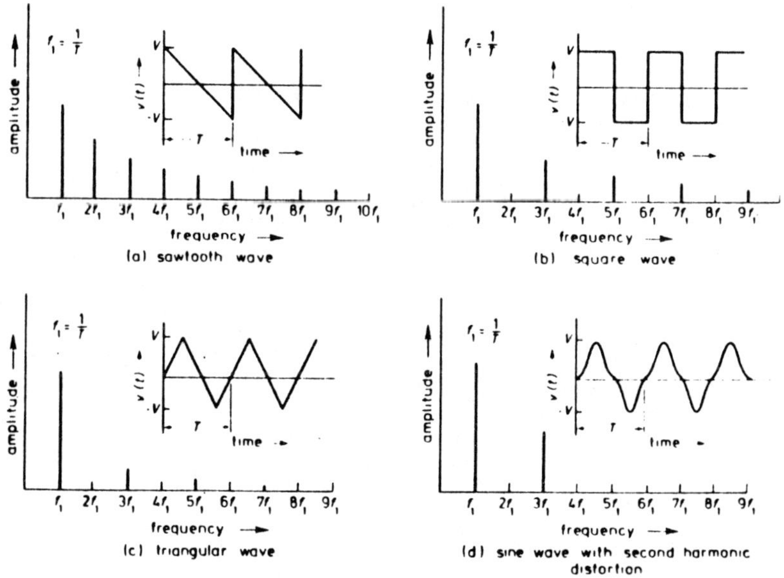

FIGURE 6.9

Frequency spectra of commonly encountered waveforms

spectrum for the continuous square-wave depicted in figure 6.7. It will be noted that although the original time-domain diagram specified both the frequency and the phase of the components, the frequency-domain spectral diagram ignores phase relationships between the various frequency components. Therefore it does not contain sufficient information to reconstruct the waveshape because the phase relationships of these frequency components are needed for this. Transformations from the time to the frequency-domain and vice versa can be effected by the techniques of the *Fourier transform*[25] which is beyond the scope of this text.

Figures 6.9 and 6.10 are of interest in that they depict the frequency spectra of various commonly-encountered repetitive waveforms – as the curve shapes become smoother from figure 6.9a to figure 6.9d so the frequency spectra become noticeably less rich in harmonic content and thus require progressively narrower bandwidths for faithful transmission.

It will be seen from figure 6.10 that half- and full-wave unsmoothed rectifier outputs contain only even harmonics and that in the latter case, apart from the d.c. component, the lowest frequency encountered is the second harmonic, there being no fundamental term.

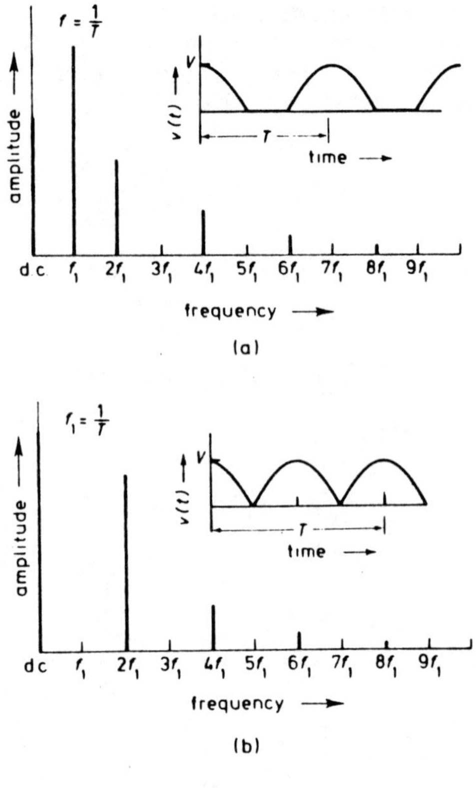

FIGURE 6.10

Spectra of waveforms encountered in power-supply design

Non-periodic Signals. So far the signals discussed have been of the repetitive or periodic type having a period T. These have exhibited frequency spectra having discrete components at multiples of the fundamental frequency $f_1 = 1/T$. The frequency spectrum of an isolated pulse may be predicted intuitively; since the period time is very long (infinite if only one pulse occurs) the fundamental frequency will approach zero and hence the harmonics will be very close together. That is, the spectrum will tend to become a continuous band instead of clearly separable discrete harmonics. The application of the Fourier transform to an isolated pulse produces a spectral distribution as shown in figure 6.12a.

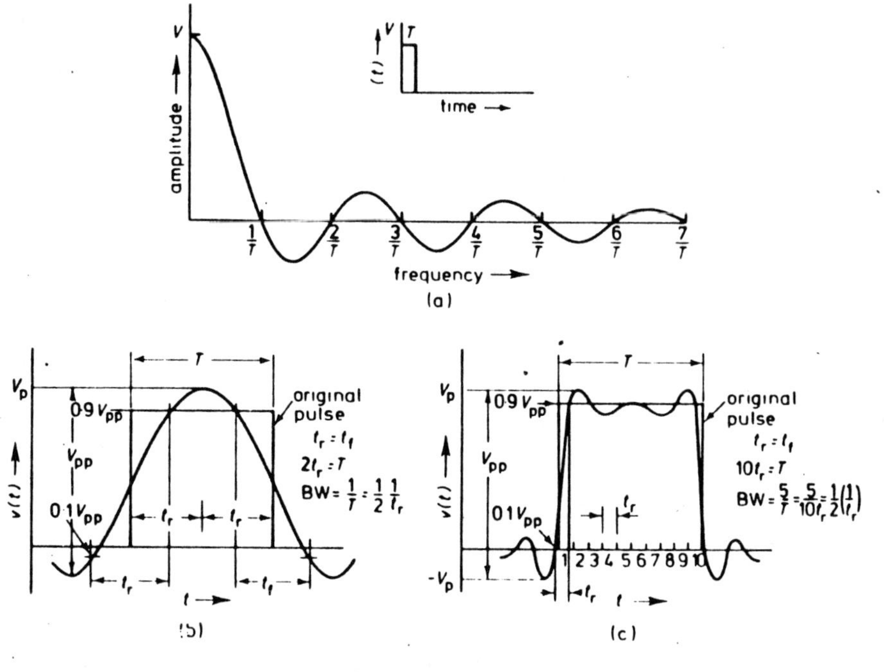

FIGURE 6.12

Isolated rectangular pulse (a) and its frequency spectrum. Its shape when reconstructed from frequencies below (b) $1/T$ only and (c) $1/5T$ only

The envelope is of the $(\sin x)/x$ form and it can be shown that the major part of the pulse power lies at frequencies below $1/T$ since the amplitude distribution is greatest here. Theoretically such a pulse requires an infinite bandwidth for faithful transmission, but reconstructions using only those frequencies below $1/T$ and $5/T$ are shown in figure 6.12b and c respectively. The response shown at (b) may be perfectly satisfactory for a pulse-code-modulated (PCM) data-link where only the presence or absence of a pulse is of interest and subsequent pulse reshaping can be carried out. Greater knowledge of the pulse duration (as in figure 6.12c) may be essential in PDM data systems (p. 193). Figures 6.12b and c illustrate the definitions of the pulse-rise time t_r and pulse-fall time t_f. These are the times needed for the pulse to rise from 0.1 to 0.9 of its peak-to-peak value, and to fall between the same limits.

It will be seen that $t_f = t_r = T/2$. Thus the required bandwidth is given for figure 6.12b by $1/T = 1/2t_r = 1/2t_f$ or for figure 6.12c by $5/T = 1/2t_r = 1/2t_f$. It is convenient to specify the required bandwidth as the reciprocal of twice the rise time since this is true of all pulse durations. Rise and fall times are easily measured directly on an oscilloscope.

Example 6.7

What is the minimum bandwidth required faithfully to transmit (i) a repetitive square wave of period 1 ms which varies (a) between 0 and + 10V (b) from − 5V to + 5V; (ii) a repetitive rectangular waveform of 100 Hz and mark-space ratio of 1:3 and (iii) what would the rise time of the pulses be after transmission?

(i) (a) (See figure 6.11a)

$$T = 1 \text{ ms} = 10^{-3} \text{ s}$$

$$f \text{ (periodic frequency)} = \frac{1}{T} = 10^3 \text{ Hz}$$

Using the approximation on p. 178 there is negligible harmonic content above $10f$, that is, 10^4 Hz. Since the waveform is asymmetrical about the x-axis, there will be a d.c. (zero frequency) component. The bandwidth is thus $10^4 - 0 = 10^4$ Hz.

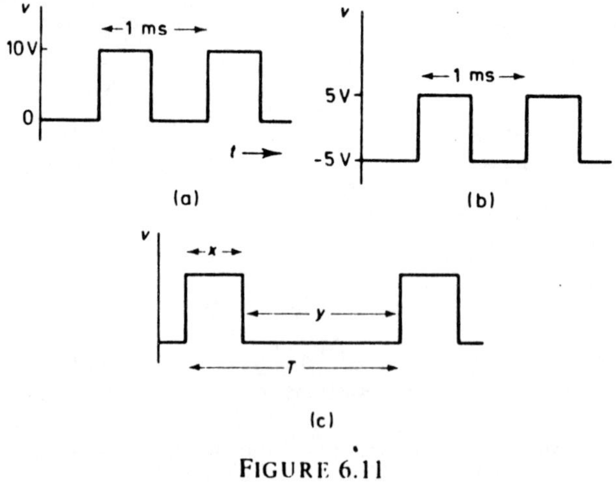

FIGURE 6.11

The waveforms of example 6.7

(b) (See figure 6.11b), since the signal is symmetrical about the x-axis there are no frequencies present less than f. Therefore the amplifier response may be limited to frequencies between f and $10f$ to reduce $1/f$ noise at low frequencies.

$$\text{Bandwidth} = 10^4 - 10^3 = 9 \times 10^3 \text{ Hz}$$

(ii) (Figure 6.11c), since the mark-space ratio is 1:3

$$\frac{x}{y} = \frac{1}{3}$$

$$T = \frac{1}{f} = \frac{1}{100} = 0.01 \text{ s}$$

$$x = \frac{0.01}{4} = 0.0025 \text{ s}$$

Therefore the required bandwidth is approximately five times the reciprocal of the pulse duration.

$$\text{Bandwidth required} = \frac{5}{T} = \frac{5}{0.0025} = 2 \text{ kHz}$$

(iii) The minimum rise-time of the pulse after transmission through this bandwidth is given by

$$\text{bandwidth} = \frac{1}{2t_r} \qquad t_r = \frac{10^{-3}}{2 \times 2} = 0.25 \text{ ms}$$

6.8 Types of Channel [26, 27]

6.8.1 Line Communication

The simplest form of channel that can carry information is a pair of wires. These will have resistance depending on their length and cross-sectional area, leakage conductance between them (depending on the properties of the insulation medium), together with capacitance and inductance determined by their geometry.

These properties affect the nature of any signal which is propagated along them. The equations of a pure sinusoid's behaviour when applied to such a line are well known[27], and it can be shown that the attenuation of the signal becomes more severe as its frequency increases. A multiple-frequency signal of the types discussed in section 6.7 will become distorted in shape due to the different attenuation of its various frequency components.

If the transmitter and receiver are close together the use of such a line is permissible up to frequencies of several megahertz. If the line must be long due to the transducer being situated in a hazardous area (in the vicinity of explosions, high temperatures or radioactivity, for example) then the upper frequency limit is greatly reduced. A simplified equivalent circuit of a transmitter of output resistance R_1 and a receiver of input resistance R_2, together with the line capacitance C is shown in figure 6.13a. As long as the reactance of C is large compared to both R_1 and R_2 there will be negligible loss of signal energy in the uppermost frequency components of the signal. Frequently however the output impedance of the transducer (as in the case of a piezoelectric accelerometer) and therefore the input impedance of the associated receiver for maximum power-transfer (section 2.5.6) are in the order of 10^8 Ω. Quite small cable capacitances would thus cause the cable losses to be unacceptable.

Further, the presence of environmental noise (p. 175) may require the use of screened cables of the form shown in figure 6.5, or the cable may have to be

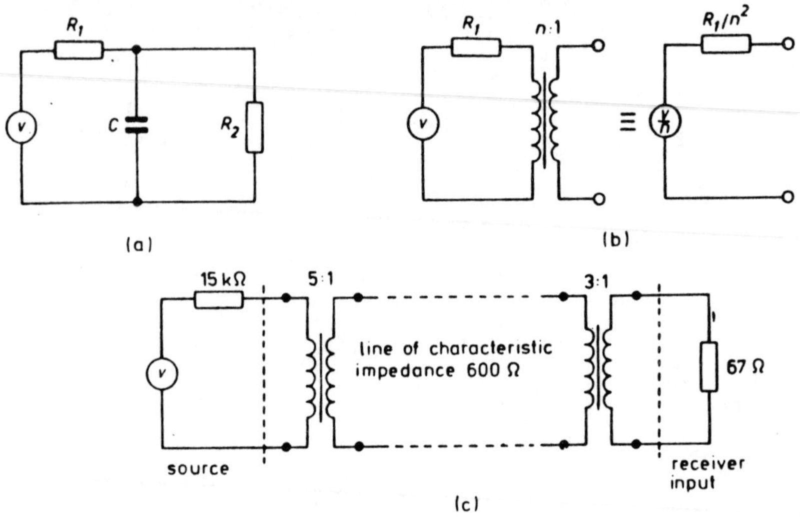

FIGURE 6.13

Impedance matching by a transformer

armoured to protect it from damage. These cables have an inherently higher self-capacitance than open wires and the impedance levels of both source and receiver must be reduced either by a transformer or by an electronic method such as the emitter-follower (p. 245) to allow a low-impedance cable to be used.

Figure 6.13b shows how the effective source impedance is modified by transformer matching.

Figure 6.13c shows a line, source and receiver coupled together with matching transformers; the use of transformers is of course limited to frequencies above about 20 Hz and if the d.c. or low-frequency components of a signal are to be retained then some electronic impedance-transformation device such as an emitter-follower must be attached directly to the transducer in a screened container. Such an arrangement is often referred to as a pre- or head-amplifier.

The equations governing the transfer of energy along transmission lines show that, in order to effect maximum power-transfer, not only must the source and receiver input impedances be matched together, but both must be matched to the channel's characteristic impedance. This has been standardised at 600 Ω for twin open-wires or 75 and 50 Ω respectively for normal and low-loss coaxial cables.

Example 6.8

A transducer has a resistive output impedance of 50 ohms. Calculate the turns ratios of transformers required to connect the transducer to an amplifier of 2 kΩ input impedance (i) directly and (ii) via a line of 600 Ω characteristic impedance.

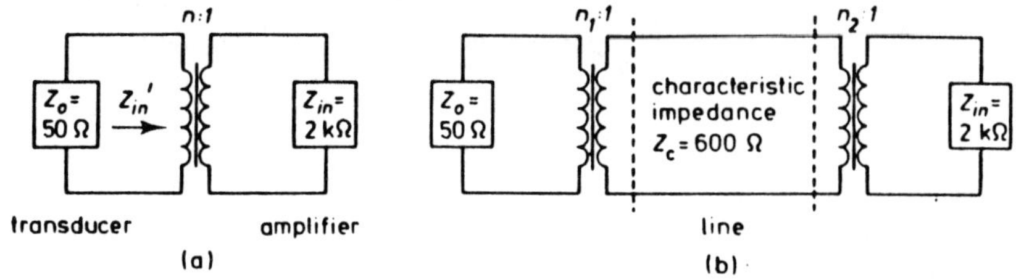

FIGURE 6.14

Example 6.8

(i) (See figure 6.14a.) For maximum power-transfer

$$Z_o = Z_{in}' = 50 \text{ ohms}$$

$$\frac{Z_{in}'}{Z_{in}} = \frac{n^2}{1}$$

therefore

$$n = \sqrt{\left(\frac{50}{2000}\right)}$$
$$= 0.158 : 1$$

(ii) (See figure 6.14b) For maximum power-transfer

$$n_1 = \sqrt{\left(\frac{50}{600}\right)} \qquad n_2 = \sqrt{\left(\frac{600}{2000}\right)}$$
$$= 0.288 : 1 \qquad = 0.0548 : 1$$

6.8.2 Radio Communication

In many systems it is necessary to use only one channel for the simultaneous transmission of more than one signal. Since many transducers produce an output which is a sine wave whose *amplitude* is proportional to the variable, *amplitude-modulated* carrier channels are often used. The signal is transported along the channel by means of a carrier frequency which must be at least twice as high as the highest frequency component of the signal to be transmitted (see p. 193). Carrier frequencies of between 100 kHz and several megahertz are often used since they are readily propagated through space as radio waves.

Frequency Bands for Radio Communication and Telemetry

The allocation of frequencies for radio transmission is closely controlled by international agreement to avoid spectrum overcrowding. The earliest attempts in radio transmission were made at low frequencies that have the advantage of providing simple long-distance coverage because of their long wavelengths. Examples of this are the Droitwich Radio 2 transmitter giving European coverage on a wavelength of 1500 m (200 kHz), or the very long wavelengths corresponding to frequencies of about 100 kHz which enable worldwide communication to be established between submerged submarines and their base.

The next band of frequencies from 0.6 to 4.0 MHz is allocated to commercial and entertainment communications. Above this frequency the transmission of scientific data or telemetry is possible because the shorter wavelengths allow aerials of practical dimensions to be used. Telemetry chains, to monitor ocean temperatures and other variables, consist of unmanned buoys operating between 4 and 23 MHz; meteorological radiosonde balloons operate at 27 MHz. Frequencies as high as 86 MHz must be employed to provide an aerial small enough to fit inside an internal-combustion-engine crankcase to monitor that environment. Frequencies up to 10 GHz (10^{10} Hz) are used for line-of-site telemetry in microwave links of approximately 3 cm wavelength.

Satellite telemetry poses an interesting problem since frequencies must be used within the spectrum of the space 'window' to which the Earth's atmosphere is transparent. This window ranges from a low limit of 100 MHz, where ionospheric reflection occurs, to 10 GHz where absorption effects from rain and atmospheric dust begin to manifest themselves.

Perhaps the most promising area for development is in the field of modulated optical-light beams, either propagated from lasers or 'piped' along bundles of fibre-optical waveguides. These operate at visible-light frequencies (33×10^{13} Hz to 67×10^{13} Hz) corresponding to wavelengths of 4500 - 9000 Å respectively (1 Å = 10^{-10} m). These enormous frequencies offer possibilities for bandwidth and therefore information-rate capacity far beyond anything currently possible.

Carrier frequencies up to tens of megahertz may be used in coaxial-line transmission so the following section on modulation is valid for line as well as for radio transmission.

6.9 Modulation Systems

Modulation modifies the measured variable signal, introducing a carrier signal and transferring the variable information to the modulated carrier which has properties more suitable for the chosen channel.

Carrier modulation takes place by either (i) continuous variation of one of the carrier parameters (amplitude, frequency or phase), or (ii) by pulse modulating the carrier (chopping it into segments).

6.9.1 Amplitude Modulation (A.M.)[28]

This is the simplest of modulation systems; a constant-frequency carrier wave has its *amplitude* varied by the signal in a linear manner. Let us take a simple time-varying signal ($P + Q \cos \omega t$), with which we wish to amplitude modulate a carrier signal of frequency ω_c. If we multiply these two signals together we will obtain an expression for the instantaneous value of the modulated carrier signal

$$
\begin{aligned}
v &= (P + Q \cos \omega t) \cos \omega_c t \\
&= P \cos \omega_c t + Q \cos \omega t \cos \omega_c t
\end{aligned}
$$

since cos A. cos B $= \frac{1}{2}$ [cos (A + B) + cos (A − B)]

$$v = P\cos\omega_c t + \frac{Q}{2}\cos(\omega_c - \omega)t + \frac{Q}{2}\cos(\omega_c + \omega)t$$

The final equation clearly shows the presence of three frequencies formed from the multiplication process. One at the carrier frequency whose amplitude is proportional to the d.c. component of the modulating signal, and two others at angular frequencies equal to the sum and difference of the modulated signal. The carrier frequency has an amplitude equal to the d.c. term of the signal and the *upper and lower side-frequencies* are proportional in amplitude to the amplitude of the time-variant signal term. If a complex signal is used to modulate the carrier, it is obvious that the *bandwidth* covered by the modulated carrier signal will be $2\omega_h$ where ω_h is the highest frequency component of the complex signal (see figure 6.15).

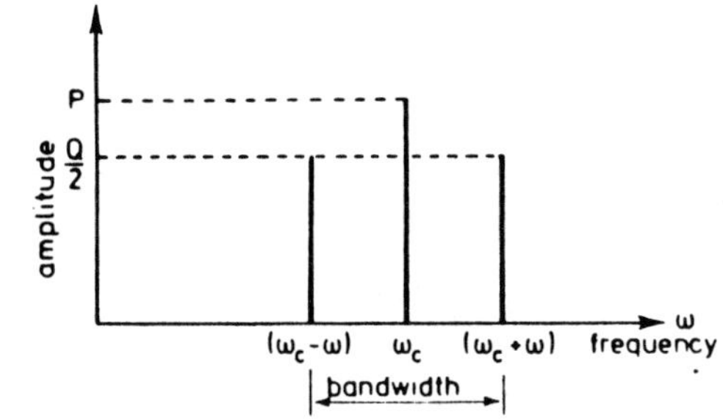

FIGURE 6.15

Frequency spectrum of an amplitude modulated (A.M.) carrier of frequency ω_c

Example 6.9

The Radio 4 Droitwich transmitter operates at a wavelength of 1500 m. Calculate the required bandwidth and frequency range of its radio-frequency amplifiers if the audio range is to be faithfully transmitted.

Assuming that frequencies up to 15 kHz are required to be present in the signal, the required bandwidth is 2 × 15 kHz = 30 kHz.

The carrier frequency is given by $f = c/\lambda$, where λ is the carrier wavelength and c the velocity of light (3×10^8 m/s)

Therefore $f = 3 \times 10^8/1500 = 200$ kHz, this gives a frequency range from 200 + 15 = 215 kHz, to 200 − 15 = 185 kHz respectively.

This type of amplitude-modulated signal is produced by passing the original signal, together with the unmodulated carrier through a *non-linear* device. Most of the devices used in electronics exhibit non-linearities to some extent. We usually operate them over the linear portions of their characteristics to give distortionless operation. Consider the shape of the characteristic curve of a diode shown in figure 6.16.

This type of gradual non-linearity is represented by a power series of the form $i = av + bv^2 + cv^3 + \ldots$etc. From the equivalent circuit of figure 6.17 we

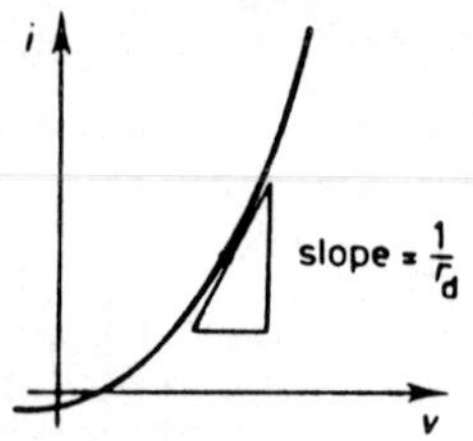

FIGURE 6.16

The non-linear characteristic of a diode

can obtain the output voltage expression as

$$v_o = aR_L\,(V_c \cos \omega_c t + V_s \cos \omega_s t) + bR_L(V_c \cos \omega_c t + V_s \cos \omega_s t)^2 + \ldots \text{etc.}$$

assuming R_L is much less than the minimum value of r_d.

Expanding, this becomes

$$v_o = R_L(aV_c \cos \omega_c t + aV_s \cos \omega_s t + bV_c^2 \cos^2 \omega_c t + 2bV_sV_c \cos \omega_s t \cos \omega_c t + bV_s^2 \cos^2 \omega_s t)$$

or

$$v_o = R_L(aV_c + 2bV_sV_c \cos \omega_s t) \cos \omega_c t \qquad (6.4)$$

plus other higher-order terms.

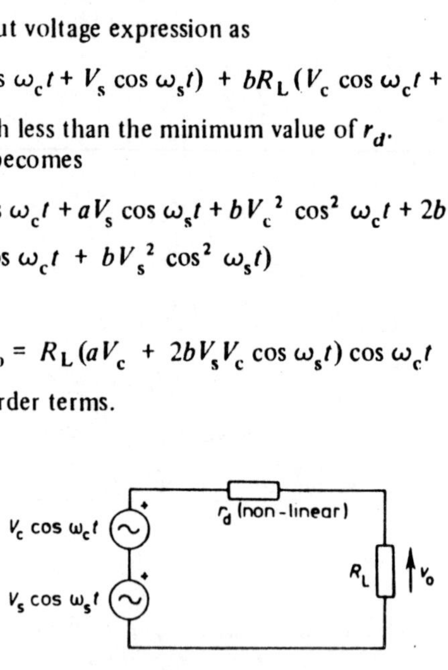

FIGURE 6.17

The equivalent circuit of a diode modulator

Equation 6.4 represents the A.M. waveform, the other terms may be rejected by a filter (usually a tuned circuit) of suitable bandwidth. The filter's bandwidth is determined by its Q (see section 2.4.3) where

$$Q = \frac{\text{centre frequency}}{\text{bandwidth}} = \frac{\omega_c}{2\omega_s}$$

When displayed on an oscilloscope a correctly amplitude-modulated waveform appears as in figure 6.18c. If the constants a and b, and the signal amplitude V_s are such that the complete signal just decreases to zero at the modulation troughs (figure 6.18e) then the modulation is said to be full or one hundred per

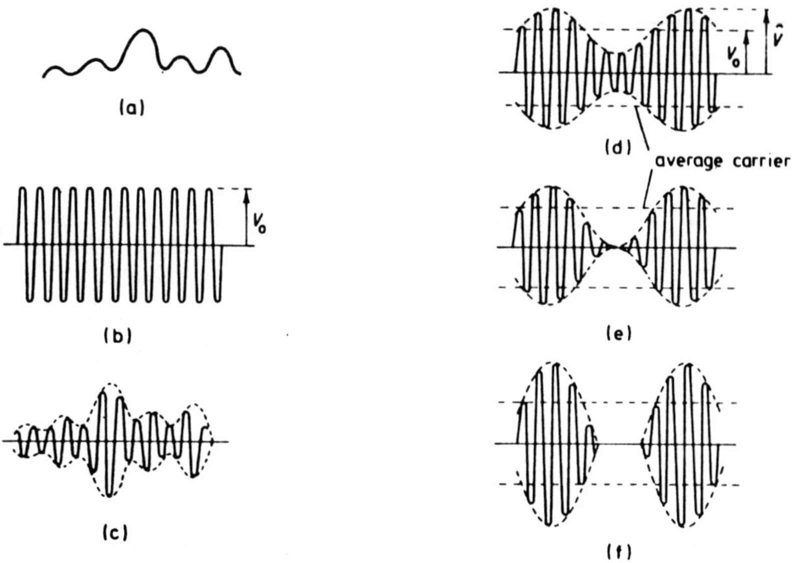

FIGURE 6.18

(a) Nature of the waveform which is to be transmitted; (b) a constant amplitude (unmodulated) carrier wave; (c) the carrier modulated by a curve (a); (d) a carrier modulated by a single low frequency; (e) full or 100 per cent modulation using this frequency; (f) overmodulation using this frequency

cent. An increase in signal-voltage amplitude produces overmodulation (figure 6.18f) which results in distortion of the signal. The percentage modulation or modulation index may be obtained from

$$\text{Modulation index } M = \frac{\begin{matrix}\text{maximum amplitude} \\ \text{of modulated signal}\end{matrix} - \begin{matrix}\text{amplitude of} \\ \text{unmodulated carrier}\end{matrix}}{\text{amplitude of unmodulated carrier}}$$

$$= \frac{\hat{V} - V_o}{V_o}$$

$$= \frac{2bV_sV_cR_L}{aV_cR_L} = \frac{2bV_s}{a} \quad \text{(from equation 6.4)}$$

Therefore

$$\text{percentage modulation} = \frac{2bV_s}{a} \times 100$$

Note that this equation is dimensionally correct because the units of a are A/V and those of b are A/V^2.

Example 6.10

An inductive transducer is used to measure a variable having a maximum frequency component of 100 Hz. It uses amplitude modulation (A.M.) with a carrier frequency of 10 kHz. Calculate the amplifier bandwidth required and the modulation index if the maximum and minimum instantaneous signal amplitudes are 0.1 V and 0.8 V.

To reduce noise problems the receiver for the above transducer is to be tuned (made frequency selective) by a single parallel resonant circuit using an inductor having a self-resistance of 12 ohms. Calculate the inductance of the coil and the capacitance required.

The bandwidth for amplitude modulation is from $f_c - f_s$ to $f_c + f_s$ or 9900 Hz to 10 100 Hz. Thus the bandwidth needed is 200 Hz.

$$\text{Modulation index} = \frac{\begin{matrix}\text{maximum amplitude of} \\ \text{modulated signal}\end{matrix} - \begin{matrix}\text{amplitude of} \\ \text{unmodulated carrier}\end{matrix}}{\text{amplitude of unmodulated carrier}}$$

$$= \frac{(0.8 - 0.1)/2}{(0.8 + 0.1)/2} = \frac{0.7}{0.9} = 0.77$$

$$Q \text{ of the tuned circuit} = \frac{\text{centre frequency}}{\text{bandwidth}} = \frac{10^4}{200} = 50$$

$$Q = 50 = \omega L/r$$

therefore

$$L = \frac{50r}{\omega} = \frac{50 \times 12}{2\pi \times 10^4}$$

$$= 0.00955 \text{ H} = 9.55 \text{ mH}$$

$$f_c = \frac{1}{2\pi\sqrt{(LC)}}$$

$$C = \frac{1}{4\pi^2 f_c^2 L} = \frac{1}{0.00955 \times 4 \times \pi^2 \times 10^8}$$

$$= 0.0265\ \mu\text{F}$$

This type of amplitude modulation is rarely used in measurement systems since attenuation along the channel and hence the amplitude of the received-signal components can vary in practice. This is caused by changes in atmospheric attenuation and reflection in the case of radio transmission, and by varying line-losses in line communication systems. The use of A.M. is chiefly confined to a.c. transducers whose output is already in amplitude-modulated form (see section 7.1.3) and the entertainment industry where the simplicity and hence the cost of receivers is of over-riding importance. Even in the latter field it is being replaced by *frequency-modulated* F.M. systems due to its inefficient use of bandwidth and vulnerability to atmospheric noise. It is inefficient because only one sideband of the three frequency components is required. Since the ear is not sensitive to frequencies below 20 Hz the ω_c term, which only carries information on the d.c. level of the signal, is redundant. The two sidebands carry identical information so that the redundancy of one can be dispensed with unless frequency-selective atmospheric attenuation is expected. These considerations have led to the the development of double-sideband-suppressed-carrier (D.S.S.C.) and single-sideband-suppressed-carrier systems (S.S.B.) respectively for use in increasingly crowded frequency channels.

Before advancing to other forms of modulation, a suppressed-carrier modulator which is finding increasing application in the instrumentation and control-engineering field merits attention because of its simplicity. This is the *double-balanced ring modulator.*

In the circuit of figure 6.19, as long as V_c is much greater than V_s, point A will become positive with respect to point B on one half-cycle, forward-biasing

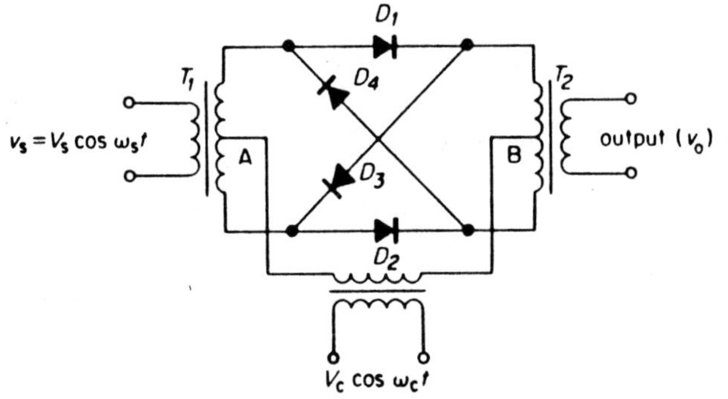

FIGURE 6.19

Double-balanced ring modulator

the diodes D_1 and D_2 and reverse-biasing D_3 and D_4. Assuming ideal diode characteristics the equivalent circuit of the centre section of figure 6.19 is now shown in figure 6.20a. Similarly on the reverse cycle, D_3 and D_4 conduct while D_1 and D_2 are cut off yielding the equivalent circuit of figure 6.20b. Providing the turns-ratios of T_1 and T_2 are the inverse of each other the effect of the circuit is to transmit the signal unchanged during one-half of the carrier period and to invert it during the remainder of the carrier cycle. The continuous signal and its inverse are shown in figure 6.21a, and the resultant waveshape produced by alternating between these in figure 6.21b. This same function is performed

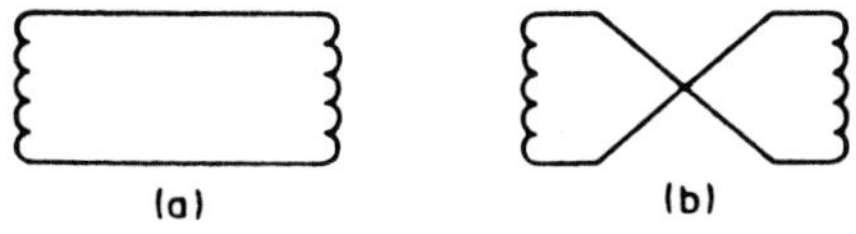

FIGURE 6.20

Equivalent circuits of figure 6.19

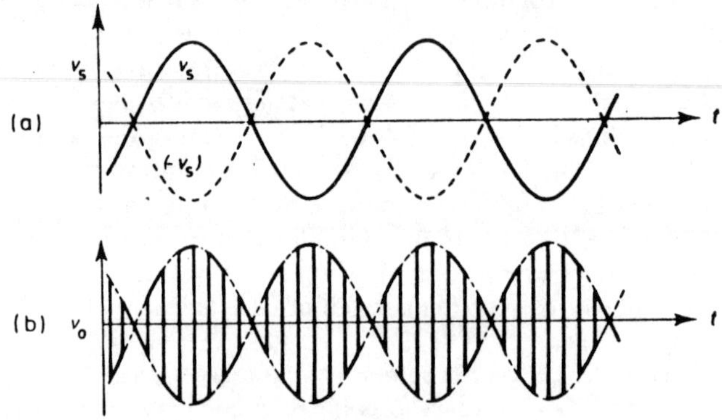

FIGURE 6.21

Waveforms of the circuit in figure 6.19

in instruments that are not required to respond to fast changes of signal by a mechanical switch (a polarised reed-relay) as shown in figure 6.22. The vibrating reed, on which is mounted a small permanent magnet, is enclosed within an evacuated glass capsule to exclude dirt and gases that may contaminate the relay contacts. The external mains-driven solenoid causes the reed to vibrate between the two fixed contacts, thus passing signal current through alternative halves of the centre-tapped transformer primary. This alternating flux induces an e.m.f. in the secondary having a waveshape very similar to figure 6.21. When the signal voltage is converted to an alternating form in this way by the circuit of figure 6.22 it is often known as a *chopper*.

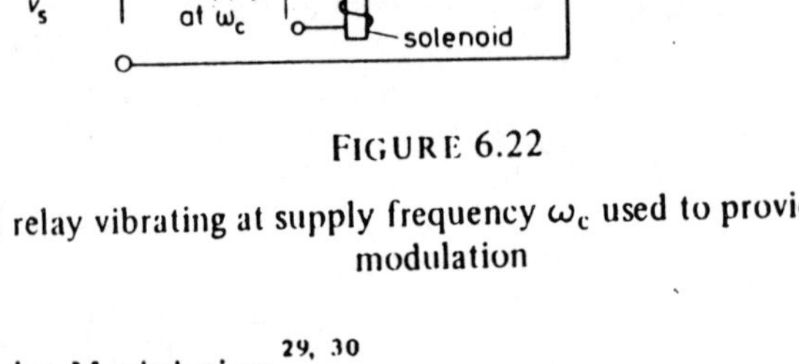

FIGURE 6.22

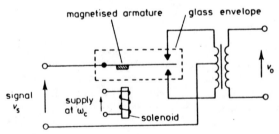

Reed relay vibrating at supply frequency ω_c used to provide balanced modulation

6.9.2 Pulse Modulation [29, 30]

Close inspection of figure 6.21 shows that it can be regarded as a series of nearly rectangular pulses whose duration is equal to the interval between them (*a mark-space ratio* of 1 : 1). Thus the *amplitude* of the pulses is a function of the measured variable thus classifying the system as *pulse-amplitude modulation*

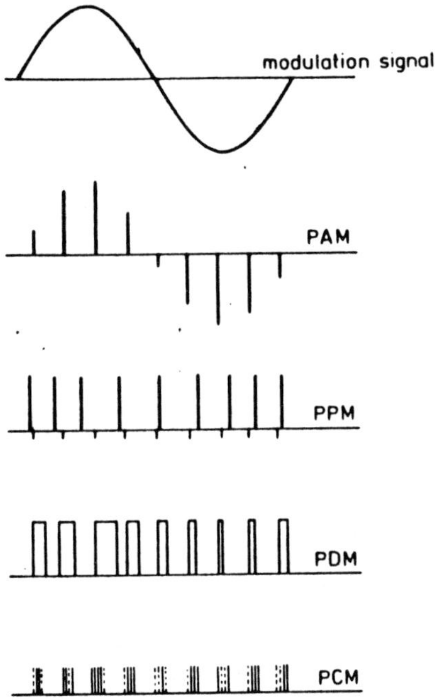

FIGURE 6.23

Differing forms of pulse modulation -- the effect of a sinusoidal signal on the output pulse shape

(*PAM*). Other pulse parameters which may be varied by the signal are its duration (PDM), its position or delay (PPM) together with pulse-code modulations, (PCM). Figure 6.23 shows the effect a sinusoidally varying signal would have upon the pulse parameters in each of the above modulation systems; for convenience the pulses have been shown of short duration except in the PDM case. In practice, the mark-space ratio of pulses is usually much less than unity since the use of short-duration pulses allows time between them for sampling other signals (time-division multiplexing p. 196).

The effect of too low a sampling rate is clearly shown in figure 6.24. It has been shown[22] that the sampling frequency must be at least twice the frequency of the highest harmonic component present in the signal for distortionless transmission.

In pulse-position modulation (PPM), the standard position of the pulse (corresponding to zero modulation) is sent to the receiver along the same channel by means of a repetitive marker pulse.

The concept of pulse-code modulation is not clear from figure 6.23 and since this form is becoming increasingly popular, brief mention of its mechanism is justified. We must first determine the span of the analogue signal and decide upon the resolution required of the system in transmitting this variable. For example, say that the pressure in a vessel is varying within the limits of zero

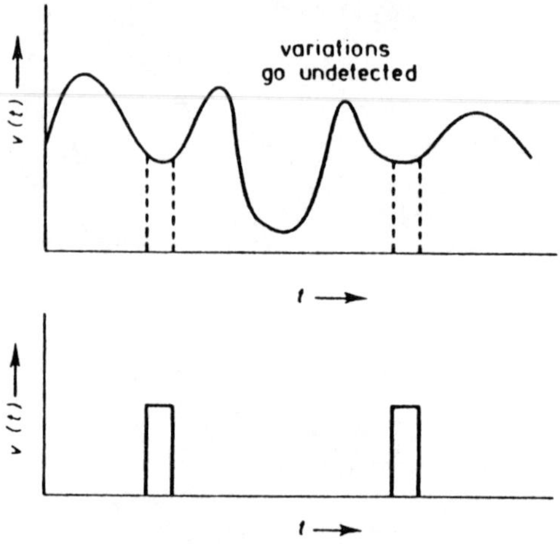

FIGURE 6.24

The effect of a rapidly varying signal being sampled too slowly

to 30 N/m^2 and the system demands an accuracy of ± 1 N/m^2. This analogue signal must be sampled at a rate appropriate to the highest significant harmonic in its fluctuation waveform, there being 30 possible values of the variable. This process of identifying the magnitude of an analogue signal by stating whether it lies between the limits 0 and 1, 2, 3 and 4 . . ., etc. is known as *quantising*. The value of the variable will now be an integer varying between 0 and 30; this

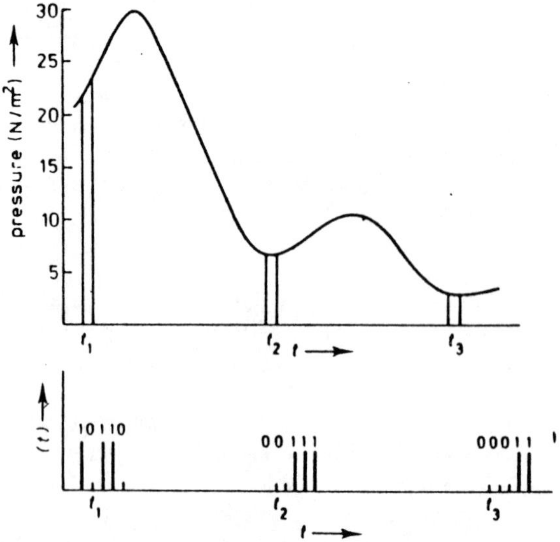

FIGURE 6.25

Basic analogue-to-digital conversion process. A pressure signal quantised and pulse coded at times t_1, t_2 and t_3

number has now to be translated from decimal to binary notation and for those readers unfamiliar with binary counting a short table of decimal and binary numbers up to 30 is depicted below. In the decimal system there are ten permitted symbols per column that is 0 to 9, after which a symbol 1 must be placed in the next column and the sequence repeated. In the binary notation there are only two permitted symbols, 0 and 1, after which a 'carry' 1 symbol must be put in the next column. This yields table 6.1.

TABLE 6.1
The decimal numbers 0 to 30 and binary equivalents

Decimal	Binary	Decimal	Binary	Decimal	Binary
01	00001	11	01011	21	10101
02	00010	12	01100	22	10110
03	00011	13	01101	23	10111
04	00100	14	01110	24	11000
05	00101	15	01111	25	11001
06	00110	16	10000	26	11010
07	00111	17	10001	27	11011
08	01000	18	10010	28	11100
09	01001	19	10011	29	11101
10	01010	20	10100	30	11110

The magnitude of the variable at times t_1, t_2 and t_3 can now be specified as the binary numbers 10110 (22 N/m^2), 00111 (7 N/m^2), and 00011 (3 N/m^2). This signal has been sampled far too infrequently but this is merely for diagram clarity. It can now be clearly seen why (on page 172) it was stated that $\log_2 n$ binary digits are required to specify a decimal number n, since the extreme right-hand binary column represents 2^0 and proceeding to the left the columns represent 2^1, 2^2, 2^3, etc. The signal magnitude at times t_1, t_2 and t_3 in binary notation may thus be transmitted as the series of pulse chains 1, 2 and 3 on figure 6.25. The simplest system would be to represent the symbol 1 by a voltage pulse of some value and to indicate the symbol 0 by zero voltage. This has the disadvantage that, should the system malfunction for a limited time and

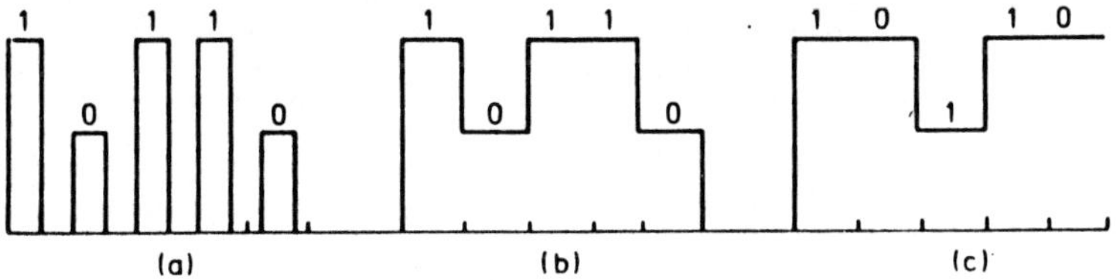

FIGURE 6.26

Pulse code modulation systems — the word 10110 sent: (a) on a return-to-zero (R.Z.) system; (b) on a non-return-to-zero-level (N.R.Z.L.) system, and (c) on a non-return-to-zero-mark (N.R.Z.M.) system

give zero signal or become submerged in noise, the receiver would interpret this as the value 0. To avoid this ambiguity the symbol 1 is represented by a positive pulse-height and the symbol 0 by a small less-positive pulse height. This is known as *positive logic* and if the symbol 1 is represented by a less-positive pulse, negative logic. There is no agreed standard between various manufacturers as to pulse heights and duration, and thus problems of system compatibility arise if one wishes to use manufacturer A's computer with manufacturer B's analogue-to-digital converter. This is partly a result of manufacturers vested interests in selling a complete system rather than one sub-unit and if items from differing sources are to be used together *interface units* must often be built or bought to make different logic systems compatible. Figure 6.26b shows the pulse train at time t_1 sent as a voltage pulse-train *word* on the non-return-to-zero-level (N.R.Z.L.) system. Figure 6.26c shows the identical word transmitted in the non-return-to-zero-mark (N.R.Z.M.) system where the symbol 1 calls for a change in level and a 0 for no change. This latter system has the advantage that less changes in level and therefore possibilities of error occur for the same information transmission. For both systems return-to-zero only occurs at the termination of a word. There is no agreement between manufacturers concerning the order of sending the word that is whether to commence with the 2^0 or the 2^n bit.

Referring back briefly to channel capacity; once the sampling rate for a given frequency has been decided, together with the number of quantisation levels, we may set a lower limit to the duration of a bit. Roughly, if the transmitted frequency band extends to zero at its lower limit we may assume that the shortest pulse capable of transmission = $2/\delta f$, where δf is the channel bandwidth. On one of the non-return-to-zero systems the word duration will thus be $2N/\delta f$ where N is the number of bits per word, equal to $\log_2 \times$ (number of quantisation levels). It may well happen that the word length only occupies a fraction of the time between successive samples (see figure 6.25) and this clearly represents an inefficient use of channel capacity. During the period

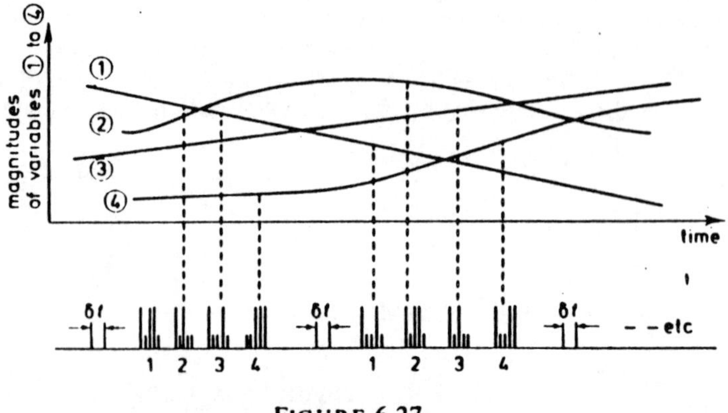

FIGURE 6.27

Time-division-multiplex. Variables 1 to 4 are sampled in rotation and synchronising pulses (two 1s separated by time δt) which the receiver can recognise are inserted in each sampling cycle

between successive words representing samples of one variable, the data channel is usually made to work to full capacity by sampling other variables in order. Thus information is being transmitted about several variables along one data channel by sampling them in rotation -- this is known as *time-division-multiplexing* and is rapidly becoming the basis for all modern forms of communication from local telephone calls to submarine intercontinental cables. A synchronising pulse is usually sent with each complete multiplex cycle in order to let the receiver decode the data into separate channels again for recording or display. Figure 6.27 shows a four-channel multiplex signal-cycle together with synchronising pulses.

Finally, concerning all forms of pulse modulation it must be stated that they can be transmitted over lines as changes in d.c. level (baseband transmission) or for wide-band or radio propagation the pulse height may be used to amplitude-modulate or frequency-modulate (see section 6.9.3) a high-frequency carrier for subsequent transmission over a radio link.

6.9.3 Frequency Modulation

Amplitude modulation suffers from the disadvantage that most external noise enters the channel in variable-amplitude form and is therefore difficult to separate from the signal. We have also seen in section 6.9.1 that unless used in single-sideband-suppressed-carrier form it is wasteful of precious bandwidth.

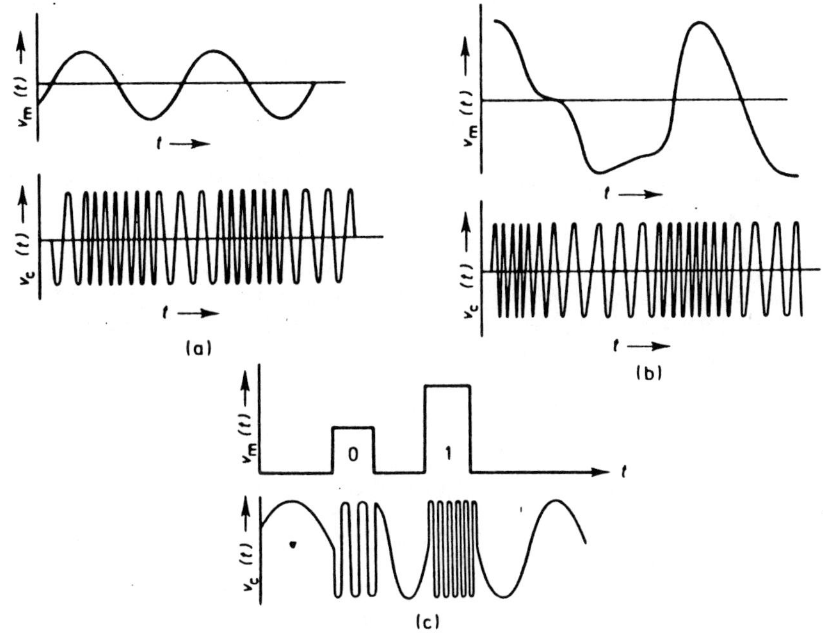

FIGURE 6.28

Frequency or angle modulation – the effects of (a) sinusoidal, (b) non-symmetrical and (c) pulse-coded variables on the amplitude-time graph of a frequency-modulated carrier wave

An alternative system by which an analogue variable may be impressed on a carrier is *angle modulation*, usually in the form of *frequency modulation*. At the time of writing this is met by most members of the public in V.H.F. national and local radio systems. Figure 6.28 illustrates the principle of the system. A constant-amplitude carrier signal is transmitted the frequency of which is varied about some fixed centre-frequency by the amplitude of the analogue variable. Below this is shown the result of using pulse-code modulated signals to frequency-modulate a carrier signal; the d.c. levels corresponding to zero, logical 1 and logical 0 will be represented by three discrete frequencies and hence this system is sometimes known as *frequency-shift modulation*. If any amplitude variations appear on the carrier when arriving at the receiver they will be caused by external noise and can be removed by an amplitude limiter before demodulation.

This type of modulation is often encountered in measurement, as opposed to entertainment, spheres when the transducer is one whose reactance (that is capacitance or inductance) varies with the measured variable. This transducer can then be employed as one element in the frequency-determining network of the oscillator producing the carrier frequency.

6.10 Problems

6.1 A picture 40 cm wide by 20 cm high is to be transmitted by a telegraphic link. Horizontal lines are scanned at a frequency of 0.5 Hz, each line consisting of square black or white dots of 0.2 mm side. Calculate the total time required for transmission of this picture and the channel bandwidth occupied if the signal-to-noise ratio of the link is 30 dB.

6.2 A transducer has an internal resistance of 50 Ω and feeds an amplifier via a 20 : 1 step-up matching transformer. If the amplifier has a noise bandwidth of 50 kHz and the amplifier's internal noise may be represented by a 5 kΩ resistor at its input, calculate the signal-to-noise ratio at the amplifier output, when the transducer terminal voltage is 1 mV. The temperature may be taken as 288 K. What would be the signal-to-noise ratio without the matching transformer?

6.3 The waveforms of the following figures are to be amplified before display.

(a) The waveform of figure 6.9a where T = 1 ms.
(b) The waveforms of figure 6.10 where T = 20 ms
(c) The isolated pulse of figure 6.12 where T = 1 ms.

Write down the minimum frequency limits of the amplification and display equipment if no change in shape is to be noticed from observation of the recorder trace.

6.4 Why are frequency modulation and pulse techniques more commonly employed in radio-telemetry equipment than amplitude-modulation methods? What advantages does amplitude modulation have?

A miniature radio transmitter is built into the piston of an experimental internal-combustion engine to detect piston flutter (transverse motion) at an expected frequency of ten times the axial piston frequency. Calculate the bandwidth required of the commercial A.M. receiver and transmitter if the engine's maximum speed is 10 000 rev/min.

6.5 What would be the minimum sampling frequency if a pulse-modulation technique were to be employed in example 6.4?

7 The Transducer and Transmitter

The transducer in a measurement system converts variations of the measured variable into an electrical signal. The transmitter then converts the signal into an electrical form suitable for transmission through the channel. Except when using electromagnetic waves (radio) as a channel, the transmitter is often an integral part of the transducer.

In this chapter attention will be confined to the measurement of non-electrical quantities by electrical means. It is assumed that the reader will be conversant with the elementary measurement of electrical quantities; single- and three-phase power measurements were discussed in chapter 2, and many standard texts[5,31,42,43] deal with the more sophisticated measurement of electrical quantities.

No attempt will be made to give exhaustive details of the many types of transducer available. The author feels that discussion of the underlying principles of operation and the inherent limitations is more appropriate at this stage.

7.1 Displacement

Displacement may be conveniently divided into two categories. Firstly the linear (translational) movement of an object and second, rotational (or angular) displacement.

A. Linear Displacement

7.1.1 The Potential Divider

This is an inexpensive transducer in which the position of the slider in a resistive potentiometer is made proportional to the displacement. The most common application is the automobile petrol gauge in which the slider position is varied by the float (figure 7.1a) thus registering the level. If a *constant-current* (high-

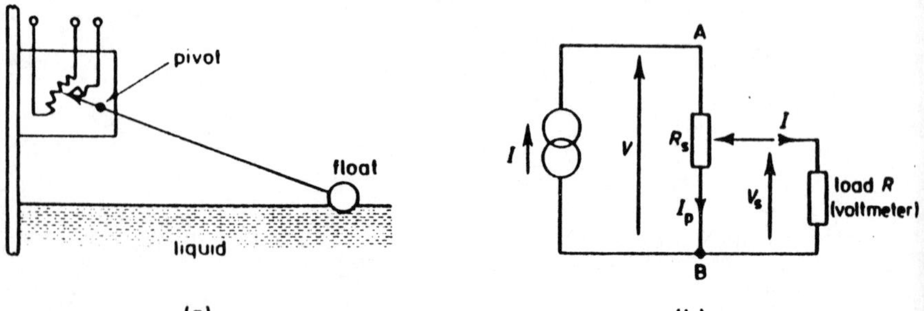

FIGURE 7.1

A potentiometric displacement transducer used as a level indicator and its circuit

impedance) source is used to energise the potentiometer (figure 7.1b), the potentiometer current I_p will be almost independent of any loading current I taken by the measuring voltmeter. The load current I may be minimised by using a high-impedance voltmeter. If one or other of these conditions is unfulfilled, *loading errors* occur. It is clear from figure 7.1b that with a *constant voltage* V applied across the potentiometer AB there will be no error when the slider is either at position A or position B. This is because the slider voltages V and 0 at these respective positions are independent of I.

Example 7.1

Calculate the minimum resistance of a voltmeter to be used with a 10 Ω potentiometer if the loading error at the mid position is not to exceed 0.01 per unit when fed from a *constant voltage* source.

Let the voltmeter resistance be R and the supply voltage V. Referring to figure 7.1b; without the voltmeter the slider voltage V_s would be $V/2$. With the voltmeter

$$V_s = I\left(\frac{R_s/2 \times R}{R_s/2 + R}\right)$$

and

$$I = V\Big/\left(\frac{R_s}{2} + \frac{R_s/2 \times R}{R_s/2 + R}\right)$$

therefore

$$V_s = V\Big/\left[\frac{R_s/2\,(R_s/2 + R)}{R_s/2 \times R} + 1\right] = V\Big/\left[\frac{R_s/2 + R}{R} + 1\right].$$

Thus the error is

$$V\Big/\left[\frac{R_s/2 + R}{R} + 1\right] - V/2 = \left[V\Big/\left(\frac{R_s}{2R} + 2\right)\right] - V/2$$

and per unit error is

$$\left[2\Big/\left(\frac{R_s}{2R} + 2\right)\right] - 1 = -\ 0.01$$

therefore

$$2.02 = \frac{R_s}{2R} + 2$$

or

$$\frac{R_s}{2R} = 0.02$$

$$\text{voltmeter resistance } R \geqslant \frac{R_s}{2} \times \frac{100}{2}$$

$$\geqslant 250\ \Omega$$

The resistive element may be wire-wound or may be a carbon or metal film. Trouble may be encountered from the slider contact which may become dirty and from pivot friction. A typical life expectancy is a million operations and the speed of response is rarely less than a quarter-second. In wire-wound versions the resolution is limited by the number of turns employed.

7.1.2 The Reluctance Transducer

This device uses the displacement to alter the geometry of a magnetic circuit, thus altering the inductance of a coil wound round the circuit. The simplest form is shown in figure 7.2a where the armature moves with respect to the E-core, thus altering the size of the airgaps. Since reluctance is equal to magneto-

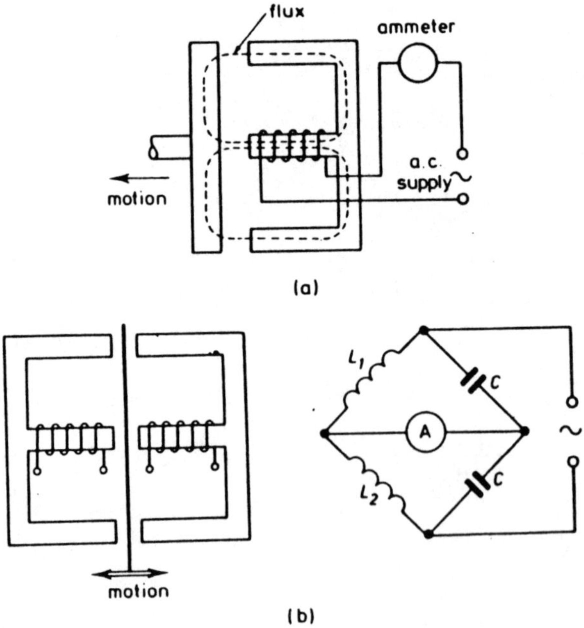

FIGURE 7.2

Single- and dual-coil reluctance transducers together with a bridge detecting circuit for the dual form

motive force /flux, a decrease in reluctance caused by narrowing the airgap results in a flux increase. The coil's inductance thus increases and the ammeter reading will fall. Unfortunately this simple arrangement is very non-linear, that is, the current is not proportional to the armature displacement. Figure 7.2b shows a two-coil transducer which gives an approximately linear detector current in the bridge circuit shown; typical maximum airgaps and hence the displacement spans are only a few millimetres but the speed of response can be up to 1 kHz.

7.1.3 The Differential Transformer (LVDT)

The limited span of the simple reluctance transducer may be extended up to 20 cm displacement by using the transformer type of magnetic device shown in figure 7.3a. With the core centralised in the bore of the former, there will be no net output voltage, V_{out}, because identical coils L_1 and L_2 are connected back to back, the e.m.f.s cancelling ($V_{AB} = V_{CB}$). The half-wave rectifier diodes will thus give equal positive outputs to points D and E, making $V_{out\ d.c.}$ equal to zero.

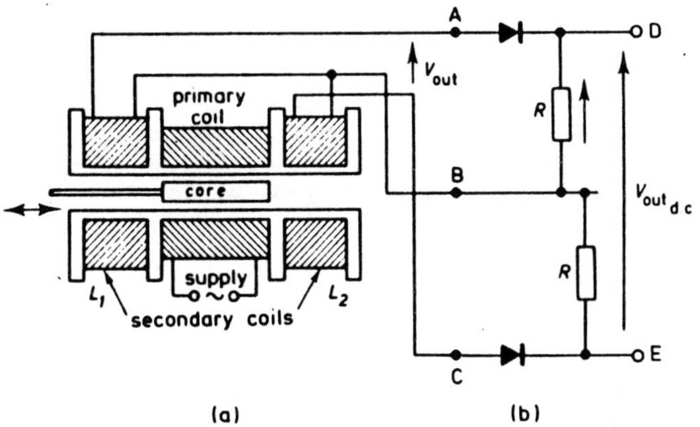

FIGURE 7.3

A linear variable differential transformer (LVDT) and its associated circuitry

As the core moves left the primary flux coupling with L_1 will exceed that coupling L_2. Thus $V_{AB} > V_{CB}$ making the point D more positive than E. $V_{out\ d.c.}$ will thus increase, having a polarity as shown. If the core moves to the right of centre, $V_{CB} > V_{AB}$ making E more positive than D, reversing the polarity of $V_{out\ d.c.}$. Sketches of the various voltages are shown in figure 7.4a and b for varying displacements. The response speed is typically up to 1 kHz.

7.1.4 The Capacitive Displacement Transducer

Because the capacitance of a parallel-plate capacitor depends upon plate area and separation together with the relative permittivity of the dielectric, variation of any of these can be converted to an electrical output. Figure 7.5 illustrates some of the more common forms of capacitive transducer; type c is an interesting variation suitable for use with liquids which are insulating and whose relative permittivity is markedly different from that of air. The disadvantage of capacitive

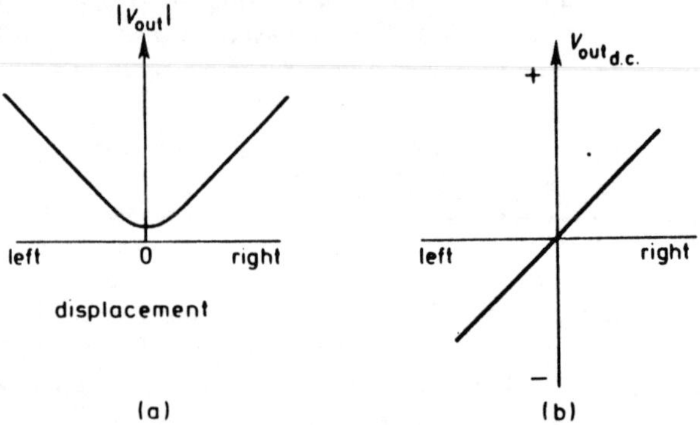

FIGURE 7.4

Variations of the voltages in figure 7.3

displacement transducers is that the necessary plate area often renders them bulky. Nevertheless the forces required to operate them are very small and therefore they can be designed to have an extended upper-frequency response.

7.1.5 Seismic Methods for Displacement

Displacement may be regarded as the second time-integral of acceleration. The output of any of the seismic accelerometers described in section 7.4 may therefore be integrated twice electronically to give a signal proportional to displacement, provided that information on the initial conditions (velocity and displacement) is available. These transducers have a wide frequency response (up to 1 MHz), the necessary integrating circuits are dealt with in section 8.3.1.

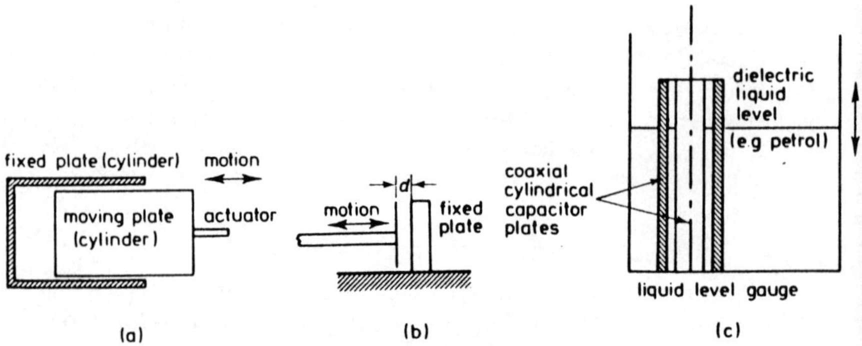

FIGURE 7.5

(a) Cylindrical variable area, (b) variable-separation and (c) cylindrical variable-dielectric level transducers

7.1.6 Digital Methods for Displacement

In its crudest form this method depends on counting the number of pulses observed by a photocell or other sensor when a striated surface moves beneath it as in figure 7.6a. The pulses are fed to an electronic counter whose display can be made to indicate the lateral position of the surface.

In practice, the dark areas are often formed by interference fringes produced by two transparent surfaces each ruled with a fine grating and placed at a small angle to each other as in figure 7.6b. This effect is known as the 'Moiré fringe' effect[32]. When light is shone through both the surfaces a pattern of light and dark fringes is seen whose movement is greatly affected by very small relative movements between the surfaces. This method has been extensively used for the automatic control of machine-tool slides.

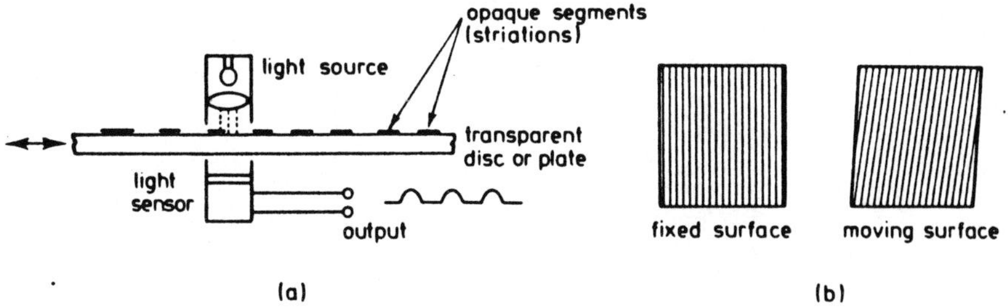

FIGURE 7.6

(a) Principle of digital displacement methods; (b) surface patterns for generating Moiré fringes

B. Angular Displacement

Distinction must be made between measurements of angular movement of a shaft with respect to a stationary datum and relative displacement between two parts of a rotating shaft. The latter measurement is complicated by the fact that the transducers must be mounted on a rotating member and connections made to the stationary display equipment. There are two main methods of achieving these connections.

(i) For many short-term measurements with low shaft speeds slip-rings and brushes can be used as in figure 7.7a. The main difficulties are caused by contact potentials and variable contact resistances between slip-rings and brushes.

(ii) For higher shaft speeds the transformer effect may be employed; one winding is mounted upon the shaft, the other enclosing it and stationary (figure 7.7b). This method is only applicable to alternating signals and it must be remembered that the frequency and voltage of the signal in the stationary secondary will depend upon the relative angular velocity between the two windings. The analysis of this situation is similar to that of the rotor e.m.f. and rotor frequency of an induction motor (section 4.5.2) because in both cases relative motion between the two windings occurs. Thus if the primary is fed with a constant frequency f and if the standstill secondary output voltage is V_o, when rotation occurs

$$\text{secondary output voltage, } V_o' = V_o\left(1 - \frac{n_{rel}}{f}\right)$$

$$\text{secondary output frequency, } f_o = f\left(1 - \frac{n_{rel}}{f}\right)$$

where n_{rel} is the speed of shaft rotation (rev/s). These equations show that the speed of shaft rotation must not approach the frequency of the alternating signal, so the frequency must be chosen accordingly.

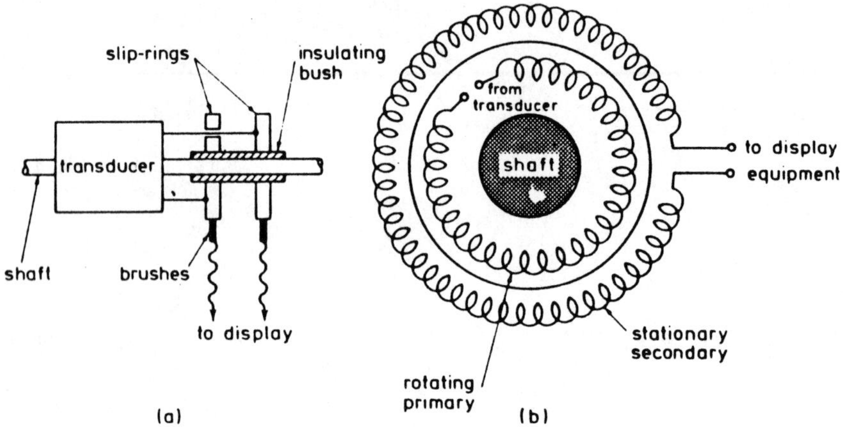

FIGURE 7.7

Two methods of transferring a signal from a rotating member to stationary equipment

Neglecting the rather crude visual inspection of shaft deformation under a stroboscopic light, there are several methods of detecting relative angular displacement between two rotating members.

7.1.7 Inductive Bridge Transducer

Figure 7.8 shows four coils L_1 to L_4 mounted upon a rotating backplate A. If shaft B moves relative to A in the direction shown, L_2 and L_4 will increase in reactance as their magnetic cores enter them more deeply; whereas L_1 and L_3 will decrease for the oppositve reason. If wired in the bridge circuit of figure 7.8b a bridge imbalance will occur if B moves relative to A in either direction. The phase of the bridge output voltage will indicate the direction of relative displacement. The detector output from the bridge is suitable for transmission to stationary equipment by either of the methods of figure 7.7.

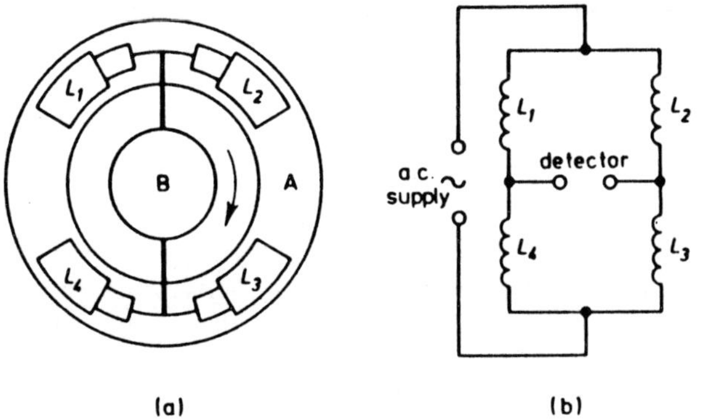

FIGURE 7.8

The inductive angular displacement bridge

7.1.8 Timing Methods

The techniques of figure 7.9 are indirect ones which assume a constant value for the rotational speed of the shaft and which infer the angular displacement from a time interval between two events. These are the passage of two projections beneath either magnetic transducers or photocells. Any misalignment in the projections will appear as a time interval between the two which may be measured using an electronic timer-counter. Because intervals of only a few microseconds can be recorded, very high shaft speeds can be catered for.

If suitable values are assumed for the elastic constants of the shaft in the methods of figure 7.8 and 7.9, the torque in the shaft may be inferred.

We will now examine methods more suitable for the measurement of displacement with respect to a stationary datum.

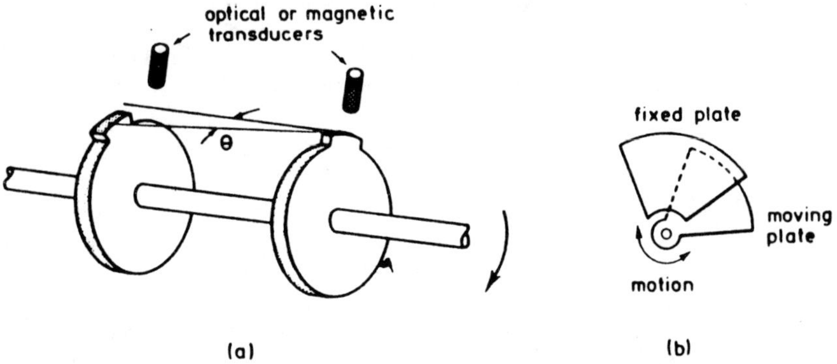

FIGURE 7.9

(a) Indirect time-interval method and (b) variable capacitance method for angular displacement

7.1.9 Capacitive Methods for Angular Displacement

Figure 7.9b illustrates the principle of a capacitive transducer used for this measurement. Many interleaved alternately fixed and moving plates may be used to give a capacitance change much greater than the standing capacitance of the connecting leads. But this causes them to be bulky (see section 7.1.4).

7.1.10 The Synchro

If two three-coil windings are connected together as shown in figure 7.10 and an a.c. flux established in rotor A from an external voltage source, the relative magnitudes of the e.m.f.s in stator A will be determined by the rotor position. These three e.m.f.s induce currents in stator B such that the

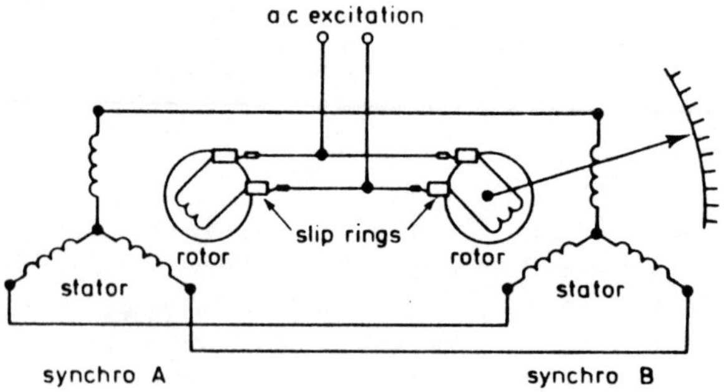

FIGURE 7.10

The principle of the synchro

resultant flux-pattern in B exactly corresponds in direction to that produced by rotor A. Rotor B thus aligns itself in this direction and will follow any movement of rotor A. Considerable torque can be exerted by a large (15 A) synchro; quite sufficient to act as a remote servo-control for a valve or other equipment.[33]

Differential and *summing* synchros are available which will take up a position proportional to the difference or sum respectively of the angular deflections of two other remote synchros.

7.1.11 Disc Encoders

This method is a simple and accurate one which provides a shaft-position signal consisting of a binary coded number. A transparent disc with opaque segments is mounted upon the shaft and light from a lamp is interrupted by the segments before falling on a series of photocells. An alternative, less accurate method is to use a series of brushes contacting conducting segments. The segment width limits the resolution to about 1° on a four-inch disc using brushes, but with photoreduction techniques opaque segments only a few microns in width can be produced for the photocell method. This enables a 14-bit binary output with its resolution of approximately one minute of arc to be obtained.

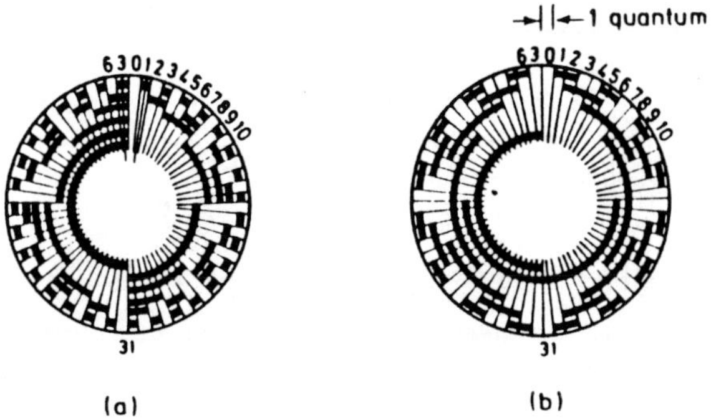

FIGURE 7.11

Shaft encoding discs (a) 8421 binary coded decimal; (b) Gray code

Figure 7.11a shows a disc employing the well known 8421 BCD code; its use, however, causes some difficulties. In moving between adjacent segments many bits often change simultaneously, the extreme case being sector 15 to sector 0 (1111 to 0000). Even with precision construction all four bits will not move simultaneously because of stagger in the brush or photocell positions. At the moment of changeover a large error can thus occur, for example 15 to 0 might give a momentary 0111 during changeover corresponding to a sector 7 signal.

TABLE 7.1
Binary Codes

Decimal number	8421 BCD	Gray code
0	0000	0000
1	0001	0001
2	0010	0011
3	0011	0010
4	0100	0110
5	0101	0111
6	0110	0101
7	0111	0100
8	1000	1100
9	1001	1101

Accordingly the Gray or *reflected-binary code* is employed (figure 7.11b) in which only one bit changes in passing between adjacent sectors giving a maximum error of one sector during changeover. Table 7.1 sets out the two codes for comparison – it is relatively easy to change from one to another by electronic means.

C. Strain Gauges

A special type of displacement transducer which deserves a separate section is the strain gauge. It is designed to detect the small fractional change in dimensions which occur when a body is stressed.

7.1.12 Metallic Strain Gauges

If a wire is stretched from length l to $(l + \delta l)$ there will be accompanying reduction in the cross-sectional area A. Since its resistance equals $\rho l/A$ where ρ is its resistivity in ohm metres (Ω m), both these effects will lead to a resistance rise. Because the resistance changes observed in actual gauges cannot be completely explained by the above effect, it is thought that structural changes also occur within the wire which have an important effect. Nevertheless, for per unit changes in length of less than 0.01 the corresponding fractional change in resistance is approximately linear.

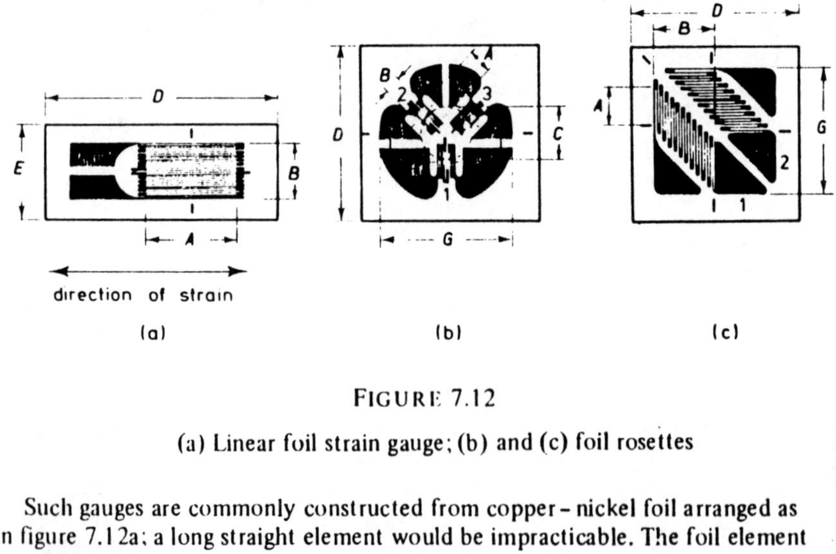

FIGURE 7.12

(a) Linear foil strain gauge; (b) and (c) foil rosettes

Such gauges are commonly constructed from copper – nickel foil arranged as in figure 7.12a; a long straight element would be impracticable. The foil element is embedded in a plastic backing strip (for electrical insulation and moisture exclusion) and the whole is bonded to the material under test using special adhesives.

The gauge must be mounted parallel to the strain direction as shown in figure 7.12a. Where this direction is unknown a compound gauge or rosette (figure 7.12b) will determine the strain in three directions from which both the magnitude and direction of the total strain may be inferred. It is possible, given the elastic constants, to calculate the torque in a shaft from knowledge of the strain in two directions each at 45° to the shaft axis. Two-element rosettes (figure 7.12c) are supplied for this purpose.

The most important drawback to the use of strain gauges is the fact that resistance and dimensional changes are also caused by temperature variations.

Accordingly gauges are often used in pairs or groups of four to compensate for temperature changes. Details of the theory and arrangement of suitable bridges for these measurements are given in section 8.6.2. Per unit strains down to a few *microstrain* can be detected (1 microstrain = one part in 10^6) and gauges are constructed with preferred values of resistance of 120 and 600 ohms. The relationship between the strain ϵ and the fractional change in resistance $\delta R/R$ is given by

$$\frac{\delta R}{R} = F\epsilon$$

where F is the *strain sensitivity* or *gauge factor* and is a constant (≈ 2) for a given gauge. The manufacturer always quotes the gauge factor and sets of gauges with closely matched gauge-factors are obtainable for use in bridges. The application of strain-gauge techniques is often quite sophisticated and the reader is referred to standard texts on two-dimensional stress[34] and strain-gauge applications[35] for further reading.

7.1.13 Semiconductor Strain Gauges

Piezoresistive materials suffer gross resistance changes when strained because of changes in the crystal structure. This allows very sensitive gauges to be constructed having gauge factors between 20 and 100 times those of metallic gauges. The resistance - strain relationship however is non-linear and they are heavily temperature-dependent. It is possible that these disadvantages will be minimised in the future. Meanwhile they must be carefully temperature-compensated and prestressed to give a more linear output.

Perhaps the only remaining disadvantage is that strain gauges can only monitor *surface strain*; for bulk strains, photoelastic techniques must be employed.

7.2 Time

Time and its reciprocal (frequency) are the variables that we can measure most accurately. For example, sophisticated laboratory instruments measuring other variables are often limited to 0.01 per unit accuracy, whereas such a crude mechanical timepiece as a long-case (grandfather) clock should be accurate to a minute per week (approximately 0.0001 per unit).

Most modern laboratory timer-counters employ an electronic clock, which is an oscillator often with a frequency of 1 MHz, followed by electronic divider circuits to reduce this frequency to the 1Hz region. Oscillators using the simple *LC* frequency-determining networks of section 5.5 are not sufficiently temperature-stable for this application. Dimensional changes caused by thermal expansion in both capacitors and inductors cause the resonant frequencies of these oscillators to suffer long-term 'drift'. The classical method of ensuring frequency stability has been to employ the mechanical resonances which occur in thin wafers of quartz vibrating in their thickness mode. The resonant frequency can be accurately adjusted by grinding the crystal to the required thickness. The

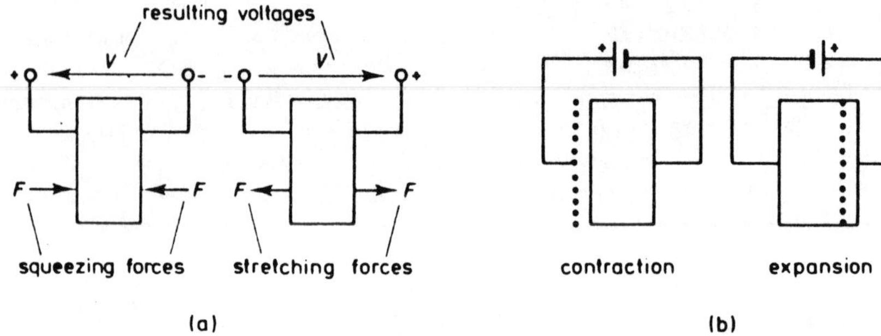

FIGURE 7.13

Piezoelectric behaviour

mechanical resonances are sustained electrically. Because quartz is piezoelectric, potentials will appear across its faces when stressed as shown in figure 7.13a and conversely it will deform mechanically if potentials are applied to it (figure 7.13b). Thus because the mechanical resonant frequency is determined by the crystal thickness, this is the only frequency at which corresponding electrical oscillations may be sustained. Figures 7.14a and b show a block diagram and a

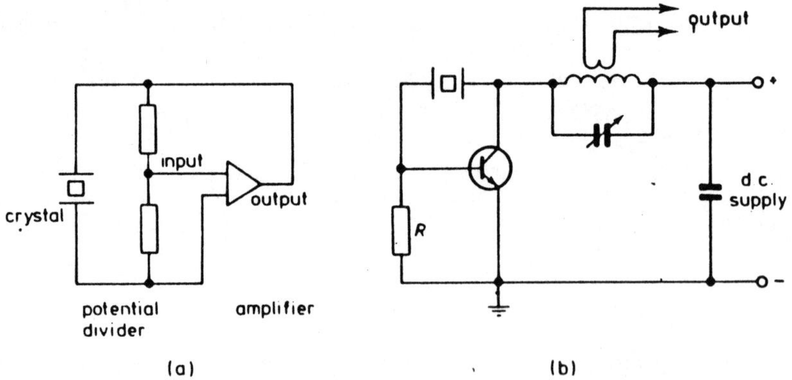

FIGURE 7.14

Block and practical diagrams of crystal-controlled oscillators

practical circuit of a crystal-controlled oscillator. Although crystal dimensions are less susceptible to thermal expansion than inductors or capacitors, some form of thermostatically controlled enclosure is often provided for the crystal if accuracies of more than 1 part in 10^6 are required. Accuracies up to 1 part in 10^8 can easily be obtained in commercial instruments employing temperature-controlled crystals.

A block diagram of a timer-counter employing an electronic clock is given in figure 7.15. Signals of unknown period are applied to input 1 and, after electronic squaring, are used to hold gate 1 open. The number of clock pulses passing through the gate to the display will thus reveal the signal period in time units (in this case milliseconds). For frequency determination the signals are applied to input 2 where they are shaped and allowed through gate 2 to the display counter for unit time. The display will show the number of cycles passing in unit time.

The accuracy of timer-counters can be increased by locking their electronic clocks to some external frequency standard. The frequency of the BBC 200 kHz (1500 m) transmitter is often employed because its accuracy is maintained to 5 parts in 10^9. The ultimate accuracies of 1 part in 10^{11} are obtained in atomic clocks in which the output of a quartz-crystal oscillator is continuously compared with the frequency of radiation emitted from a transition between two electron states in the emission spectrum of the caesium atom.[36]

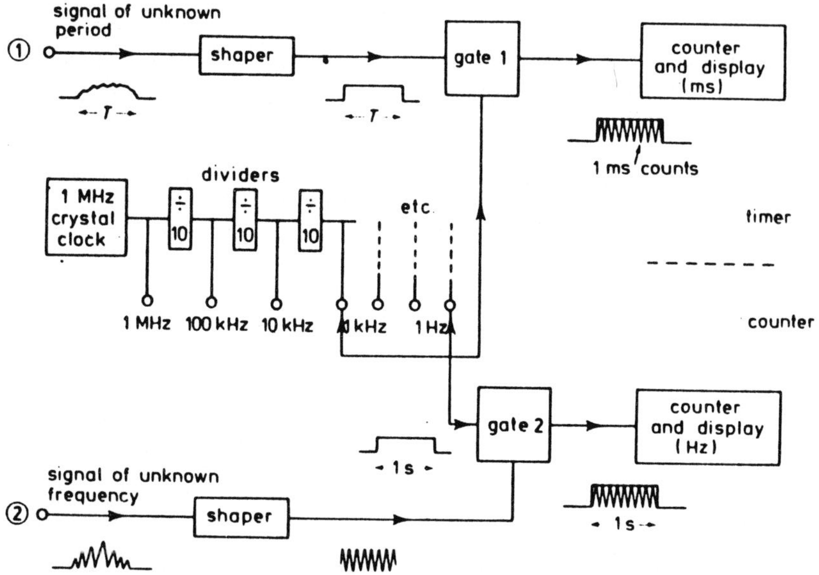

FIGURE 7.15

Block diagram of an electronic timer/counter

7.3 Velocity

Linear Velocity

In applications where the linear velocity of a uniformly moving object is to be determined the combination of an electronic timer with photocells or an inductive magnetic pickup is often employed. Figure 7.16 illustrates these two simple methods.

In applications where the instantaneous value of a constantly changing velocity must be determined the electromagnetic transducer of figure 7.17 is employed.

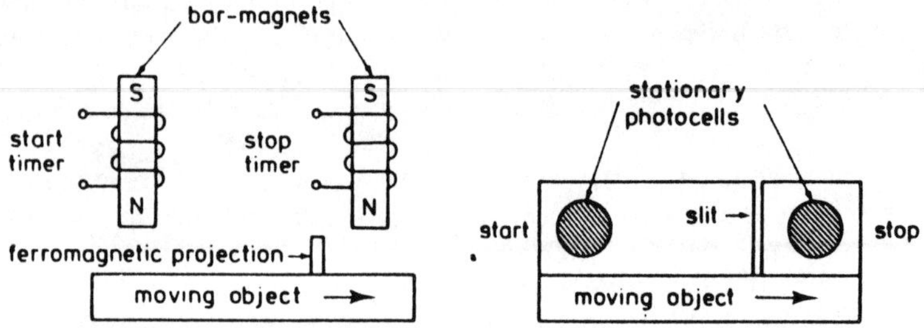

FIGURE 7.16

Principles of velocity measurements employing timing

Because the induced e.m.f. is proportional to the rate of flux cutting an oscilloscope placed across the coil will monitor the instantaneous velocity of the magnetic plunger.

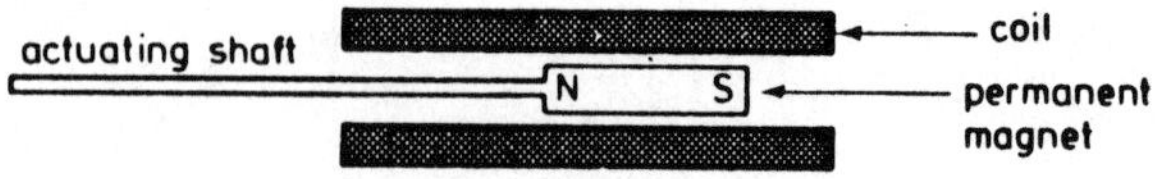

FIGURE 7.17

Electromagnetic velocity sensor

Angular Velocity

7.3.1 Drag-cup Tachometers

This instrument is simple and reliable, being commonly encountered in automobile instrumentation as a speedometer and tachometer. Figure 7.18a shows the constructional features while figure 7.18b demonstrates the flux pattern produced within the aluminium drag-cup C. The surrounding soft-iron cylinder B provides a low reluctance path for the magnetic flux from the bar magnet A which rotates at the speed to be measured. Eddy currents caused within the cup tend to rotate the cup in the same direction of motion against the return spring S. The angle of deflection of the pointer is thus proportional to the angular velocity of the magnet.

The disadvantage of this mechanism is that it does not provide a remote indication unless the drive to the input shaft is taken through some form of flexible cables. The length of the latter is limited to a few metres and they have poor reliability. Figure 7.19 shows an adaptation of the drag-cup principle to give an a.c. tachogenerator whose output voltage is proportional to the angular velocity of the rotating drag-cup. When the cup is stationary there will be no e.m.f. induced in L_2 by flux from L_1 because their axes are perpendicular. As the cup rotates it distorts the flux pattern as shown; the faster the rotation, the more

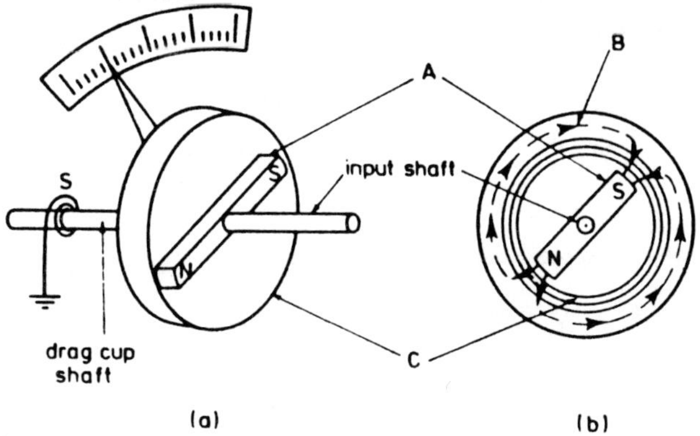

FIGURE 7.18

Mechanical drag-cup tachometer

flux is swept into the axis of L_2 giving an increased e.m.f. induced in L_2. The output voltage is thus velocity-dependent but the frequency of the output voltage is fixed at the supply frequency. The chief disadvantage is that reversal of the direction of drive only causes the phase of the output voltage to be reversed which is often difficult to detect and gives an ambiguous output.

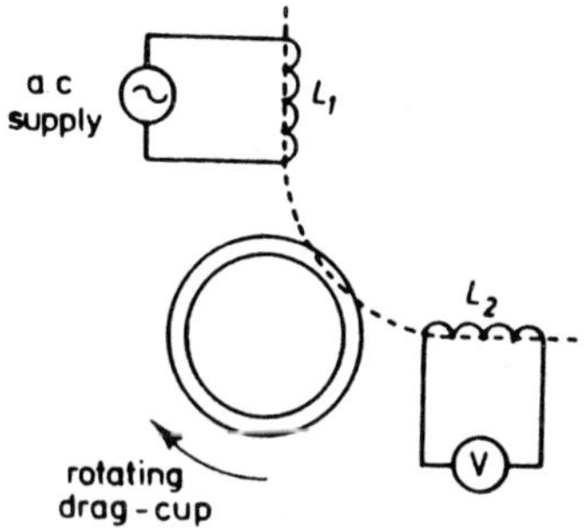

FIGURE 7.19

The drag-cup tachogenerator

7.3.2 Tachogenerators

A simple tachometer can be made by monitoring the output voltage of a small d.c. or a.c. generator driven at the speed to be measured. These are usually two-pole d.c. machines with a simple commutator and permanent-magnet field giving about 5 V output per 1000 rev/min. The output polarity, of course, clearly demonstrates the direction of drive.

The complication and hence poor reliability of brushgear can be avoided by employing an a.c. generator with a permanent-magnet rotor and fixed field-coils

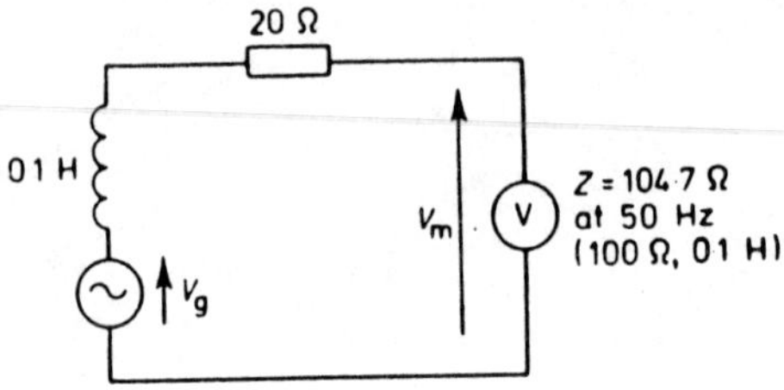

FIGURE 7.20
Example 7.2

as shown in figure 7.20. The output voltage is proportional to speed but unfortunately the output frequency ($f = np$) is also speed-dependent. This causes the tachogenerator coils to have higher reactance at higher speeds giving errors unless the generator output is monitored by a high-impedance voltmeter.

Example 7.2

A two-pole permanent magnet alternator is used as a tachogenerator and is calibrated by a digital voltmeter of 2 MΩ input resistance at 1500 rev/min. If the resistance and inductance of the field coils are 20 Ω and 0.1 H respectively, calculate the error if the output is read at 3000 rev/min with a moving-iron voltmeter of 100 Ω resistance and 0.1 H inductance.

At 1500 rev/min

$$f = np$$
$$= \frac{1500 \times 1}{60} = 25 \text{ Hz}$$

Field-coil impedance

$$Z = \sqrt{[R^2 + (2\pi f L)^2]}$$
$$= \sqrt{[20^2 + (2\pi \times 25 \times 0.1)^2]}$$
$$= \sqrt{[400 + 246]} = 25.4 \ \Omega$$

This is negligible compared with the digital voltmeter's impedance of 2 MΩ.

At 3000 rev/min

$$f = np = 50 \text{ Hz, thus field coil impedance}$$
$$Z = \sqrt{[20^2 + (2\pi \times 50 \times 0.1)^2]}$$
$$= \sqrt{[400 + 987]} = 37.2 \ \Omega$$

This is not negligible compared with the meter impedance

$$Z_m = \sqrt{[R_m{}^2 + (2\pi f L_m)^2]}$$

$$= \sqrt{[10^4 + (2\pi \times 50 \times 0.1)^2]}$$
$$= \sqrt{[10^4 + 987]} = 104.7\ \Omega$$

Referring to figure 7.20

$$V_m = V_g \times \frac{Z_m}{Z_{total}} = V_g \times \frac{104.7}{\sqrt{[120^2 + (2\pi \times 50 \times 0.2)^2]}}$$
$$= V_g \times \frac{104.7}{\sqrt{[14400 + 3940]}} = V_g \times \frac{104.7}{135}$$
$$= 0.77 V_g$$

Thus per unit error is $1 - 0.77 = 0.23$.

7.3.3 Digital Methods

All these methods depend on some form of projection or slit being attached to the shaft under investigation. Figure 7.21 shows magnetic and optical transducers being operated by a toothed wheel and a slit disc respectively. The frequency of the output can be measured by an electronic timer-counter set in the counter mode. If more than one tooth or slit passes the transducer per revolution, the final reading must be divided by this number.

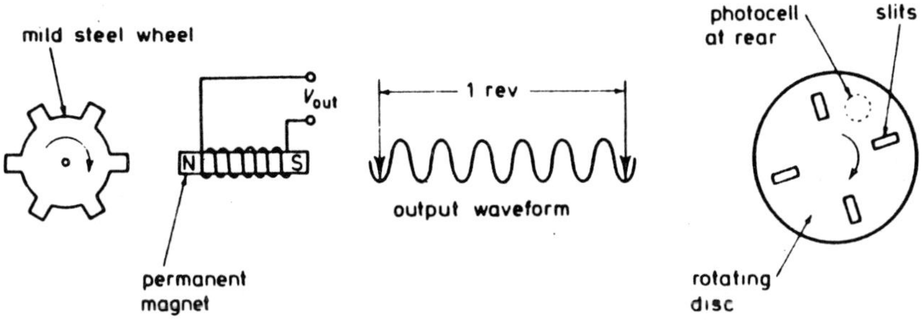

FIGURE 7.21

Digital methods for angular velocity

7.3.4 Stroboscopic Methods

If the angular velocity to be measured is constant for a reasonable period (10 seconds or so) it may be measured with limited accuracy by a stroboscope. The stroboscope is basically a high-intensity gas-discharge lamp emitting short flashes and powered by a variable-frequency oscillator. The oscillator's frequency is adjusted until some mark or feature on the rotating object appears stationary – this is caused by the persistence of vision. The oscillator frequency will now be equal to a sub-multiple of the shaft speed. The apparent number of marks has to be examined carefully to find the actual speed (and thus frequency) at which the number of actual and apparent marks coincide. Nevertheless provided this possibility of error is allowed for the method is convenient and requires no modification to existing plant.

7.4 Acceleration and Vibration

Because acceleration is the second time-differential of displacement, the output of a displacement transducer may be differentiated twice by electronic means and the result used to provide an acceleration signal. Unfortunately, in the majority of cases, large displacements are encountered rendering this method impracticable.

7.4.1 Seismic Methods

These methods employ Newton's law by transmitting the acceleration or vibration forces from the test object to a small mass via a force transducer. Figure 7.22 shows three examples of typical seismic accelerometers. Type (a) is applicable to low-frequency investigations where the relatively low stiffness of the springs is not a disadvantage. The mass is a magnet which moves along the axis of a cylindrical coil producing e.m.f.s proportional to its acceleration. For higher-frequency investigations pattern (b) is preferred; the acceleration is transmitted from the transducer body to the mass by a high-stiffness piezoelectric crystal.

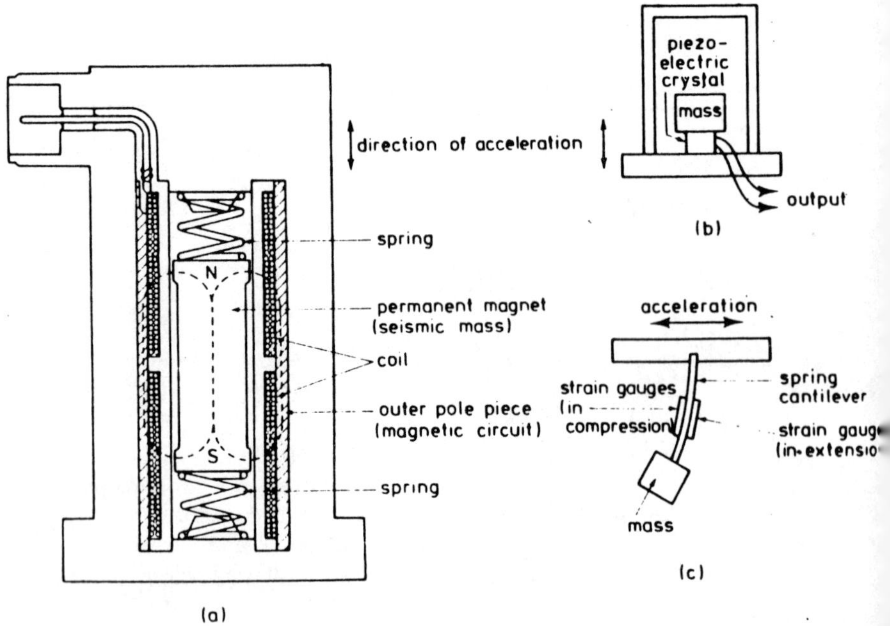

FIGURE 7.22

Types of seismic accelerometer

The voltages set up across the crystal are proportional to the accelerating force and hence to the acceleration. This transducer suffers from the disadvantage that, with slowly varying accelerations, the electric charge produced across the crystal by the piezoelectric effect tends to leak away. This necessitates the use of a charge amplifier of extremely high input impedance (section 8.3.2) to minimise leakage of this charge. The third type (figure 7.22c) is convenient for use in low- and medium-frequency ranges, the accelerating forces being transmitted to the mass by a spring cantilever blade. Strain gauges (usually in pairs) are mounted on opposite sides of the blade giving suitable outputs for incorporation in a four-active-arm bridge (section 8.6.2).

7.4.2 Production of Vibrations

In many structural engineering investigations the response of a complex structure to forced vibration over a wide frequency spectrum is sought. Such vibrations can conveniently be excited by electromagnetic means, employing the principle of the moving-coil loudspeaker. Figure 7.23 depicts a typical transducer in

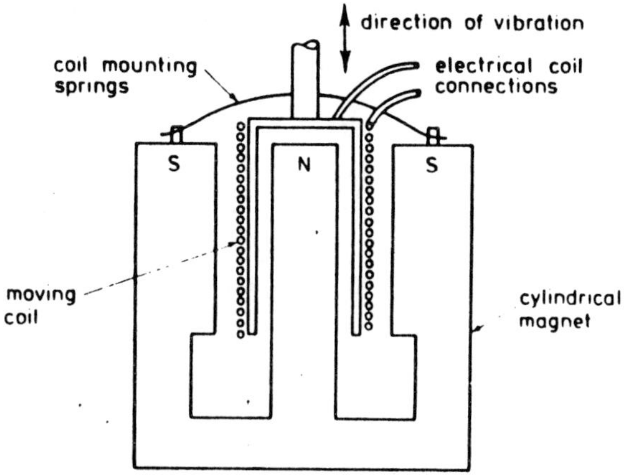

FIGURE 7.23

Moving-coil vibrator

which a cylindrical coil is mounted within the airgap of a powerful cylindrical permanent magnet. The application of an alternating e.m.f. to the coil results in a reciprocating movement along the coil axis. The frequency range of such a vibrator may extend from a fraction of a hertz to several kilohertz which is ample for almost all investigations.

7.5 Pressure

Most pressure transducers employ the pressure differential between the measured pressure and atmospheric pressure to distort some elastic structure. The *displace-*

ment which occurs in the structure is then measured using one of the methods of section 7.1.

7.5.1 Bellows-type Transducers

Figure 7.24a shows a transducer which uses a bellows to convert pressure changes to displacement variations. The relatively large displacement of the

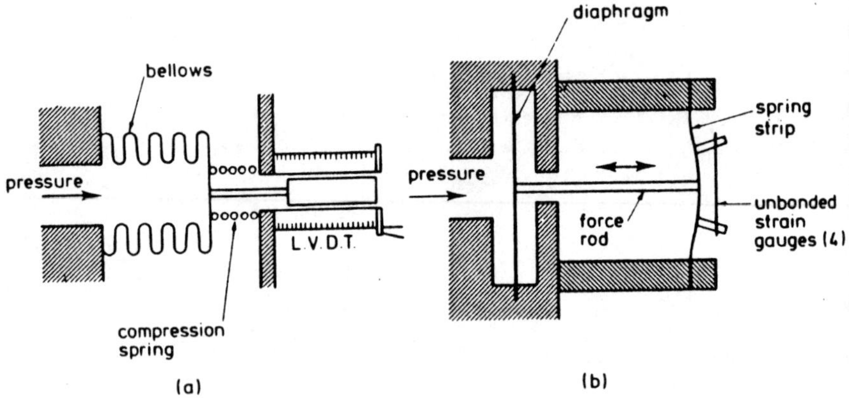

FIGURE 7.24

Electrical pressure transducers

bellows makes a linear variable differential transformer (LVDT) more suitable for providing an electrical output.

7.5.2 Diaphragm-type Transducers

Figure 7.24b illustrates the principle of the diaphragm transducer whose relatively small displacements lie within the span of strain gauges. These gauges are unbonded, that is, the wires are stretched directly between insulating supports mounted on a flexible spring strip and not bonded to a surface. Four gauges are usually employed to form all four arms of the active bridge.

7.5.3 Quick-response Pressure Transducers

Although bellows and diaphragm transducers are sensitive, their speed of response is limited by the high inertia of their moving parts. For rapidly changing pressures such as those within the cylinder of an internal-combustion engine, piezoelectric transducers are often employed. The pressure variations are made to compress a piezoelectric crystal or crystals (section 7.4.1) and the resultant voltage observed. Unfortunately such transducers are relatively insensitive (rarely less than 100 kN/m^2 span) but this suffices for the above example. In addition, because of charge leakage they are unsuitable for static pressure measurements. Transducers with higher sensitivity (up to 1 kN/m^2 span) which retain low inertia are available (figure 7.25); they convert the pressure fluctuations into small

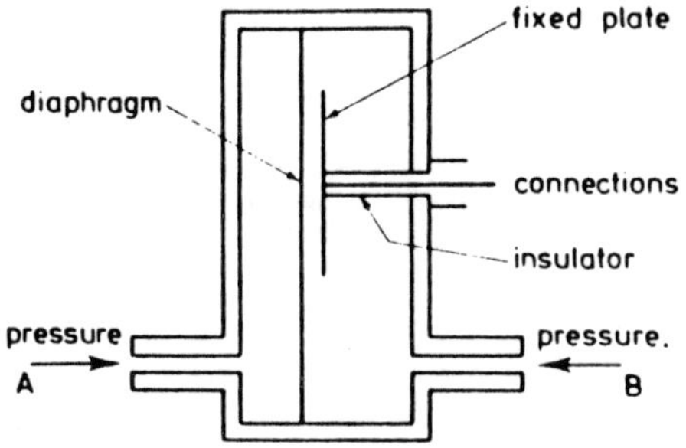

FIGURE 7.25

Capacitive differential-pressure transducer

movements of one of the two plates of a capacitor. Because the capacitance depends upon plate separation, its variation is pressure-dependent.

7.6 Flow

The word flowmeter is often used loosely to describe both *rate-of-flow* meters (m^3/s or kg/s) and *integrating* flowmeters (m^3 or kg) which are used to measure the total volume (or mass) of fluid delivered. Integrating flowmeters are usually of the positive-displacement type in which rotary vanes, or reciprocating pistons or bellows, monitor the amount of fluid passing.

Electrical methods are commonly employed for rate-of-flow meters.

7.6.1 Hot-wire Flowmeters

A current-carrying wire is placed in the fluid stream and its fall in temperature, as disclosed by its electrical resistance, indicates the rate of flow past it. Figure 7.26a illustrates the construction and an elementary circuit for its operation. This circuit suffers from the disadvantage that its output voltage which is proportional to electrical resistance and therefore hot-wire temperature is unlikely to be linearly proportional to the rate of flow. Figure 7.26b illustrates a more sophisticated closed-loop control circuit in which any changes in hot-wire resistance unbalance the bridge whose output voltage is fed back to the control unit. The latter adjusts the input voltage to the bridge until the hot-wire returns to its original temperature through electrical heating thus restoring the bridge balance. The output voltage required to restore balance in this constant-temperature method is more nearly proportional to the rate of flow of liquid.

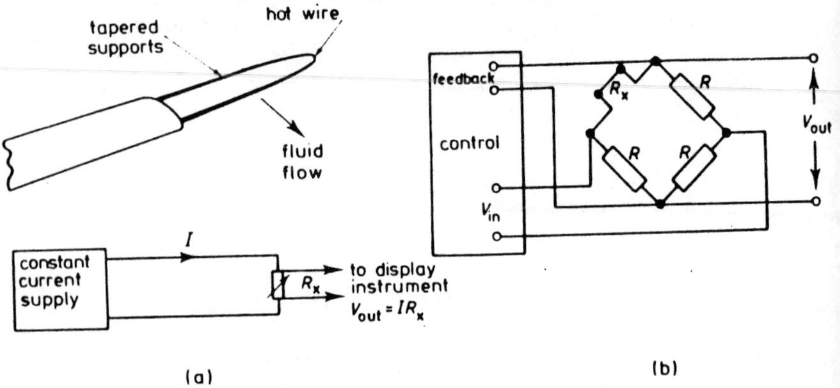

FIGURE 7.26

The hot-wire flowmeter

7.6.2 Turbine Flowmeters

Figure 7.27 shows that, in this type, the liquid flows past a turbine or impeller, revolving it at a speed proportional to the flow rate. Measurement of this speed is often performed by the magnetic method illustrated to avoid sealing problems in the pipe wall. The frequency of electrical voltage impulses from the pickup coil discloses the speed of turbine rotation.

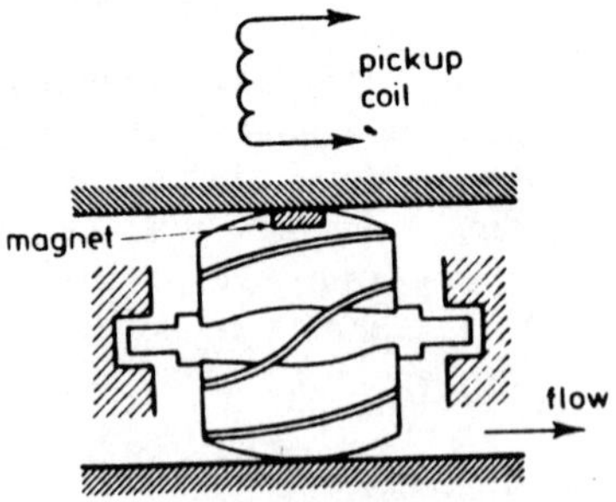

FIGURE 7.27

Turbine flowmeter

7.6.3 Electromagnetic Flowmeters

This method, which can be used for electrically conducting liquids, monitors the induced e.m.f. produced when a moving conductor (the liquid) moves across a magnetic field. An approximately uniform field from a permanent magnet passes through a section of non-conducting plastic pipe into the walls of which two metallic electrodes are set flush with the surface (figure 7.28). Because the directions of fluid flow and magnetic field are perpendicular an e.m.f. is generated between the electrodes proportional to the fluid velocity.

Example 7.3

Calculate the e.m.f. induced between the electrodes set in opposite walls of an electromagnetic flowmeter whose pipe is 1 cm diameter, if the magnetic flux density is 1 T and the flow rate is 1 m^3/min. Assume that the velocity distribution across the pipe diameter is uniform.

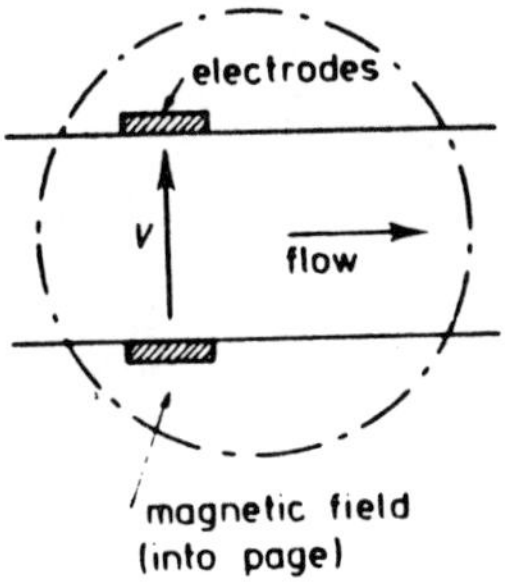

FIGURE 7.28

The electromagnetic flowmeter

Fluid velocity v = volume per sec/area

$$= \frac{1}{60} \times \frac{1}{\pi \times r^2}$$

$$= \frac{1}{60} \times \frac{1}{\pi(0.5 \times 10^{-2})^2}$$

$$= \frac{10^4}{60\pi \times 0.25} = 212 \text{ m/s}$$

e.m.f. = flux density × conductor length × velocity

$$= 1 \times 10^{-2} \times 212$$

$$= 2.12 \text{ V}$$

In practice, flow rates and flux densities much lower than the above are often encountered and e.m.f.s below 1 mV are common.

7.6.4 Differential-pressure Flowmeters

A common method of flow measurement is to pass the liquid through an orifice plate or venturi as shown in figures 7.29 a and b.

Bernoulli's equation states that

$$Q = \frac{a}{\sqrt{[1 - (a/A)^2]}} \frac{2\Delta_p}{\rho} (\times C_d)$$

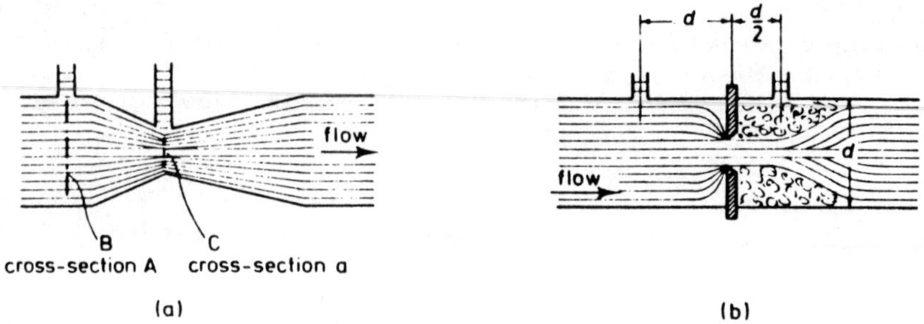

FIGURE 7.29

Venturi-tube and orifice-plate flowmeters

where Q = volumetric rate of flow; a = tube cross-sectional area at C; A = tube cross-sectional area at B; ρ = density of the fluid; Δ_p = pressure difference between B and C.

In practice, energy losses and non-uniform flow distributions occur which require that the right-hand side of the equation be multiplied by an experimentally determined coefficient C_d as shown. Nevertheless the volumetric flow is proportional to the pressure difference across BC. The differential pressure between points B and C may be conveniently measured electrically by the capacitive differential-pressure transducer shown in figure 7.25.

7.7 Temperature

7.7.1 Electrical Resistance Thermometers

Both metallic and semiconductor materials exhibit changes in electrical resistance with temperature. The amount of crystal-lattice vibration increases with temperature rise which produces in metals a rise in resistance because the conduction electrons are hindered in their flow by collisions with the vibrating atoms. As explained in section 3.4.1, the resistance of semiconductors falls rapidly with temperature and the two contrasting effects are compared in figure 7.30.

The most accurate temperature measurements are obtained by using a platinum resistance element whose resistance varies according to the equation

$$R_t = R_0 (1 + \alpha t + \beta t^2 + \text{etc.})$$

where R_t = resistance at temperature t; R_0 = resistance at temperature 0 °C; α and β are constants. The relative magnitudes of α and β are such that for temperatures below 300 °C the expression becomes

$$R_t \approx R_0 (1 + \alpha t)$$

yielding an almost linear resistance – temperature relationship as in figure 7.30. The value of α for platinum is only 0.00391/°C giving only a small fractional change in resistance compared with semiconductors. Because of their linearity

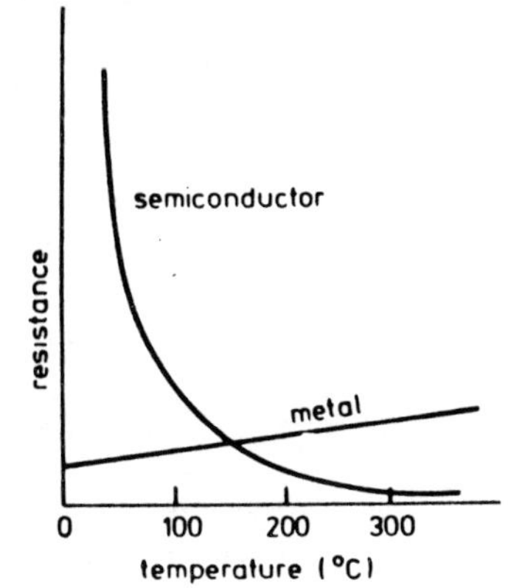

FIGURE 7.30

Resistance changes with temperature

and accuracy, however, they are the preferred method for laboratory and many industrial applications and if used with Smith[37] or Müller[38] bridges which eliminate the effects of connecting-lead resistance they are capable of resolving to 0.0001 °C.

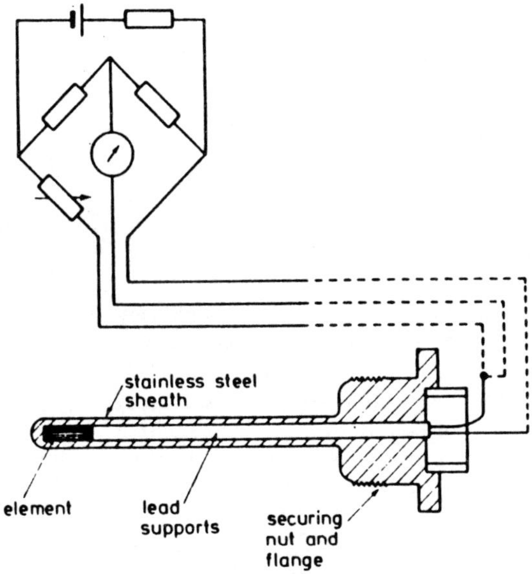

FIGURE 7.31

Resistance thermometer and three-lead compensation

Where the non-linearity of *thermistors* (semiconductor resistance elements) is acceptable their increased sensitivity can prove useful. All types of resistance thermometer however are fragile, expensive and have long response-times compared with thermocouples. For ordinary industrial applications they are used in both balanced and unbalanced bridge circuits (see section 8.6.2). Figure 7.31 illustrates a simple three-wire bridge connection to minimise inaccuracies caused by lead resistance. The standard methods for measurement appear in BS 1041 : 1969.

7.7.2 Thermocouples

Wherever a junction between two dissimilar metals occurs a small e.m.f. is observed dependent in magnitude upon the metals used and the junction temperature. Figure 7.32a shows two junctions J_1 and J_2 between metals A and B. If

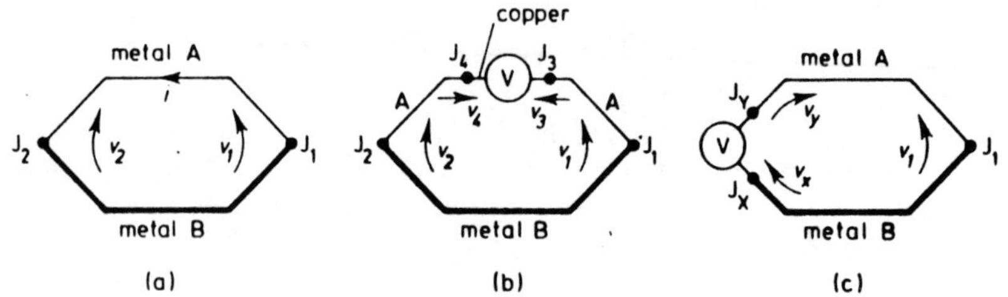

FIGURE 7.32

Thermocouple circuit and intermediate junctions

the two junction temperatures are equal, the opposing e.m.f.s v_1 and v_2 are equal and no current flows round the circuit. If J_1 becomes hotter than J_2, v_1 is greater than v_2 and a current flows in the direction shown. For many thermojunctions the relationship between e.m.f. and temperature is approximately linear so that if J_2 is held at some reference temperature the e.m.f. $(v_1 - v_2)$ will be approximately proportional to the temperature difference between J_1 and J_2.

To measure this e.m.f., or the current produced by it, a measuring device which is nearly always of some third metal (often copper) must be inserted into the circuit. Figure 7.32b shows a meter constructed with conductors of copper introduced into one-half of the thermocouple circuit. It will be seen that two new junctions J_3 and J_4 with their associated e.m.f.s have been formed. Provided J_3 and J_4 are at the same temperature v_3 and v_4 will exactly cancel each other, the circuit current being identical to that in figure 7.32a. More usually the measuring device is inserted as shown in figure 7.32c. The *law of intermediate metals* states that the junction J_2 has merely been split and that provided J_x and J_y are at the same temperature $v_x + v_y$ is identical to v_2 of figure 7.32a.

Table 7.2 illustrates the magnitude and permissible working temperatures of commonly encountered pairs of thermocouple materials. It will be seen that temperature differences between junctions of a hundred degrees will only produce an e.m.f. of a few millivolts. This fact combined with the relatively high resistance of junctions and leads makes direct measurement with a moving-coil instrument inaccurate. The preferred method is to employ a manual or self-balancing potentiometer (see section 9.3.3) which draws no current at balance and thus is unaffected by thermocouple resistance.

In industrial applications, economics may dictate the use of a simple moving-coil instrument in which case a large *swamping resistance* is often incorporated in series with the meter movement. Any small changes in the thermocouple resistance caused by temperature changes in the leads are now negligible compared with the large total-circuit resistance. The manufacturer will calibrate the instrument to be used with a given external resistance and any departure from this, caused by variation in the thermocouple length, can be compensated by adjusting the series resistor R_x inside the meter case (figure 7.33a).

TABLE 7.2
Thermocouple pairs

Material	Approximate sensitivity (mV/°C)	Working range (°C)
Copper - constantan	0.05	0 - 370
Iron - constantan	0.05	0 - 760
Nickel - chromium - nickel - aluminium (Chromel - alumel)	0.04	0 - 1260
Platinum - platinum - rhodium	0.01	up to 1750

Reference-junction Compensation

In use the junction J_1 of figure 7.32 is allowed to reach the unknown temperature whereas junctions J_2 or J_x, J_y are held at some reference temperature (often 0 °C) for which e.m.f. tables[39] have been prepared.

It is often impracticable in plant situations to obtain this reference temperature and reference-junction compensation must be applied arithmetically or automatically.

Because the temperature - e.m.f. relationship is slightly non-linear, each e.m.f. corresponds to a particular junction temperature not to a temperature difference between J_1 and J_2. If junctions J_1 and J_2 are at temperatures T_1 and T_2 corresponding to e.m.f.s v_1 and v_2 respectively and if e.m.f. v_0 corresponds to the correct reference temperature T_0, then

$$\text{Measured e.m.f. } (v_1 - v_2) = (v_1 - v_0) - (v_2 - v_0)$$

Rearranging

$$(v_1 - v_0) = (v_1 - v_2) + (v_2 - v_0)$$

or

Correct e.m.f. for measured temperature = measured e.m.f. + e.m.f. for reference-junction temperature from tables

Because of the non-linearity mentioned at the start of this paragraph, care must be taken never to obtain a temperature difference based upon the measured e.m.f. from the tables and to add this to the reference-junction temperature. It is the e.m.f.s which must be added and only finally are the tables used for conversion to temperature.

Many industrial thermocouple meters and recorders incorporate automatic reference-junction correction by adding an internally produced e.m.f. corresponding to $(v_2 - v_0)$ to the thermocouple output. An alternative method is to employ a bimetallic strip to offset the zero position of the meter.

Thermocouple Extension (Compensating) Leads

The distances between the thermocouple measuring point and the meter or recorder often make it uneconomic to extend the thermocouple metals, A and B, right back to the meter. In some applications this connection can be made with copper leads as in figure 7.33b. This method transfers the reference junction to

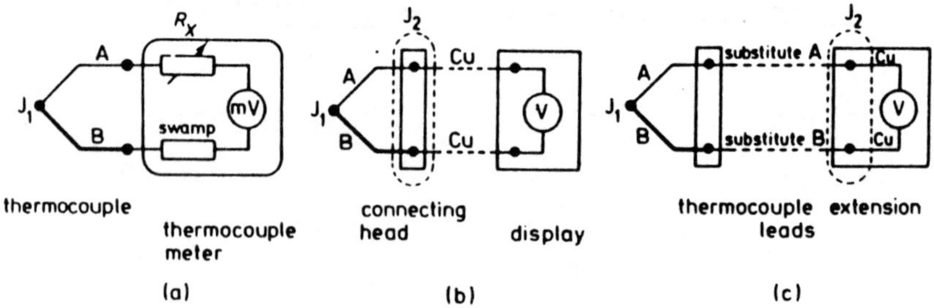

FIGURE 7.33

Thermocouple connection circuits

the connecting head whose temperature may differ from that of the compensating network in the display, producing errors. One method which is often used where many thermocouples are employed is to bring all their connections to a common connecting head situated in a constant-temperature enclosure. The display compensating circuits are then set to a constant value corresponding to the enclosure temperature.

More usually the circuit between the connecting head and display is made with cheaper substitute metals which are thermoelectrically identical to A and B, thus returning the reference junction to the display whose compensating circuits are now able to eliminate errors (figure 7.33c).

7.7.3 Radiation Pyrometry

This method is especially useful when, because of high temperatures or inaccessibility, it is impossible to place a transducer in direct contact with the object whose temperature is to be determined. Figure 7.34 illustrates the main features of the more common pyrometers. Figure 7.34a is a total-radiation device which accepts energy of all wavelengths (infrared to ultraviolet) emitted from the object and focuses them through a calibration stop on to a *thermopile.* A thermopile

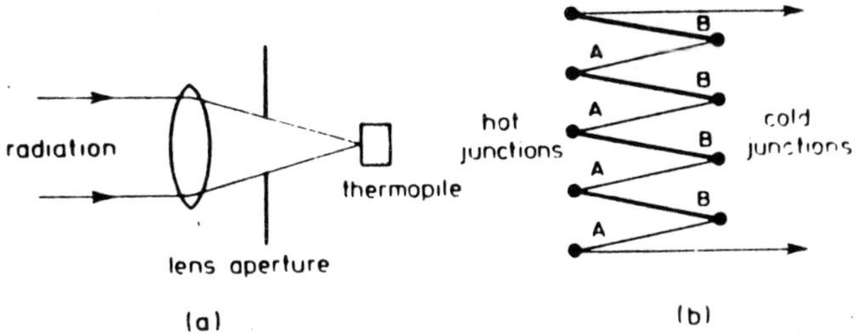

FIGURE 7.34

Thermopile total radiation pyrometer

(figure 7.34b) is a collection of thermocouple junctions placed end to end in series, the radiation being allowed to fall upon all the hot junctions. The cold junctions are placed in a position shielded from the radiation and the thermopile gives an electrical output proportional to the fourth power of the temperature measured. It is thus very non-linear and has a slow response time of several seconds until the thermopile reaches temperature equilibrium.

Much faster response (down to a few milliseconds) may be obtained by replacing the thermopile with a *photocell* (usually of the photovoltaic pattern). Radiation from a limited bandwidth (usually infrared) falls upon the cell producing an output voltage.

Perhaps the most common pyrometer is the disappearing-filament pattern in which the colour of an electrically-heated filament (figure 7.35) is compared

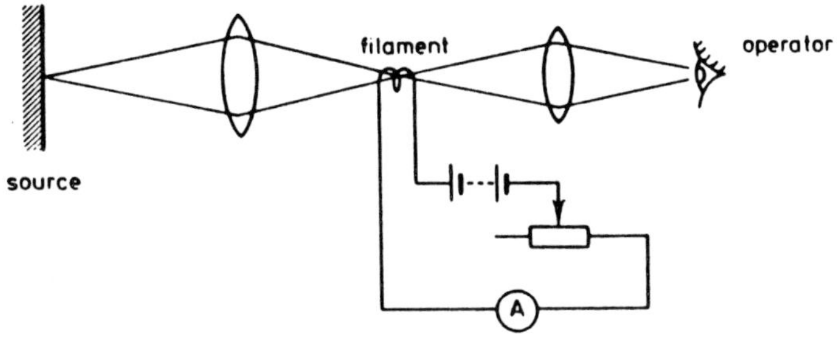

FIGURE 7.35

Disappearing filament pyrometer

visually with the radiation colour of the body. The ammeter, A, can be calibrated directly in temperature units. These devices are not automatic, however, and are confined to temperatures above 750 °C where visible radiation is emitted.

7.7.4 Piezoelectric Thermometers

Section 7.2 showed that it is possible to construct an electronic oscillator in which the frequency is controlled by the mechanical resonances of a quartz crystal. Synthetic-quartz crystals are produced which have a linear frequency-temperature relationship between −40 °C and 330 °C and can change frequency by as much as 1 kHz/°C. The output of such a thermometric oscillator may be accurately displayed on a digital counter.[40]

7.8 Sound

7.8.1 Units of measurement[41]

The sensitivity of the ear varies widely with frequency, being greatest at medium frequencies. The frequency spectrum extends in children from 20 Hz to 20 kHz, but in the elderly the upper limit can fall below 10 kHz. At 1 kHz the usual oscillatory sound pressure at the lower limit of detectability of sound is approximately 2×10^{-5} N/m² while the upper limit where pain and damage occur is roughly 20 N/m².

TABLE 7.3
Sound levels

Noise	dB with respect to lower threshold of audibility
Threshold of hearing	0
Quiet country evening	20
Quiet suburban evening	40
Normal conversation at 30 cm	60
Busy office or light machine shop	80
Noisy machine or heavy engineering shop	100
Near aero engine; threshold of pain	120

This range of 1 to 10^6 in sound pressure ratios is inconveniently large for normal use, so the logarithmic decibel scale (section 2.3.5) is adopted. Two sound powers P_1 and P_2 differ by $10 \log_{10} (P_1/P_2)$ decibels. Since power is proportional to the square of the sound pressure, two sound pressures Pr_1 and Pr_2 differ by $20 \log_{10} (Pr_1/Pr_2)$. The audio pressure range of 2×10^{-5} N/m² to 20 N/m² thus becomes 0 to 120 dB with respect to the lower limit.

A rough indication of the values of some common noises are given in table 7.3.

7.8.2 Microphones

(a) The Moving-coil (or Dynamic) Type is like a miniaturised version of the moving-coil loudspeaker operating in reverse (figure 7.36a). Sound pressure waves fall upon the diaphragm causing it to vibrate axially in the permanent-magnet field. E.M.F.s are induced in the coil proportional to the diaphragm velocity. This type of microphone is very sensitive to stray alternating fields and requires careful screening. It should not be used near electrical machines or transformers.

(b) The Piezoelectric (or Crystal) Type has a diaphragm attached to a small piezoelectric pressure transducer (section 7.4.1). These microphones (see figure 7.36b) are sensitive but mechanically somewhat fragile. Their sensitivity is temperature-dependent which makes them unsuitable for accurate measurements.

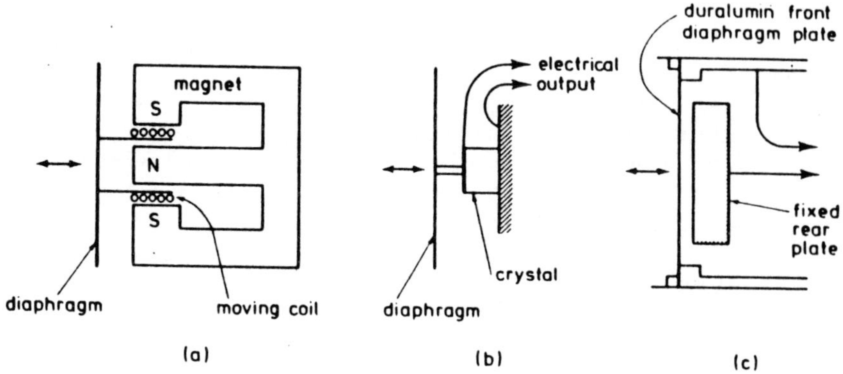

FIGURE 7.36

(a) Moving-coil; (b) piezoelectric; and (c) condenser type microphones

(c) The Condenser (Capacitor) Microphone (figure 7.36c) is almost universally employed for accurate measurements. It is less sensitive than the other types and expensive but it is capable of consistent results over long periods of use. The rear capacitor plate is a thick, highly polished and insulated surface. This is maintained at a high d.c. potential with respect to the front plate which is made from 0.03 mm duralumin sheet. This is suspended about 0.03 mm from the back plate. Pressure fluctuations cause the dielectric airgap to change, altering the capacitance. As the plates are held at a constant voltage, the charge must vary and hence currents enter and leave the microphone. These currents are detected and subsequently amplified.

7.9 Problems

7.1 A voltmeter of 10 000 Ω input resistance is used to measure the output from a potentiometric displacement transducer. If the error at the midpoint of the transducer's travel is not to exceed 0.01 per unit, calculate the maximum permissible resistance of the potentiometer winding.

7.2 A variable inductance level transducer has negligible resistance and an inductance which varies between 5 and 35 mH. The transducer together with three 630 Ω resistors form the arms of an a.c. bridge energised from a 5 V, 5 kHz supply. Calculate the maximum and minimum voltages which would appear across a detector of infinite impedance used with the bridge. How could one discriminate between a high- and a low-level voltage output?

7.3 The position of a lathe crosshead is measured by an LVDT type of transducer employing the circuit of figure 7.3. When the core is positioned 10 cm from the centre the root mean square voltages are $V_{AB} = 6$ V, $V_{CB} = 2$ V. Assuming the diodes to be ideal, calculate the d.c. output voltage V_{DE} stating its polarity. Assuming the transducer to be truly linear what would be the magnitude and polarity of V_{DE} with the core 7 cm on the opposite side of the central position.

7.4 A strain-gauge bridge has one active arm consisting of a 60 Ω gauge of gauge factor 2.2. A similar gauge is used for temperature correction and the remaining two arms are 60 Ω fixed resistors. Calculate the maximum permissible bridge supply voltage if the gauge current must not exceed 50 mA. Using this supply voltage what would be the voltage across a bridge detector of infinite impedance if the active gauge was subjected to 500 microstrain. What value of resistor, placed in parallel with the active gauge when unstrained would give a bridge output corresponding to 100 microstrain?

7.5 Describe how the fact that air insulation breaks down at an approximate electric field strength of 20 kV/cm imposes a limit on the sensitivity of air-dielectric transducers.

7.6 A barium titanate piezoelectric transducer has a sensitivity of 10^{-4} C/cm of displacement. It has a self-capacitance of 800 pF and is connected to an oscilloscope of input capacitance 50 pF by a cable of capacitance 300 pF. Express the sensitivities of (a) the transducer alone and (b) the complete system, in V/cm.

7.7 An accelerometer uses the barium titanate piezoelectric element of example 7.6 together with an inertial mass of 2 g. If the modulus of elasticity of the element is 12×10^{10} N/m^2, its area is 1 cm^2 and its thickness 5 mm, calculate the output voltage when the device is accelerated at 20g.

8 The Receiver/Amplifier

The purpose of this section of the measurement system is to receive the signal after transmission through the channel, to amplify it if necessary and to perform any signal conditioning needed before data display. We must first examine the basic amplifier circuits in common use.

8.1 The Transistor as an Amplifier

Section 3.4.4 has shown that a transistor is a current-amplifying device. Let us examine its characteristics in some detail so that we may predict its performance and design suitable circuits for its use. We have seen that variations ΔI_b in base current cause much larger variations ΔI_c in the collector current. The relationship between these two is given by the equation

$$\Delta I_c = h_{fe}\,(\Delta I_b) \tag{8.1}$$

when the collector - emitter voltage V_{ce} is constant and where h_{fe} is the common-emitter current gain of the transistor. In order to cause the base current I_b to vary, a variable input voltage V_{be} must be applied across the base - emitter junction (figure 8.1a). Any variations in output current I_c will be monitored between the emitter and the collector. It is the fact that the emitter terminal is common to both the input and output circuits which explains the name of amplifiers using transistors in this way.

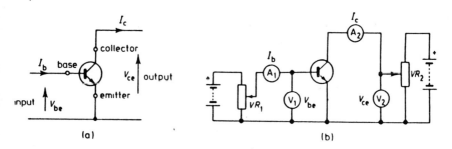

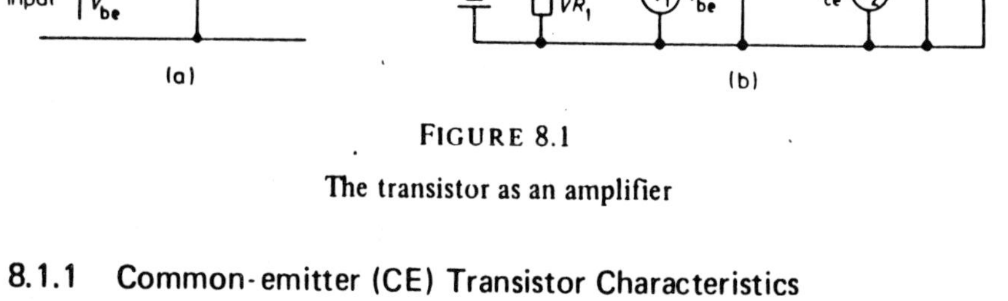

FIGURE 8.1

The transistor as an amplifier

8.1.1 Common-emitter (CE) Transistor Characteristics

If the transistor has its voltages and currents monitored in the manner of figure 8.1b we may draw its common-emitter electrical characteristics. An *npn* transistor is illustrated but the circuit is equally suitable for *pnp* devices if the supply polarities and meter senses are reversed. If I_b is held at zero by observing ammeter A_1 and V_{ce} is gradually increased by raising the slider of VR_2, increases of I_c will occur. These may be plotted as the $I_b = 0$ curve on figure 8.2a. After an initial small rise the current I_c is almost independent of V_{ce}.

If I_b is now held at constant values of 25, 50, 75 and 100 μA the curves shown may be produced by a typical transistor. In all cases, after the first volt

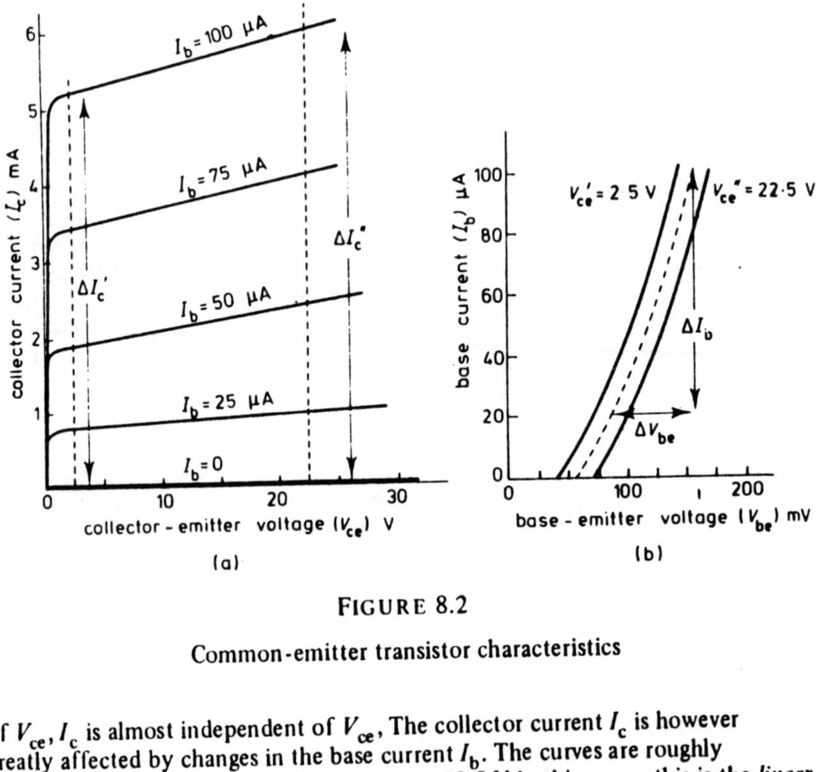

FIGURE 8.2

Common-emitter transistor characteristics

of V_{ce}, I_c is almost independent of V_{ce}. The collector current I_c is however greatly affected by changes in the base current I_b. The curves are roughly straight-line between $V_{ce} = 2.5$ V and $V_{ce} = 22.5$ V in this case – this is the *linear operating region* bounded by the dotted lines. At the extremities of this region values of

$$h_{fe} = \left| \frac{\Delta I_c}{\Delta I_b} \right|_{V_{ce} = \text{const.}}$$

are, at $V_{ce} = 2.5$ V

$$h_{fe} = \left| \frac{\Delta I_c'}{\Delta I_b'} \right|_{V_{ce} = 2.5\text{ V}}$$

$$= \frac{(5.2 - 0.05) \times 10^{-3}}{(100 - 0) \times 10^{-6}}$$

$$= 51.5$$

at V_{ce} = 22.5 V

$$h_{fe} = \left|\frac{\Delta I''_c}{\Delta I''_b}\right|_{V_{ce} = 22.5 \text{ V}}$$

$$= \frac{(6.0 - 0.05) \times 10^{-3}}{(100 - 0) \times 10^{-6}}$$

$$= 59.5$$

Note that h_{fe} is dimensionless, being the ratio of two currents. That is, there is a slight variation in h_{fe} across the operating region because the curves, though nearly linear, are not parallel.

This family of curves is known as the *common-emitter output characteristic* since it connects the two output quantities I_c and V_{ce}. It gives us no information however about the effect of V_{be} on I_b – this requires the *common-emitter input characteristic* (figure 8.2b). These two characteristics together can give us almost all the information we need to predict the operation of this transistor. Different types of transistor will exhibit different values on their characteristics but all will have the same general shape. Even transistors of the same type will exhibit some differences because of production tolerances.

Example 8.1

The base – emitter voltage of the transistor whose characteristics are shown in figure 8.2 is varied from 60 mV to 160 mV at a constant collector – emitter voltage of 12.5 V. Calculate (i) the resulting change in base current and (ii) the corresponding values of collector current.

(i) From figure 8.2b we may interpolate a curve for V_{ce} = 12.5 V centrally between the given curves. A change in V_{be} from 60 to 160 mV will correspond to values of I_b = 0 and I_b = 100 μA on the interpolated curve.

(ii) At V_{ce} = 12.5 V when

$$I_b = 0, \quad I_c = 0.05 \text{ mA}$$

and when

$$I_b = 100\ \mu\text{A}, \quad I_c = 5.6 \text{ mA}$$

8.1.2 Load-line Analysis of CE Amplifiers

The foregoing section has shown that changes in the input (base - emitter) voltage applied to a transistor in the CE configuration can produce large changes in I_c (collector current). Many amplifiers are required to convert small voltage changes into large voltage changes however so that a method must be found of converting changes in I_c into an output voltage.

The simplest method of converting a varying current into a varying voltage is to pass it through a resistor (R_L in figure 8.3a). This *load resistor* R_L will now present us with a voltage across it which is proportional to I_c. Because we prefer one terminal of an amplifier's output to be at zero (or earth) potential it is customary to take the output voltage between collector and emitter (or earth) as shown in figure 8.3a. The output voltage

$$V_{ce} = V_{cc} - I_c R_L \tag{8.2}$$

and since V_{cc} is the constant power supply or battery voltage, any change in V_{ce} will be proportional to changes in I_c. We refer to the transistor, load resistor and its associated circuitry as an amplifier *stage.* Practical amplifiers usually employ many such stages end to end.

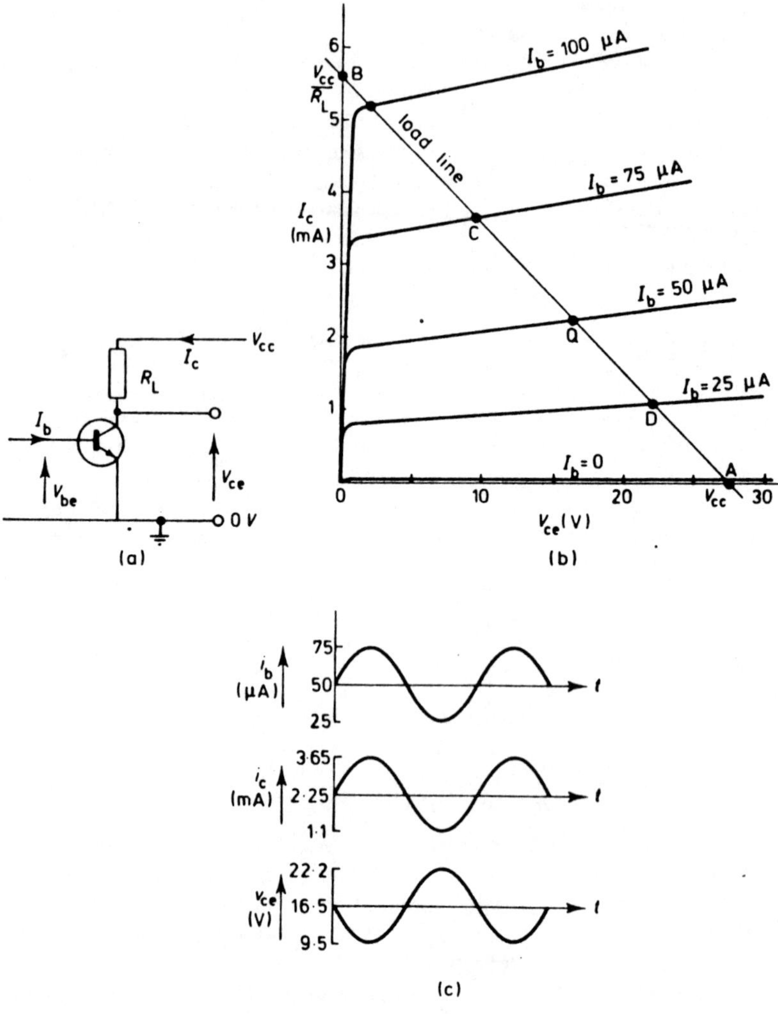

FIGURE 8.3

The common-emitter transistor amplifier

Because the output characteristics of figure 8.2 refer to the transistor alone, we cannot use them to predict the operation of this transistor-resistor combination. However, we can superimpose on the output characteristics another curve representing the behaviour of the resistor alone, *the load line.* The points at which the load line and the characteristics intersect will give the conditions for joint operation of the transistor and resistor.

Plotting a voltage-current curve for a resistor must yield a straight line (Ohm's law). Two hypothetical points could be (i) where $I_c = 0$ and (ii) where $V_{ce} = 0$. From equation 8.2

$$V_{ce} = V_{cc} - I_c R_L$$

therefore when $I_c = 0$ $\quad V_{ce} = V_{cc}$ point A

when $\quad V_{ce} = 0, I_c R_L = V_{cc}$, or

$$V_{ce} = 0, I_c = \frac{V_{cc}}{R_L} \quad \text{point B}$$

Plotting these two points on the output characteristics (figure 8.3b) gives points A and B. A straight line connecting A and B is the load line. This particular load line has been plotted for $V_{cc} = 27.5$ V and $R_L = 4.9$ kΩ giving $V_{cc}/R_L =$ 5.6 mA. Note that the slope of the load line is always $-1/R_L$ which allows us alternatively to plot it through one point only ($V_{cc} = V_{ce}$) if we know R_L.

As stated previously, the combination of the load line and characteristic curves allows us to predict the operation of the transistor-resistor combination at the given supply voltage V_{cc}. For instance if we supply the base with 50 μA the operating point will be at Q with $I_c = 2.25$ mA and $V_{ce} = 16.5$ V. Similarly if $I_b = 75$ μA or 25 μA the combination will operate at points C or D respectively.

To operate the transistor as an amplifier of a.c. waveforms we must first supply the base with a constant d.c. current called the *bias.* Let us choose a base bias-current of 50 μA. Let us now supply an alternating voltage to the base to vary the base current sinusoidally between limits of say 25 μA and 75 μA as in figure 8.3c. The bias operating point will be at Q and the transistor will make excursions to points C and D during the cycle. If we plot the associated instantaneous values of I_c and V_{ce} which are i_c and v_{ce} we will obtain graphs as shown in figure 8.3c.

Notice from the values that i_c and v_{ce} are not truly sinusoidal – this is because the characteristic curves are not exactly equidistant or parallel. Many amplifiers only use small excursions along the load lines in which case this *distortion* is negligible. It is always present however in large-signal power amplifiers.

It will be seen that the peak-to-peak value of the collector current ΔI_c is much greater than that of the base current ΔI_b. This shows that amplification has occurred as measured by the current gain A_i of the stage.

$$A_i = \frac{\Delta I_c}{\Delta I_b} \tag{8.3}$$

that is the ratio of the changes in output and input current. In this case

$$A_i = \frac{(3.65 - 1.1) \times 10^{-3}}{(75 - 25) \times 10^{-6}}$$

$$= 2.55 \times \frac{10^3}{50}$$

$$= 51.0$$

We may also speak of the voltage gain A_v of an amplifier stage, the ratio of output-to-input voltage changes.

$$A_v = \frac{\Delta V_{ce}}{\Delta V_{be}} \tag{8.4}$$

The input voltage change (ΔV_{be}) is not obtainable from figure 8.3b but we may obtain its approximate value from the input characteristic (figure 8.2b). The mean value of V_{ce} over the cycle is about 16.5 V so that we can interpolate an input characteristic for this value(dashed line). This shows that the values of base--emitter voltage V_{be} for base currents of 75 and 25 μA are 132 mV and 75 mV respectively. Thus

$$\Delta V_{be} = (132 - 75) \times 10^{-3}$$

$$= 57 \text{ mV}$$

and

$$A_v = \frac{(22.2 - 9.5)}{57 \times 10^{-3}} = \frac{12\,700}{57}$$

$$= 223$$

For completeness we may also speak of the power gain A_p of an amplifier stage. Power gain equals the product of voltage and current gains

$$A_p = A_v \times A_i$$

$$= 223 \times 51 = 11\,400 \tag{8.5}$$

We often express this in decibels

$$A_p = 10 \log_{10} (11\,400)$$

$$A_p \approx 40.6 \text{ dB}$$

Note that the output voltage waveform v_{ce} is in antiphase to the base current waveform i_b. It is therefore in antiphase to the base - emitter voltage. The phase shift of 180° between the input and output voltages is a characteristic of common-emitter amplifiers at normal frequencies.

Another limiting figure often quoted by transistor manufacturers is the maximum a.c. power which the device is recommended to handle. For a resistive load

a.c. power = r.m.s. voltage × r.m.s. current

$$= \frac{\textit{peak to peak voltage}}{2\sqrt{2}} \times \frac{\textit{peak to peak current}}{2\sqrt{2}}$$

$$= \frac{\Delta V_{ce}}{2\sqrt{2}} \times \frac{\Delta I_c}{2\sqrt{2}}$$

$$= \frac{(\Delta V_{ce} \times \Delta I_c)}{8}$$

8.1.3 Common-emitter Amplifier Design

The manufacturers specify for each transistor a maximum collector-dissipation mean power P_D below which there is no chance of thermal runaway occurring. It is advisable to ensure that the instantaneous collector power never exceeds this value. For a resistive device such as a transistor, power is the product of voltage and current hence

$$P_D = V_{ce} \times I_c$$

By calculation for a few arbitrary values we can superimpose a curve of this equation upon the characteristic curves. This has been done for a power of 60 mW on figure 8.4a. For the transistor to operate within the permitted power levels the whole of the load line must lie comfortably to the left of, and below, this power curve.

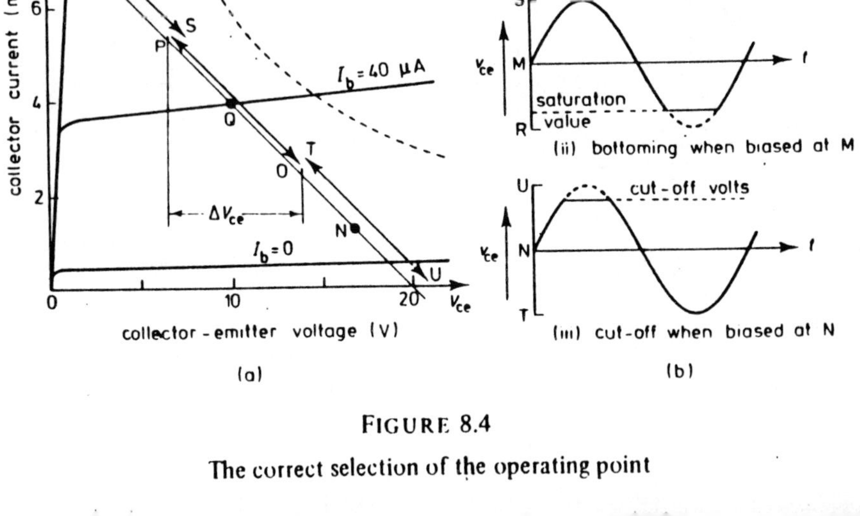

FIGURE 8.4

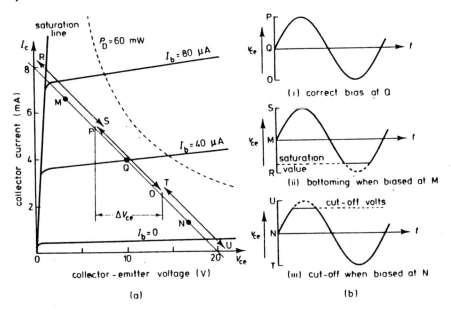

The correct selection of the operating point

The next step is to assess what size of swing or excursion the collector current and voltage will experience. If the stage is towards the output end of the amplifier these swings may represent a good proportion of the supply voltage V_{cc}. Figure 8.4a shows a swing ΔV_{ce} of several volts. In such cases the positioning of the operating point is critical. With I_b adjusted to 40 μA the operating point is central on the load line at Q, allowing maximum swing in either direction. An undistorted waveform as shown in figure 8.4b(i) is likely to result.

If the operating point is made too high (point M) the transistor will bottom or saturate in attempting to swing between points R and S resulting in bottoming (the clipped waveform of figure 8.4b (ii)). Too low an operating point results in cutoff when swinging between T and U with the accompanying distortion of figure 8.4c. If either of these conditions is suspected in an amplifier, an examination of the collector-voltage waveform with an oscilloscope will reveal their presence or otherwise.

Note that, even with a centrally situated operating point, peak-to-peak collector-voltage swings must always be slightly less than the supply voltage to ensure freedom from cutoff or bottoming. If larger swings must be accommodated a larger supply voltage must be used which may necessitate a transistor of higher power rating. In most cases it is advisable to centralise the operating point by biasing the transistor so that $V_{ce} \approx V_{cc}/2$.

We must next consider the components required to supply this bias to the transistor. Figure 8.5a shows one of the best biasing arrangements in which R_1 and R_2 act as a potentiometer across the supply voltage V_{cc} to feed the required base current to the transistor. The base current I_b will be approximately I_c/h_{fe},

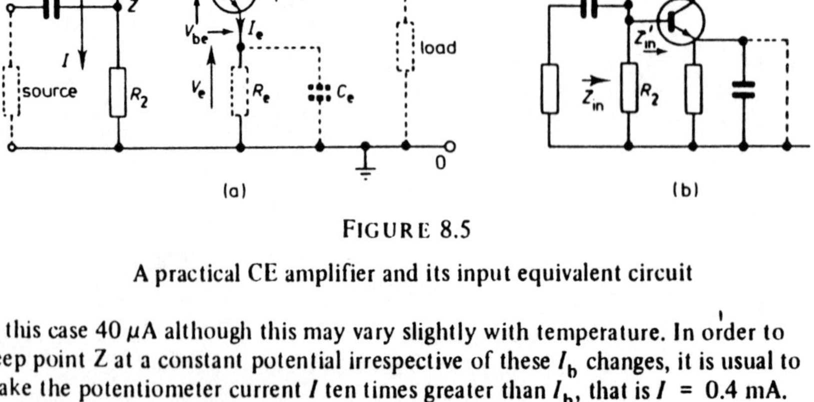

FIGURE 8.5

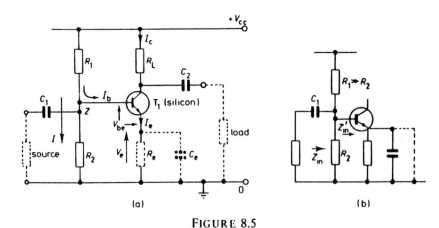

A practical CE amplifier and its input equivalent circuit

in this case 40 μA although this may vary slightly with temperature. In order to keep point Z at a constant potential irrespective of these I_b changes, it is usual to make the potentiometer current I ten times greater than I_b, that is I = 0.4 mA.

To determine the values of R_1 and R_2 separately we must consider the voltage required at point Z. A good approximation for silicon transistors is that V_{be} = 0.7 V and that V_{be} = 0.3 V for germanium transistors.

If we omit the temperature-stabilisation components R_e and C_e (dashed) then the voltage across R_2 must be 0.7 V. Hence

$$R_2 = \frac{0.7}{0.4 \times 10^{-3}} = 1.75 \text{ k}\Omega$$

$$R_1 = \frac{20 - 0.7}{0.44 \times 10^{-3}}$$

$$= 43.\dot{8} \text{ k}\Omega$$

In order to keep the operating point fixed irrespective of temperature fluctuations an emitter resistor R_e is commonly employed. Suppose that, because of temperature rise, the collector current increases. The voltage across R_e, V_e, will also increase raising the emitter voltage with respect to earth. Since point Z is held at a constant potential by R_1 and R_2, V_{be} will decrease, decreasing I_b and returning the transistor to its original operating point. The value of R_e is chosen to make V_e approximately 10 per cent of V_{cc}. We may assume $I_e \approx I_c$.

Hence

$$R_e = \frac{0.1 \times V_{cc}}{I_c}$$

in this case

$$R_e = \frac{2}{4 \times 10^{-3}} = 500 \ \Omega$$

The emitter capacitor C_e is chosen to be an effective short circuit between emitter and earth at the lowest frequencies which the amplifier will encounter. This is to prevent V_e varying when the applied signal varies I_c. R_e is effectively in parallel with the output impedance of the transistor (Z_o) looking in between emitter and earth. The expression for this is derived on p.246 for an emitter follower. With typical values of $h_{ie} = 2000 \ \Omega$ and $h_{fe} = 60$, $Z_o = 2000/60 \approx 33 \ \Omega$. Then $R_e // Z_o$ is $500 \ \Omega // 33 \ \Omega \approx 30 \ \Omega$. Usually X_{C_e} is made one-tenth of this, $X_{C_e} = 3 \ \Omega$. Therefore at the lowest frequency of, say, 100 Hz

$$X_{Ce} = \frac{1}{2\pi f C_e} = 3 \ \Omega$$

or

$$C_e = \frac{1}{2\pi \times 100 \times 3}$$

$$C_e = 500 \ \mu\text{F}$$

The inclusion of R_e necessitates recalculation of R_1 and R_2 in order to keep V_{be}

at 0.7 V for silicon transistors or 0.3 V for germanium transistors. The voltage across R_2 must now be $(V_{be} + V_e)$ for the same current I. In this case

$$R_2 = \frac{0.7 + 2}{0.4 \times 10^{-3}} = 6.75 \text{ k}\Omega$$

$$R_1 = \frac{20 - 2.7}{0.44 \times 10^{-3}}$$

$$= 39.3 \text{ k}\Omega$$

In order to take the alternating signal to and from the stage without disturbing the d.c. voltages on base and collector, coupling capacitors C_1 and C_2 are provided to isolate the d.c. from the source and load. Their reactances at the lowest signal frequencies encountered must be much less than the stage-input impedance and the load impedance respectively. We cannot know the latter without more information, but the former may be estimated from the characteristics and circuit. Figure 8.5b shows the equivalent circuit of the amplifier input where Z_{in} is the input impedance of the transistor and biasing resistors and Z_{in}' is the input impedance of the base with respect to earth. Because C_e has been designed to be effectively a short-circuit at the signal frequencies, Z_{in}' is the impedance between the base and emitter terminals alone. This may be expressed as

$$Z_{in}' = h_{ie} = \frac{\Delta V_{be}}{\Delta I_b} \tag{8.6}$$

Sometimes h_{ie} is quoted by the transistor manufacturers but in this case we must work from the input characteristics; let us assume these are as in figure 8.2b: $V_{be} = 60$ mV when $I_b = 80\ \mu$A, therefore

$$h_{ie} = \frac{60 \times 10^{-3}}{80 \times 10^{-6}}$$

$$= 750\ \Omega$$

Thus

$$Z_{in} = h_{ie} \text{ in parallel with } R_2$$

since $R_1 > R_2$. So

$$Z_{in} = 750//6750$$

$$Z_{in} = \frac{750 \times 6750}{7500} = 675\ \Omega$$

C_1 should therefore have a reactance of about 67 Ω at the lowest signal frequency of 100 Hz.

$$X_{C_1} = \frac{1}{2\pi f C_1} = 68\ \Omega$$

$$C_1 = \frac{1}{68} \times 200\pi$$

$$= 23.4\ \mu F$$

A similar calculation could be performed for C_2 if R_L were known, that is

$$X_{C_2} \approx \frac{R_L}{10}$$

Summary of CE Amplifier Design

- i Determine the output voltage and current swings required.
- ii Select $V_{cc} > 2 \times$ voltage swing.
- iii Select a transistor capable of withstanding this V_{cc} and of handling the a.c. power needed.
- iv Using the value of V_{cc} from (ii) select R_L to give the maximum use of the permissible portion of the linear characteristics.
- v Select the operating point to avoid cutoff or bottoming. Determine the value of I_c at the operating point.
- vi Calculate R_e (if used) to give $V_e \approx 0.1\ V_{cc}$ at the I_c chosen. Calculate C_e so that $X_{ce} \approx 0.1\ R_e$ at the minimum signal frequency.
- vii Determine the value of I_b at the operating point and calculate $I \approx 10\ I_b$.
- viii Calculate the value of R_2 to give $V_{be} = 0.3$ V or 0.7 V for a current of I. (Do not forget V_e if R_e is used).
- ix Calculate the value of R_1.
- x Calculate the value of C_1 so that $X_{C_1} < 0.1\ Z_{in}$.
- xi Calculate the value of C_2 if the load is known.

Example 8.2

A small-signal voltage amplifier is to use a germanium transistor having $h_{fe} = 100$ and $h_{ie} = 2000\ \Omega$ on a 9 V d.c. supply. Assuming that the amplifier is to work into a 1 kΩ load and is to handle signals down to 100Hz, calculate the component values for a suitable common-emitter stage. The mean collector current should be 1 mA.

This example does not supply the characteristic curves but since it is to be for small signals we may assume small swings and therefore little likelihood of cutoff or bottoming if we make $V_{ce} \approx 0.5\ V_{cc}$. Likewise there is little chance of exceeding the maximum collector dissipation but this can be checked at the operating point. Using the circuit of figure 8.5a

$$V_{ce} = 9/2 = 4.5\ \text{V}$$

The collector dissipation is

$$P_D = V_{ce} \times I_c$$

$$= 4.5 \times 10^{-3}$$

$$= 4.5\ \text{mW}$$

We can check that this does not exceed the maker's maximum rating. The load resistor R_L must drop 4.5 V at 1 mA.

$$R_L = \frac{4.5}{10^{-3}} = 4.5\ \text{k}\Omega$$

R_e must give $V_e = 9 \times 0.1 = 0.9$ V at 1 mA.

$$R_e = \frac{0.9}{10^{-3}} = 900\ \Omega$$

$Z_o = h_{ie}/h_{fe} = 200/100 = 200\ \Omega$ therefore $Z_o//R_e = 900\ \Omega//20\ \Omega \approx 20\ \Omega$

$$X_{Ce} = 2\ \Omega = \frac{1}{2\pi f C_e}$$

$$C_e = \frac{1}{2 \times 200\pi} = 800\ \mu\text{F}$$

$$I_b \approx \frac{I_c}{h_{fe}} \approx \frac{10^{-3}}{100} \approx 10\ \mu\text{A}$$

Thus

$$I = 10 I_b = 100\ \mu\text{A}$$

For a germanium transistor $V_{be} = 0.3$ V therefore the voltage across R_2 is (0.3 + 0.9) V.

$$R_2 = \frac{1.2}{10^{-4}} = 12\ \text{k}\Omega$$

$$R_1 = \frac{9 - 1.2}{1.1 \times 10^{-4}}$$

$$= \frac{7.8}{1.1 \times 10^{-4}} = 70.9\ \text{k}\Omega$$

No information is given on the input resistance h_{ie} of the device or its input characteristics, but h_{ie} rarely exceeds 2 kΩ. Therefore

$$Z_{in} = R_2//h_{ie} = \frac{2 \times 12}{2 + 12} \times 10^3$$

$$\approx 1.7\ \text{k}\Omega$$

Thus

$$X_{C_1} \approx 170\ \Omega = \frac{1}{2\pi f C_1}$$

and

$$C_1 = \frac{1}{170 \times 20\pi}$$

$$= 93\ \mu\text{F}$$

$$X_{C_2} \approx 0.1 \times \text{load} = 100\ \Omega$$

$$C_2 = \frac{1}{100 \times 20\pi}$$

$$= 159\ \mu F$$

In conclusion therefore the common-emitter amplifier is capable of giving voltage and current gains approaching 100 with an input impedance of a few kilohms. At normal frequencies there is a phase shift of 180° between the input and output voltages.

8.1.4 The Emitter Follower

For many applications the low input impedance of a CE transistor amplifier is insufficient because it loads the source too heavily. Consider the emitter-follower circuit of figure 8.6a – an increase in the input voltage V_{in} causes I_b to increase.

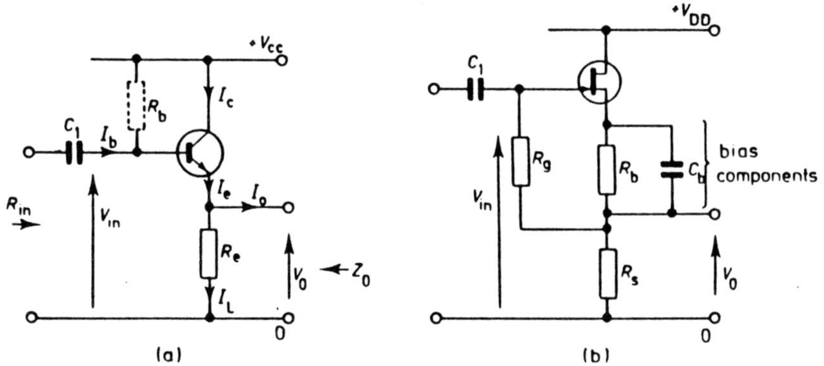

FIGURE 8.6

Emitter and source follower circuits

This causes I_c to increase which produces a greater voltage drop across R_e. Thus V_o rises *in phase* with V_{in}. The emitter voltage *follows* the input voltage. Quantitatively

$$I_b = \frac{(V_{in} - V_o)}{h_{ie}}$$

$$I_e = I_c + I_b = (1 + h_{fe})I_b$$

and

$$V_o = (I_e - I_o)R_e$$

Combining these gives

$$V_o = \frac{V_{in}(1 + h_{fe})R_e/h_{ie}}{1 + (1 + h_{fe})R_e/h_{ie}}$$

$$= \frac{I_o R_e}{1 + (1 + h_{fe})R_e/h_{ie}}$$

The voltage gain is thus

$$A_v = \frac{V_o}{V_{in}} = \frac{(1 + h_{fe})R_e/h_{ie}}{1 + (1 + h_{fe})R_e/h_{ie}} \quad (8.7)$$

Inserting the typical values of $h_{fe} = 200$, $h_{ie} = 1.5$ kΩ and $R_e = 4.7$ kΩ

$$A_v = \frac{(1 + 200) \times 4.7 \times 10^3/(1.5 \times 10^3)}{1 + (1 + 200) \times 4.7 \times 10^3/(1.5 \times 10^3)}$$

$$= \frac{630}{631} \approx 1$$

The amplifier voltage gain is thus nearly unity and the phase shift is zero. Because the unity term in the denominator can be ignored the output impedance Z_o reduces to

$$Z_o = \frac{-V_o}{I_o} \approx \frac{R_e}{(1 + h_{fe})R_e/h_{ie}}$$

the minus sign is because I_o flows outwards whereas we are looking inwards.

$$Z_o \approx h_{ie}/h_{fe} = 7.5\ \Omega$$

because $h_{fe} \gg 1$

This calculation has assumed an input voltage from a zero-impedance source. If the source has a resistive impedance R_s, Z_o becomes

$$Z_o = \frac{(R_s + h_{ie})}{h_{fe}} \quad (8.8)$$

which will still have a very low value.

The purpose of this circuit is to present a high input impedance R_{in} to the source. Let us examine the size of this.

$$V_o = I_L R_e$$

Which, if I_o is small is

$$V_o \approx I_e R_e = (1 + h_{fe})I_b R_e$$

Now

$$V_{in} = h_{ie}I_b + V_o$$

$$= h_{ie}I_b + (1 + h_{fe})I_b R_e$$

$$= I_b [h_{ie} + (1 + h_{fe})R_e]$$

The input resistance is

$$R_{in} = \frac{V_{in}}{I_b} = h_{ie} + (1 + h_{fe})R_e$$

$$\approx h_{ie} + h_{fe}R_e$$

Therefore

$$R_{in} \approx h_{fe}R_e \tag{8.9}$$

and with our typical values

$$R_{in} \approx 200 \times 4.7 \times 10^3$$

$$\approx 9.4 \times 10^5 \ \Omega$$

which is very high.

Summary of Emitter-follower Properties

(i) Near unity voltage gain
(ii) Non-inverting
(iii) Low output impedance
(iv) High input impedance

All these factors combine to make the emitter follower an ideal amplifier to interpose between a high-impedance source such as a photocell and a low-impedance load such as a pen recorder.

The purpose of R_b is merely to act as a bias resistor. Its value can be computed because the base to earth voltage will be the d.c. voltage across R_e, that is I_cR_e plus either 0.3 V or 0.7 V dependent upon the material of the device. Its upper end is at voltage V_{cc} and it is required to carry I_b which is approximately I_c/h_{fe}. Ohm's law thus gives a rapid solution.

Field-effect transistors (FETs) are often used in this circuit because of their higher values of input impedance h_{ie}. In this case the circuit is called a *source follower* and is arranged as in figure 8.6b. It is identical to the emitter follower except that the FET requires different bias conditions resulting in R_b and R_g being placed differently.

More experienced readers will recognise the similarities between the circuits for an FET and those for thermionic valves (the cathode follower)!

8.2 Amplifier Frequency Response

We have so far considered the operation of amplifiers at normal medium frequencies only, perhaps represented by a frequency spectrum from 50 Hz to 10 kHz. Outside these limits the electrical equivalents of inertia in the amplifier circuits modify their behaviour. The electrical equivalent of inertia is reactance, consequently we must consider the various capacitances in the circuit even though their effect is negligible at medium frequencies.

8.2.1 The Effects of Amplifier-circuit Capacitances

We have seen that the transistor stage is basically a current amplifier and hence we may represent it by the current generators $h_{fe}i_b$ in the equivalent circuit of figure 8.7b. The capacitances appearing in these two cascaded stages are of two kinds. First, the capacitors included in the design C_1 and C_2; second the internal capacitances of the transistor, chief of which is the capacitance across the reverse-biased collector-base junction (section 3.4.4). This can be shown[44]

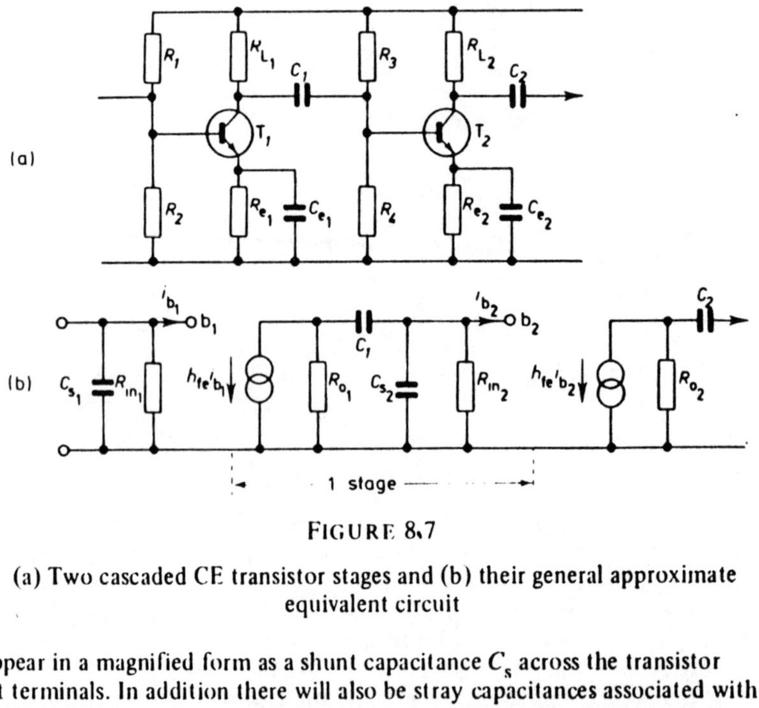

FIGURE 8.7

(a) Two cascaded CE transistor stages and (b) their general approximate equivalent circuit

to appear in a magnified form as a shunt capacitance C_s across the transistor input terminals. In addition there will also be stray capacitances associated with the wiring which enlarge the effective value of C_s. The emitter bypass capacitors C_e are ignored since they are designed to have low reactance at normal frequencies and are themselves shunted by low resistances R_e. The resistances R_{o_1} and R_{o_2} are demanded by Norton's theorem (section 2.5.2) and represent the output impedances (resistances) of the two transistors in parallel with their load resistors R_{L_1} and R_{L_2}

We may define one stage as the circuitry between any two equivalent components in the two stages. For simplicity let us divide the circuit as shown by the dashed lines. As mentioned at first, all the capacitances may be ignored at medium frequencies when the equivalent circuit simplifies to that of figure 8.8b. The output voltage at these medium frequencies will be

$$V_{OM} = -h_{fe}i_b \frac{R_{o_1}R_{in_2}}{R_{o_1} + R_{in_2}}$$

$$= -h_{fe}i_bR' \qquad (8.10)$$

where R_{in_2} is the input resistance of the second amplifier stage (T_2 and biasing components), R_{O_1} is the output resistance of the first amplifier stage (T_1 and R_{L_1}) and R' is R_{in_2} in parallel with R_{O_1}. Note that the minus sign represents the phase inversion of a common-emitter stage.

At low frequencies the reactance of a capacitor becomes high, therefore C_1 will tend to impede the signal flow and must be taken into consideration. Very little current will flow through C_s and it may therefore be neglected. This yields the low-frequency equivalent circuit of figure 8.8a. By the current division rule (section 2.5.3)

$$i_1 = h_{fe} i_b \times \frac{R_{O_1}}{[R_{O_1} + (R_{in_2} - j/\omega C_1)]}$$

$$V_{O_L} = -h_{fe} i_b R' \left(\frac{R_{O_1} + R_{in_2}}{R_{O_1} + R_{in_2} - j/\omega C_1} \right)$$

where, as before, $R' = R_{O_1} // R_{in_2}$. From equation 8.10

$$V_{O_L} = \frac{V_{O_M}}{[1 - j/\omega C_1 (R_{O_1} + R_{in_2})]}$$

$$V_{O_L} = \frac{V_{O_M}}{(1 - j/\omega \tau_L)} \qquad (8.11)$$

where τ_L is the low-frequency time-constant $C_1(R_{O_1} + R_{in_2})$.

At high frequencies the coupling capacitor C_1 has negligible reactance and may be neglected. However the stray capacitance C_{s_2} now shunts increasing amounts of current from R', yielding the high-frequency equivalent circuit of figure 8.8c.

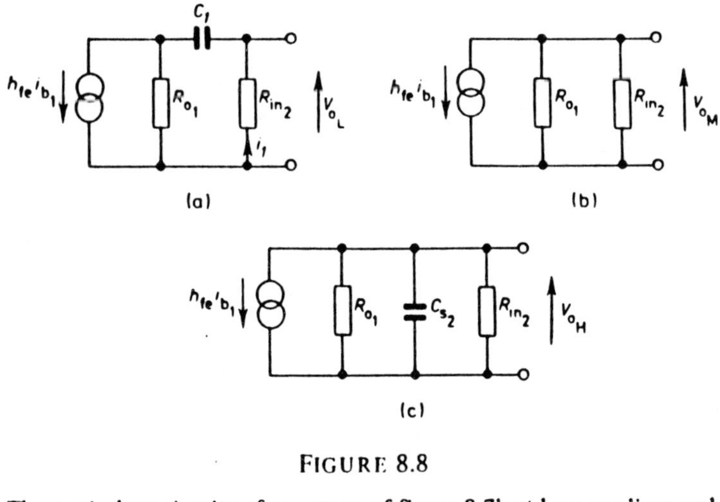

FIGURE 8.8

The equivalent circuits of one stage of figure 8.7b at low, medium and high frequencies

Taking R' again as the parallel combination of R_{O_1} and R_{in_2}, the current through this combination will be, by current division

$$i_2 = \frac{h_{fe} i_b(-j/\omega C_{S_2})}{R_e - j/\omega C_{S_2}}$$

therefore

$$V_{O_H} = \frac{-h_{fe} i_b R'}{1 + jR'\omega C_{S_2}}$$

From equation 8.10

$$V_{O_H} = \frac{V_{OM}}{(1 + j\omega\tau_H)} \qquad (8.12)$$

where τ_H is the high-frequency time-constant $R'C_{S_2}$.

Summarising therefore, at high and low frequencies the output voltage and therefore the voltage gain falls from its mid-frequency value. Also the presence of imaginary terms in equations 8.11 and 8.12 reveals a phase shift additional to the 180° mid-frequency shift.

8.2.2 Bandwidth and the Bode Diagram[45]

Let us examine the amplifier's performance at frequencies

$$\omega_L = \frac{1}{\tau_L} = \frac{1}{C_1(R_{O_1} + R_{in_2})}$$

and

$$\omega_H = \frac{1}{\tau_H} = \frac{1}{C_{S_2} R'}$$

At $\omega_L = 1/\tau_L$

$$V_{O_L} = \frac{V_{OM}}{1 - j}$$

so that

$$|V_{O_L}| = \frac{|V_{OM}|}{\sqrt{2}}$$

in addition, after rationalisation

$$V_{O_L} = \frac{V_{OM}}{2}(1 + j)$$

which implies an additional phase advance of arctan 1 = 45°, that is, from

180° to 225°. Conversely, at

$$\omega_H = 1/\tau_H$$

$$V_{O_H} = \frac{V_{OM}}{1 + j}$$

so that

$$|V_{O_H}| = \frac{|V_{OM}|}{\sqrt{2}}$$

also, after normalisation

$$V_{O_H} = \frac{V_{OM}}{2}(1 - j)$$

which implies an additional phase retardation of arctan 1 = 45°, that is, from 180° to 135°.

At both these frequencies therefore the magnitude of the output voltage has fallen by √2 and therefore the power (proportional to voltage squared) has fallen to one-half of its mid-frequency value. These frequencies are called the *half-power points* or alternatively the – *3 dB points* (see section 2.3.5). These frequencies are often quoted to define the usable range or bandwidth of amplifiers.

In the high-frequency band, from equation 8.12

$$V_{O_H} = \frac{V_{OM}}{(1 + j\omega\tau_H)}$$

At the highest frequencies $j\omega\tau_H \gg 1$ therefore

$$V_{O_H} \approx \frac{V_{OM}}{j\omega\tau_H} \tag{8.13}$$

The voltage output and therefore the voltage gain A_v is thus *inversely* proportional to frequency. If the frequency rises by one decade (a factor of 10) the voltage gain will fall by 10, or in decibel notation – 20 dB. The slope of a graph of decibel voltage gain against the logarithm of frequency will thus be a straight line having a slope of – 20 dB/decade.

Similarly in the low-frequency band from equation 8.11

$$V_{O_L} = \frac{V_{OM}}{(1 - j/\omega\tau_L)}$$

at the lowest frequencies $j/\omega\tau_L \gg 1$ therefore

$$V_{O_L} \approx V_{OM}\, j\omega\tau_L \tag{8.14}$$

The voltage output and therefore the voltage gain is thus *directly* proportional to frequency. By the previous argument a slope of + 20 dB/decade will be seen in the decibel-gain – logarithmic-frequency graph.

Plotting this graph and inserting the half-power points and phase-shift curve

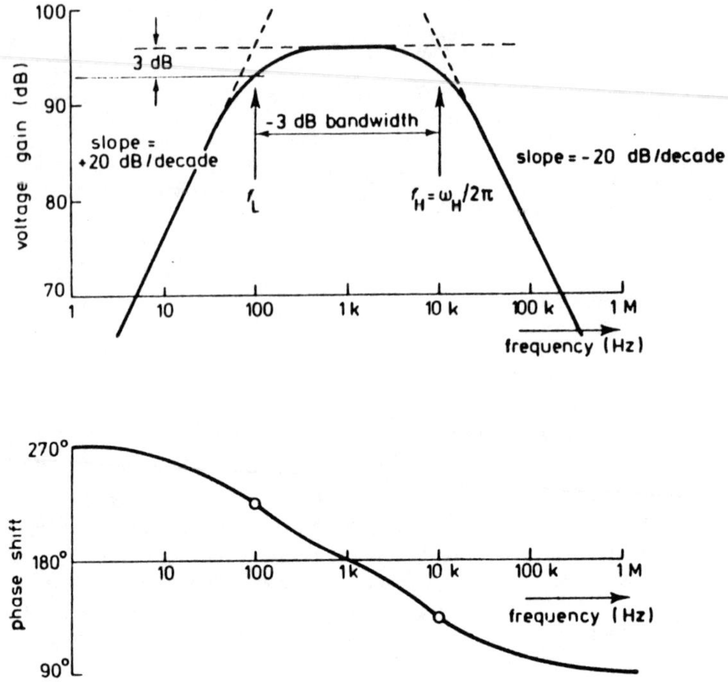

FIGURE 8.9

The Bode diagram

gives figure 8.9. This is called a Bode diagram, the three straight-lines are asymptotes to the frequency-response curve. If a little rounding is done at f_L and f_H which are the half-power or – 3 dB frequencies, a very good approximation results. Because of this diagram f_L and f_H are sometimes known as the *corner frequencies* in control engineering texts.

Equations 8.13 and 8.14 show that the phase shift at very high and very low frequencies will be 90° retarded and 90° advanced respectively compared to the mid-band phase shift of 180°. This gives a phase shift of 90° and 270° respectively. In addition, we know that at the half-power frequencies the phase shift is 45°. In this example, the half-power frequencies are 100 Hz and 10 kHz.

Bode's analytical techniques[45] were first used to determine the electrical stability of electronic amplifiers. It has since become an important tool with which to investigate any system, electrical, mechanical or fluid, and will prove useful to the reader.

8.2.3 Gain - Bandwidth Product

Section 6.7 showed that most signals consist of many frequency components. For the signal to emerge undistorted after amplification, all the components must be amplified equally. Even the best amplifiers do not exhibit a perfectly flat frequency response in the mid-frequency range. One criterion originally employed in gramophone amplifiers was the half-power or − 3 dB bandwidth. This was because a power change of − 3 dB (2 : 1) was the smallest detectable by the human ear. This criterion has been carried over to describe the performance of most electronic amplifiers used for instrumentation purposes.

For relatively undistorted output all the frequency components of the signal must lie within the − 3 dB bandwidth of the amplifier.[44] Let us examine the factors affecting the bandwidth of an amplifier. The low-frequency bandwidth may be extended by raising the values of the coupling capacitors C_1 and C_2. This lowers their reactance and minimises the signal voltage lost across them. Similarly to extend the high-frequency response C_{s_2} must be kept as low as possible by choice of transistors and minimisation of circuit stray capacitance. Both these processes may result in a more sophisticated and therefore more expensive amplifier.

One criterion of an amplifier's merit is its gain-bandwidth product. Usually f_H is much greater than f_L therefore the bandwidth is

$$B = f_H - f_L \approx f_H$$

and

$$f_H = 2\pi\omega_H = \frac{2\pi}{C_{s_2} R'}$$

From equation 8.10 the mid-band voltage gain is proportional to $h_{fe}R'$. Therefore the gain-bandwidth product is proportional to

$$f_H \times h_{fe} R'$$

$$= \frac{2\pi}{C_{s_2} R'} \times h_{fe} R'$$

$$= \text{constant} \times \frac{h_{fe}}{C_{s_2}} \qquad (8.15)$$

The gain-bandwidth product is therefore *a constant* for any particular amplifier, being dependent upon the current gain of the transistor and the value of the shunt capacitance. The word product implies that for a given amplifier we may choose between a high-gain-low-bandwidth combination or vice versa. This is so; the methods of varying the relative sizes of gain and bandwidth are discussed in section 8.3.1.

Nearly all electronic equipment employs amplifiers and the cost of such equipment depends greatly upon the gain-bandwidth product required. The gain-bandwidth product may not be directly stated but if the reader orders an oscilloscope that has both high sensitivity and wide bandwidth the invoice will clearly demonstrate the truth of this statement! Accordingly, electronic equipment should be purchased with a clear understanding of its future application if economic waste is to be avoided.

Clearly an amplifier whose response falls away at low frequencies in the manner of figure 8.7 is unsuitable for the amplification of signals from thermocouples or strain gauges which are only cycling slowly. In thermocouple measurements especially the cycle time may be many hours. Amplifiers with level frequency-response down to zero frequency are available and are called *direct coupled* (d.c.) amplifiers. This response is obtained through eliminating coupling capacitors by suitable circuit design. In return for this wider frequency-response the user must tolerate poorer temperature stability. The output voltage of the amplifier will *'drift'* with time. In the normal amplifier these slowly moving thermal effects are blocked by the coupling capacitors.

One example of the use of d.c. amplifiers is the operational amplifier (section 8.3.1).

8.3 The Integrated Circuit Amplifier

The amplifier circuits considered in sections 8.1 and 8.2 have been constructed from *discrete components.* Each transistor, resistor and capacitor was physically identifiable. Modern constructional methods have made it possible to fabricate a complete amplifier circuit containing say 15 transistors and 20 resistors on one chip of semiconductor perhaps 1 mm square. Apart from the saving in material, significant advances in reliability, performance and cost are possible. The internal circuitry of such amplifiers is beyond the scope of this text but we can consider the applications of such devices when they are considered as a 'black box'.

8.3.1 The Operational Amplifier (Op-amp)[46, 47, 48]

Let us briefly examine the properties required of a versatile voltage amplifier. Assume that the amplifier is a black box having an internal circuit whose details need not concern us. Thévenin's theorem (section 2.5.1) states that the output may be represented by a voltage generator in series with the output impedance (figure 8.10). The value of the voltage generator is $A_v V_{in}$ where V_{in} is the voltage across the input terminals which themselves have an input impedance Z_{in}.

$$A_v = \text{voltage gain of the black box on no load } (Z_L = \infty)$$

$$= \frac{V_o}{V_{in}} \tag{8.16}$$

input impedance of the black box

$$Z_{in} = \frac{V_{in}}{I_{in}} \tag{8.17}$$

output impedance of the black box

$$Z_o = \left| \frac{V_o}{I_o} \right|_{V_{in}=0} \tag{8.18}$$

Usually the amplifier will be driven from a voltage source having an internal impedance Z_s. Let us insert the following typical values and calculate the output voltage in each case.

Example 8.3

The amplifier of figure 8.10 has the following values: $Z_{in} = 10\ \text{k}\Omega$, $Z_L = 1\ \text{k}\Omega$, $V_s = 1$ mV and $A_v = 1000$. Calculate the output voltage if

(a) $Z_s = 0$ and $Z_o = 50\ \Omega$
(b) $Z_s = 0$ and $Z_o = 100\ \Omega$
(c) $Z_s = 1\ \text{k}\Omega$ and $Z_o = 100\ \Omega$

(a) Since $Z_s = 0$, $V_{in} = V_s = 1$ mV. Therefore

$$A_v V_{in} = 1000 \times 10^{-3} = 1\ \text{V}$$

$$V_o = \frac{A_v V_{in} Z_L}{Z_o + Z_L}$$

$$= \frac{1 \times 1000}{50 + 1000} = 0.95\ \text{V}$$

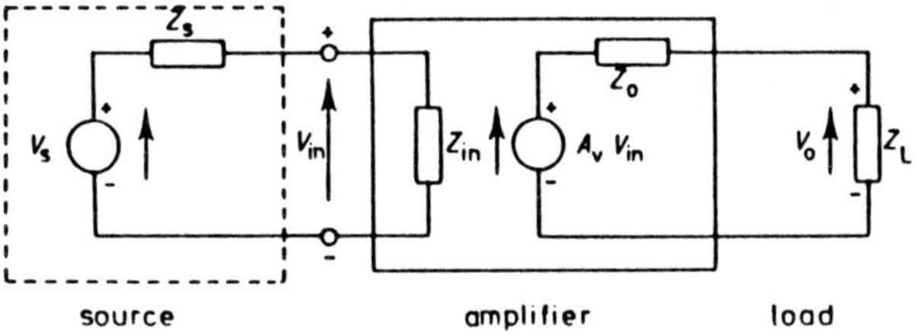

FIGURE 8.10

The black-box circuit of an amplifier

(b) As before $V_s = V_{in} = 1$ mV and $A_v V_{in} = 1$ V

$$V_o = \frac{A_v V_{in} Z_L}{Z_o + Z_L}$$

$$= \frac{1 \times 1000}{100 + 1000} = 0.91\ \text{V}$$

(c) Since $Z_s \neq 0$, $V_{in} \neq V_s$.

$$V_{in} = \frac{V_s Z_{in}}{Z_o + Z_{in}} = \frac{10^{-3} \times 10^4}{10^3 + 10^4}$$

$$= 0.91\ \text{mV}$$

Therefore

$$A_v V_{in} = 1000 \times 0.91 \times 10^{-3} = 0.91\ \text{V}$$

$$V_o = \frac{A_v V_{in} Z_L}{Z_o + Z_L}$$

$$= \frac{0.91 \times 1000}{100 + 1000} = 0.83 \text{ V}$$

Therefore for a high output voltage: from part (a) A_v must be large; from part (b) Z_o must be small compared with Z_L; from part (c) Z_{in} must be large compared with Z_s.

The first of these criteria is easily met in integrated-circuit operational amplifiers. The voltage gain at low frequencies is often about 50 000. The output impedance may be made lower and the input impedance higher by the application of negative feedback as described later in this section. Figure 8.11a shows the external circuit diagram of a typical operational amplifier (Fairchild μA 709).

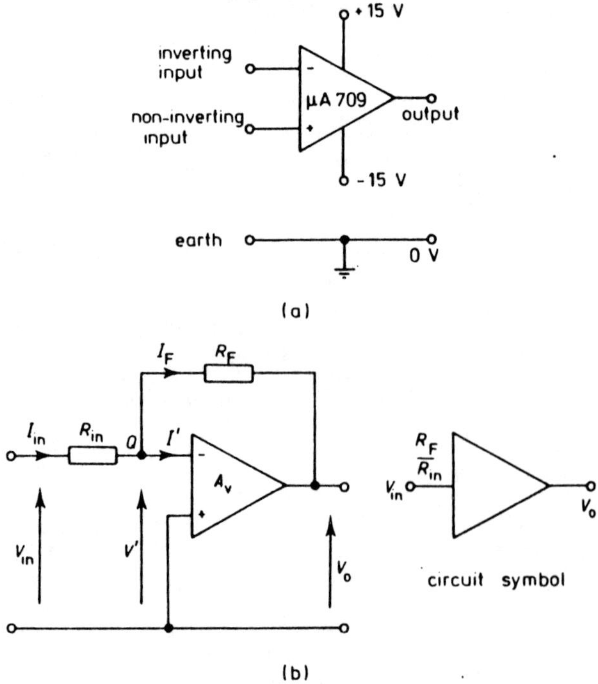

FIGURE 8.11

An operational amplifier used as a scaling circuit

A signal applied between the inverting input and earth will appear amplified and phase-shifted by 180° at the output. A signal applied between the non-inverting input and earth is also amplified but does not suffer any phase change. There are also two further terminals on the device which are not shown because they are for the attachment of external components to limit the amplifier's frequency response.

Such amplifiers may be used as the building blocks of analogue computers in the following ways.

Scaling Amplifiers

Consider the amplifier connected to two resistors as shown in figure 8.11b. R_{in} is an input resistor, R_F is a feedback resistor. If the voltage gain of the amplifier is high, V' will be much less than V_o. V' may be regarded as practically zero which makes the point Q a *virtual earth*. Thus I' is virtually zero and

$$I_{in} = I_F$$

Therefore

$$\frac{(V_{in} - V')}{R_{in}} = \frac{(V' - V_o)}{R_F}$$

If $V' \rightarrow 0$

$$\frac{V_o}{V_{in}} = -\frac{R_F}{R_{in}}$$

$$V_o = -V_{in}\frac{R_F}{R_{in}} \qquad (8.19)$$

We have an amplifier which multiplies or scales the input by a factor equal to the ratio between the feedback and input resistors together with an over-all sign inversion.

Summing Amplifiers

Figure 8.12 represents a circuit capable of summing several input voltages and

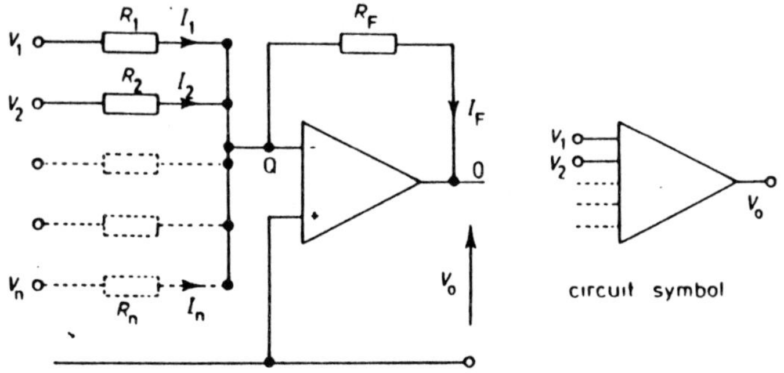

FIGURE 8.12

A summing and scaling amplifier

scaling them simultaneously. Again point Q is a virtual earth and

$$I_F = I_1 + I_2 + \ldots$$

$$\frac{(V' - V_o)}{R_F} = \frac{(V_1 - V')}{R_1} + \frac{(V_2 - V')}{R_2} + \ldots \tag{8.20}$$

$$V_o = -\left[V_1 \frac{R_F}{R_1} + V_2 \frac{R_F}{R_2} + \ldots\right]$$

If $R_1 = R_2 = \ldots = R_n = R_F$ we have an amplifier which adds together V_1, V_2, etc. and inverts the sign of their sum. If $R_1 \neq R_2 \neq R_n \neq R_F$ we are able,to sum V_1 to V_n after multiplying each by a different scaling factor.

Integrating Amplifiers

In figure 8.13, again because of the virtual earth

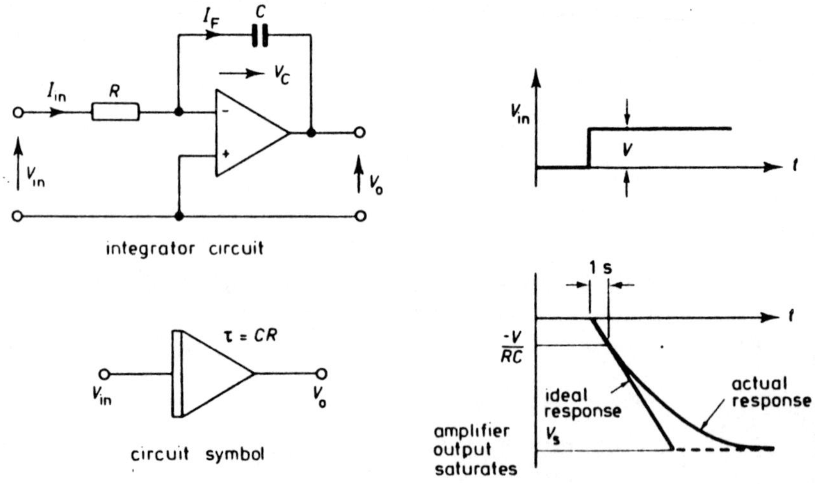

FIGURE 8.13

The integration circuit using an op. amp.

$$I_{in} = I_F \text{ and } V_c = V_o$$

so

$$\frac{V_{in}}{R} = -C\frac{dV_c}{dt}$$

or

$$V_o = \frac{-1}{RC}\int V_{in}\, dt \tag{8.21}$$

That is, the output voltage is equal to the time integral of the input voltage multiplied by a constant $-1/RC$. The constant is the reciprocal of the time-constant RC with a negative sign. Accordingly if the input voltage varies as shown in figure 8.11 the output should follow the ideal curve shown until the amplifier output saturates at the supply voltage V_s. The equation of this curve will be (from equation 8.21)

$$V_o = \frac{-1}{RC} \times Vt$$

therefore the slope will be $-V/RC$ V/s as shown.

Unfortunately because the voltage gain of the operational amplifier is not infinite and as the capacitor has some leakage resistance the curve will actually be non-linear as shown. If the integrator's use is restricted to the first portion a reasonable approximation to true integration can be obtained. In practice, there is usually a switch to allow the initial voltage of the capacitor to be set to represent the initial conditions of the integration.

8.3.2 Elements of Analogue Computation

An analogue computer consists of an interconnection of operational amplifiers of various types to perform the mathematical solution of an integro-differential equation. It can thus be used to model the response of a real system to various stimuli providing the behaviour equations of the system may be written down. Perhaps the simplest way to illustrate this is by an example.

Example 8.4

Draw the interconnections of amplifiers required to solve the behaviour equation of a moving-coil galvanometer

$$\frac{1}{\omega_n^2}\frac{d^2\theta}{dt^2} + \frac{2\xi}{\omega_n}\frac{d\theta}{dt} + \theta = \theta_F$$

Where ω_n, ξ and θ_F are constants.

The first step is to rewrite the equation so that the highest-order differential is separated with a coefficient of unity.

$$\frac{d^2\theta}{dt^2} = -2\xi\omega_n\frac{d\theta}{dt} - \omega_n^2\theta + \omega_n^2\theta_F$$

Starting with this highest-order differential we next draw the arrangement of integrators required to generate the lower order terms. These are then passed through scalers to obtain the correct coefficients and inverters (scalers of coefficient -1) as necessary.

The terms are then brought together in a 'summer' and equated to the highest order term by closing the loop at AB. Application of the input signal θ_F will now cause the network to behave according to the equation. The behaviour of the angular acceleration, angular velocity and deflection may now be observed with an oscilloscope or voltmeter attached to the appropriate point in the circuit.

A final check on the circuit may be made by counting the number of amplifiers in series round any closed path – the number must *always be odd.* This is because each amplifier contributes 180° phase shift and an even number would yield a loop phase-shift of 0. This loop would then oscillate due to positive feedback (see sections 8.4 and 8.5).

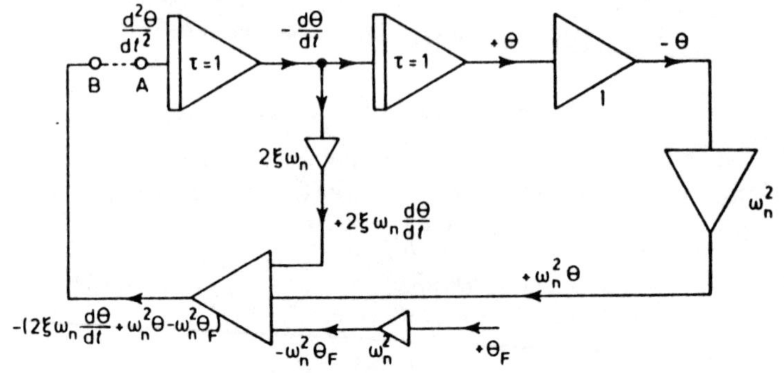

FIGURE 8.14

Analogue computation – example 8.4

The circuit as shown in figure 8.14 can be simplified by remembering that the input resistors of the integrators and the summer may be adjusted to give simultaneous scaling. This can result in a decrease in the total number of amplifiers used for this equation (figure 8.15). These minimisation procedures are important when a small analogue computer is used to represent a complex system. Again a check of the number of amplifiers in series around each loop can help to indicate errors in drawing the circuit.

The above techniques are merely the basis of analogue computation. Real computers employ techniques such as *amplitude scaling* and *time scaling* which

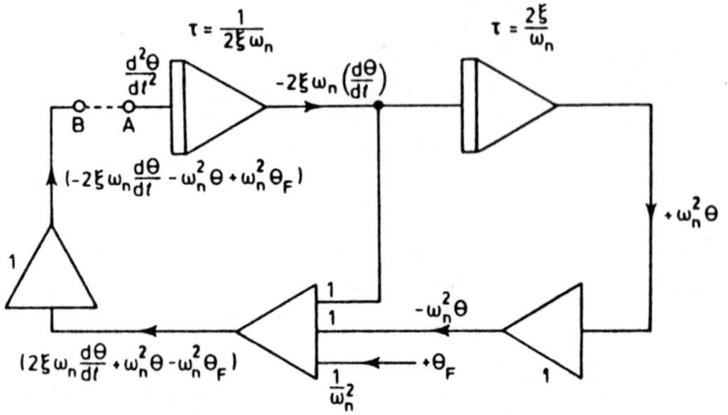

FIGURE 8.15

A minimised version of figure 8.14

are outside the scope of this text. For a fuller exposition one of the standard texts on analogue computation should be consulted[48].

The Charge Amplifier Many instrumentation transducers such as piezo-electric accelerometers (section 7.4.1) exhibit very high capacitive source impedances. The input impedance of a normal amplifier would load them excessively resulting in little output voltage. The charge amplifier of figure 8.16 is an operational amplifier in which both the input and the feedback impedances are capacitors C_1 and C_2. The fact that C_1 is incorporated in the source and not the amplifier is irrelevant to the circuit's operation. From equation 8.19

$$V_0 = -v\frac{Z_F}{Z_{in}}$$

$$= -v\frac{1/(2\pi f C_2)}{1/(2\pi f C_1)}$$

$$= -v\frac{C_1}{C_2}$$

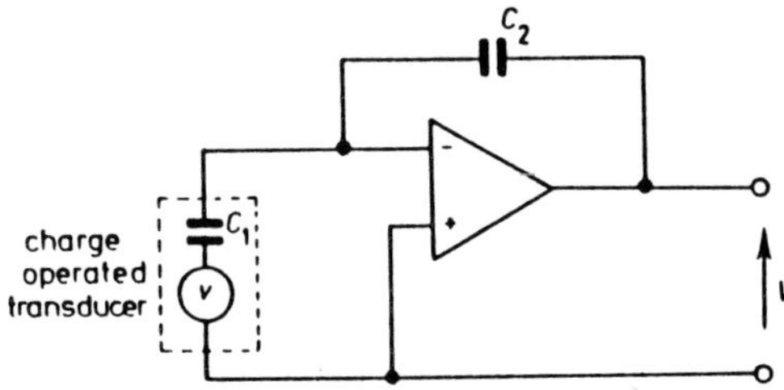

FIGURE 8.16

The charge amplifier

As stated before C_1 and C_2 are usually equal so that the amplifier's voltage gain is unity and negative. Its chief advantage however lies in its ability to transform the high impedance level of the transducer down to the low levels required to drive indicating display instruments.

8.4 Feedback

In the section 8.3.1 we saw how an amplifier having an original (or *open-loop*) voltage gain of several thousand could have its gain reduced drastically by the addition of two resistors. The effect of these was to feed back a proportion of the output voltage V_0 in such a sense as to oppose the input voltage V_{in}. This was a particular example of negative feedback, a topic that we must examine more generally. It will be found to yield significant advantages which often offset the loss in gain.

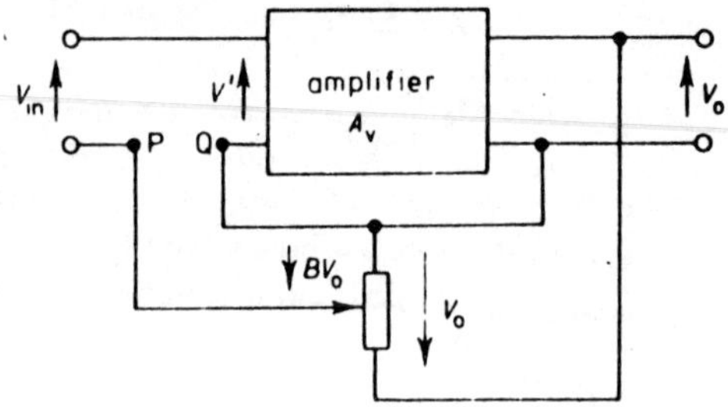

FIGURE 8.17

Series voltage feedback

Figure 8.17 shows a 'black box' amplifier in which a proportion B of the output is returned in series with the input by the potentiometer set at B per unit.

$$V' = V_{in} + BV_o$$

and

$$V_o = A_v V'$$

Thus

$$V_o = A_v V_{in} + A_v B V_o$$

$$V_o (1 - A_v B) = A_v V_{in}$$

The over-all gain of the amplifier and its feedback loop is A_v' where

$$A_v' = \frac{V_o}{V_{in}}$$

$$= \frac{A_v}{1 - BA_v} \qquad (8.22)$$

This is the fundamental equation for the study of feedback effects. In particular the denominator $(1 - BA_v)$ reappears in expressions for nearly all the feedback amplifier's properties. In figure 8.17 the output from the potentiometer is fed into the input in such a way as to assist the original *input* voltage. If the connections from the potentiometer to terminals P and Q were reversed, B would become negative – this is called negative feedback. Let us examine its effects on the main amplifier parameters.

Over-all Gain

Equation 8.22 shows that the over-all gain of the amplifier has been changed from A_v to $A_v/(1 - BA_v)$. If BA_v is negative this is a *reduction in gain*. But the gain is far more immune from changes caused by component ageing (see example 8.5).

Bandwidth

Section 8.2.3 showed that the product of gain and bandwidth for a particular amplifier is constant. If the gain is reduced by negative feedback therefore the *bandwidth is increased* by the same ratio $(1 - BA_v)$. We are now in a position to understand why in section 8.2.3 it was stated that techniques are available to exploit *either* the bandwidth *or* the gain potential of a given amplifier.

If increased bandwidth is desired it may be obtained at the *expense of gain* by the application of negative feedback.

Input Impedance

It can be shown[49] that negative feedback raises the input impedance so that

$$R_{in}' = R_{in}(1 - BA_v)$$

where BA_v is negative. This is usually an advantage because the amplifier will be less likely to load its source excessively.

Output Impedance

Similarly it can be shown[49] that the output impedance is lowered by negative feedback to

$$R_o' = \frac{R_o}{1 - BA_v}$$

This is also an advantage because it allows the amplifier to drive the following circuits more effectively.

Distortion and Noise

These internal effects of the amplifier are similarly reduced. It is this reason together with the increased bandwidth that accounts for the wide use of negative feedback in high quality audio equipment.

In conclusion therefore, although the application of negative feedback results in some loss in voltage gain, practically every other amplifier parameter is improved.

Example 8.5

An amplifier has a nominal gain of 10 000 subject to a reduction of 10 per cent caused by ageing to 9 000. Its input impedance, output impedance and bandwidth are 10 kΩ, 100 Ω and 10 kHz respectively.

Calculate the percentage gain stability when 0.02 of the output is fed back in opposition to the input

$$A_v' = \frac{A_v}{1 - BA_v}$$

Because we have negative feedback (opposition)

$$B = -0.02$$

Originally

$$A_v' = \frac{10^4}{1 + 0.02 \times 10^4}$$

$$= \frac{10^4}{201} = 49.75$$

Finally

$$A_v' = \frac{9 \times 10^3}{1 + 0.02 \times 9 \times 10^3}$$

$$= \frac{9 \times 10^3}{181} = 49.7$$

The percentage gain stability is thus

$$\frac{49.75 - 49.7}{49.75} \times 100$$

$$= 1.005 \text{ per cent}$$

This is a considerable improvement on the original 10 per cent variation.

$$R_{in}' = R_{in}(1 - BA_v)$$

$$= 10^4(1 + 0.02 \times 10^4)$$

$$= 2.01 \times 10^6 \ \Omega$$

and

$$R_o' = \frac{R_o}{1 - BA_v}$$

$$= \frac{100}{1 + 0.02 \times 10^4}$$

$$= 0.4975 \ \Omega$$

The bandwidth is increased to

$$BW' = BW(1 - BA_v)$$

$$= 10^4(1 + 200)$$

$$= 2 \times 10^6 \text{ Hz}$$

8.5 Oscillators

Consider the effect on equation 8.22 of making BA_v unity and positive. The over-all gain would be

$$A_v' = \frac{A_v}{1 - 1} = \text{infinity}$$

Since the over-all gain is the ratio of output to input voltage, this implies a finite output for zero input

$$A_v' = \frac{V_o}{V_{in}} = \frac{V_o}{0} = \text{infinity}$$

An amplifier which yields an output signal without a corresponding input signal is said to oscillate. The energy for this output voltage is obtained from the amplifier power supplies so we are not infringing the laws of energy conservation! Let us examine the significance of A_vB being unity positive.

$$A_vB = +1$$

or

$$A_v = \frac{1}{B}$$

In words – making the closed-loop gain A_vB equal to unity means exactly balancing the open-loop voltage gain of the amplifier A_v to the loss in gain $1/B$ of the feedback network. In addition the positive sign indicates that there is zero phase-shift around the loop.

Some practical circuits for oscillators have already been considered in section 5.5.3 and 5.5.4 but any standard text on electronics[44] will give circuits encompassing many output waveshapes.

8.6 Bridges

The inclusion of this topic here may seem at first sight inconsistent because bridges are often used as a complete measurement system. In instrumentation applications however the unknown arm is often separated from the rest of the bridge by some distance over which the information is transmitted by two or more conductors (a channel). The larger portion of the bridge therefore fulfils the role of receiver in the measurement system.

8.6.1 The Simple Wheatstone Bridge

This is the simplest d.c. bridge (figure 8.18a). The unknown resistor R_x is inserted and R_3 adjusted until the current through the detector is zero when *at balance*

$$V_2 = V_1$$

$$\frac{VR_2}{R_1 + R_2} = \frac{VR_3}{R_x + R_3}$$

$$1 + \frac{R_x}{R_3} = 1 + \frac{R_1}{R_2}$$

$$\frac{R_x}{R_3} = \frac{R_1}{R_2} \qquad (8.23)$$

If the values of R_1, R_2 and R_3 are known, R_x can be calculated. R_1 and R_2 are known as the ratio arms, they are often equal or in decade ratio, that is, 10 : 1 or 100 : 1. This allows for greater accuracy because R_3 usually consists of a four- or five-decade resistor. If R_x is greatly different from the total value of R_3 accuracy is lost unless R_1/R_2 is adjusted to make use of all the R_3 decades.

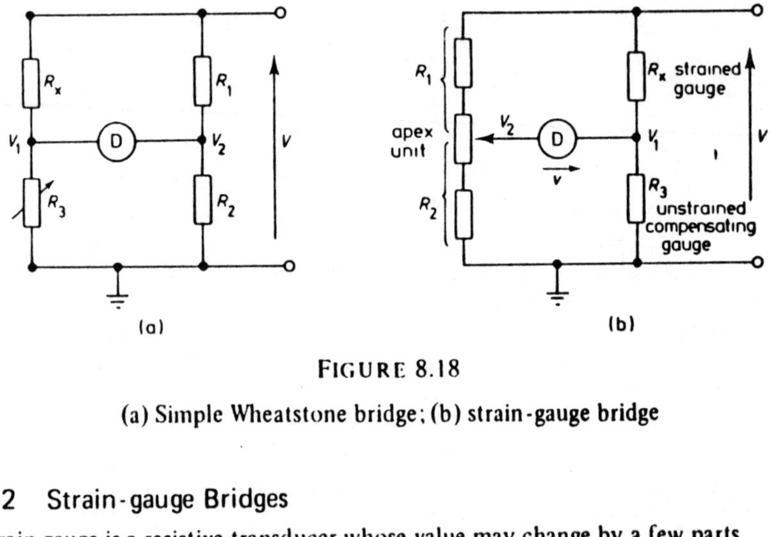

FIGURE 8.18

(a) Simple Wheatstone bridge; (b) strain-gauge bridge

8.6.2 Strain-gauge Bridges

A strain gauge is a resistive transducer whose value may change by a few parts per million when strain is applied (sections 7.1.12 and 7.1.13). Measurement of these small changes by the simple Wheatstone method would need a resistor R_3 of six or seven decades and great sophistication to prevent thermal e.m.f. errors. Accordingly an *unbalanced* Wheatstone bridge is often used similar to figure 8.18a except that no attempt is made to vary R_3 for balance. R_3 is usually an identical strain gauge to R_x but placed in an unstrained position. Since R_x and R_3 are not exactly equal because of manufacturing tolerances an apex unit (figure 8.18b) has to be used to balance the bridge when the strain is zero. Let R_x change its value to $R_x + \delta R_x$ after straining, assume $R_1 = R_2$, $R_3 = R_x$ and that the detector has infinite impedance.

$$V_2 = \frac{V}{2} \quad \text{and} \quad V_1 = \frac{VR_3}{R_x + R_3}$$

Thus the detector voltage v

$$= \frac{V}{2} - \frac{VR_x}{2R_x + \delta R_x}$$

$$= \frac{V\delta R_x}{4R_x + 2\delta R_x}$$

There is thus a nonlinear relationship between v and δR_x. However, if $\delta R_x \ll R_x$ which is usual (typically 1 : 10 000) then

$$v \approx \frac{V}{4} \times \frac{\delta R_x}{R_x}$$

which is linear. The fraction $\delta R_x/R_x$ is called the electrical strain and is related to the mechanical strain ϵ by the equations of section 7.1.12. Figure 8.18b is known as a single-arm bridge since there is only one active strained arm. Temperature compensation is provided by the unstrained gauge R_3. If both R_3 and R_x rise in value due to an increase in temperature this does not affect the value of V_1.

Greater sensitivity can be obtained in the study of some mechanical structures if R_x and R_3 can both be placed in positions having equal and opposing strains. If R_x is in tension its new value will be $R_x + \delta R_x$ and if R_3 is in compression its new value will be $R_3 - \delta R_3$. The bridge output voltage, if $R_3 = R_x$ will now be (for two active arms)

$$v \approx \frac{V}{2} \times \frac{\delta R_x}{R_x}$$

Temperature compensation is still present because temperature changes do not change the resistances of R_x and R_3 in *opposite* directions as does strain.

For the greatest sensitivity with temperature compensation we can use a four-active-arm bridge in which all four resistors are strained gauges. Diagonally opposed arms must be similarly strained. That is R_x and R_2 in tension, R_3 and R_1 in compression. If the magnitude of the strain on all four gauges is equal

$$v \approx V \times \frac{\delta R_x}{R_x}$$

an arrangement four times as sensitive as the single-arm bridge.

Example 8.6

A strain-gauge bridge consists of two strain gauges of 100 Ω unstrained resistance and gauge factor 2.2 together with two fixed 100 Ω resistors fed from a 4 V supply. If one of the gauges is strained to 101 Ω, calculate the bridge sensitivity in mV/Ω across an infinite impedance detector.

A 100 kΩ calibration resistor is placed in parallel with one of the two gauges when both are unstrained, calculate the mechanical strain that would produce the same bridge output.

Referring to the diagram of figure 8.18b, $R_1 = R_2 = R_3 = 100\Omega$ and $R_x = 101\Omega$ when strained.

$$V_2 = 4 \times 100/200 = 2 \text{ V}$$
$$V_1 = 4 \times 100/201 = 1.99 \text{ V}$$

The detector voltage = $V_2 - V_1$ = 10 mV. Thus the bridge sensitivity is 10mV/Ω.

A 100 kΩ resistor in parallel with a 100 Ω gauge gives an effective resistance of

$$R_e = \frac{10^2 \times 10^5}{10^5 + 10^2} = \frac{10^7}{100\,100}$$

$$= 99.88\ \Omega$$

This represents an electrical strain of $\delta R_x / R_x = 0.12/100 = 12 \times 10^{-4}$.

$$\text{Mechanical strain} = \frac{\text{electrical strain}}{\text{gauge factor}}$$

$$\epsilon = \frac{12 \times 10^{-4}}{2.2}$$

$$= 540\ \mu\ \text{strain}$$

8.6.3 A.C. Bridges

Until recently a.c. bridges were used to measure the value of electrical components other than pure resistors. In this role the classical bridges have largely been supplanted by the transformer ratio bridge.[55]

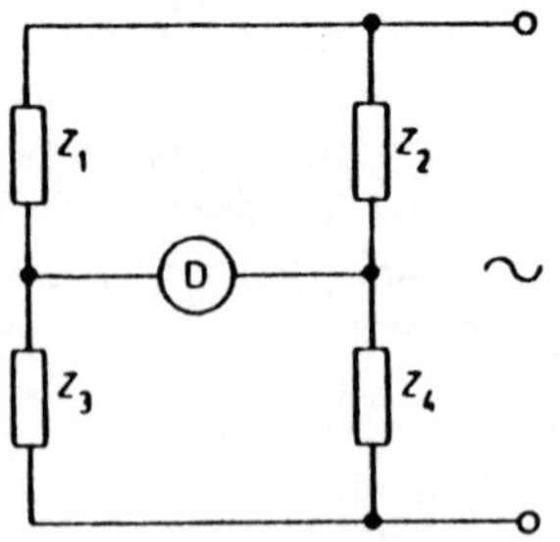

FIGURE 8.19

Generalised a.c. bridge

For completeness, however, the general method for their classical solution[1] is included as it gives insight into the use of the j operator. Referring to figure 8.19, the balance equation may be written

$$\frac{Z_1}{Z_3} = \frac{Z_2}{Z_4}$$

provided that the impedances are expressed in complex form. Real and imaginary terms on each side of the equation may be equated to give the balance conditions.

Example 8.7

In the Schering bridge of figure 8.20, balance is obtained by adjusting the variable

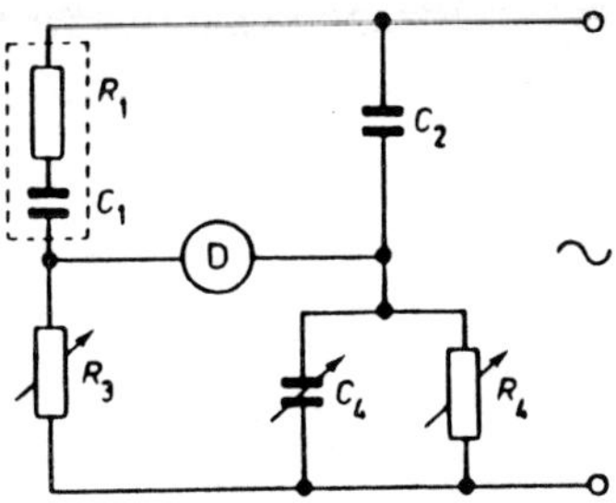

FIGURE 8.20

The Schering bridge of example 8.7

components. Obtain expressions for the series capacitance and resistance of a lossy capacitor Z_1 in terms of the other components.

$$Z_1 = R_1 + \frac{1}{j\omega C_1}, \qquad Z_2 = \frac{1}{j\omega C_2}, \qquad Z_3 = R_3$$

and

$$Z_4 = \frac{R_4}{j\omega C_4} \Big/ \left(R_4 + \frac{1}{j\omega C_4}\right)$$

$$\frac{Z_1}{Z_2} = \frac{Z_3}{Z_4}$$

$$\left(R_1 + \frac{1}{j\omega C_1}\right) j\omega C_2 = R_3 \left[j\omega C_4 \left(R_4 + \frac{1}{j\omega C_4}\right)\right] \Big/ R_4$$

$$j\omega C_2 R_1 + \frac{C_2}{C_1} = j\omega R_3 C_4 + \frac{R_3}{R_4}$$

Real terms

$$\frac{C_2}{C_1} = \frac{R_3}{R_4}$$

or

$$C_1 = \frac{C_2 R_4}{R_3}$$

Imaginary terms

$$j\omega C_2 R_1 = j\omega R_3 C_4$$

or

$$R_1 = \frac{R_3 C_4}{C_2}$$

8.7 Problems

8.1 The transistor whose characteristics are shown in figure 8.2 is used as a common-emitter amplifier with a load resistor of 4.2 kΩ and a supply voltage of 25 V. Draw an appropriate load line.

(a) What value of base - emitter voltage would cause the base current to be 50 μA and what would be the corresponding values of collector voltage and current?

(b) What powers are dissipated in the load and the transistor in (a)?

(c) What is the d.c. input resistance of the transistor under these conditions?

(d) What r.m.s. voltage must be applied to the base - emitter junction to cause the base current to vary between limits of 25 μA and 75 μA?

(e) The voltage calculated in part (d) is applied to the base - emitter junction in addition to 50 μA d.c. bias. Between what limits do the collector current and the collector - emitter voltage vary?

(f) Calculate the current and voltage gains of the stage (transistor plus load) when operating as in (e)?

(g) What would be the reading indicated by a simple a.c. voltmeter connected via a low-reactance capacitor between collector and emitter?

8.2 Design an amplifier stage similar to that of figure 8.5 using a germanium transistor having an h_{fe} of 100. Let the collector current be 1 mA at the operating point and allow 1 V drop across the emitter resistor for each 15 V of the 20 V voltage supply V_{CC}.

8.3 An emitter follower uses a silicon transistor having h_{ie} = 2 kΩ and h_{fe} = 100. If the emitter load resistor is 100 Ω calculate the voltage gain, output resistance and input resistance of the stage. Calculate the value of the single bias resistor required if the transistor is to operate at 10 mA from a 10 V d.c. supply.

8.4 The amplifier of figure 8.7a uses two identical stages where $R_1 = R_3$ = 200 kΩ, $R_2 = R_4$ = 20 kΩ, $R_{L_1} = R_{L_2}$ = 10 kΩ and $R_{e_1} = R_{e_2}$ = 1.2 kΩ. For both transistors h_{fe} = 100 and h_{ie} = 2.5 kΩ. The coupling capacitor C_1 = 100 μF and C_{e_1}, C_{e_2} and C_2 may be assumed infinite. The total shunt capacitances may be represented by a 15 000 pF capacitor between the base and emitter of T_2. The output resistances of the transistors alone may be assumed much greater than 10 kΩ. Calculate the mid-frequency voltage gain in decibels, the upper and lower half-power frequencies, the half-power bandwidth and the gain - bandwidth product. What would be the gain in decibels and the phase shift at 0.013 Hz and 5.89 MHz.

T_2.

8.5 A two-input summing amplifier uses an 'ideal' operational amplifier with a 1 MΩ feedback resistor. If 2 V are applied to the 10 MΩ input resistor and 50 mV are applied to the 100 kΩ input resistor, calculate the output voltage.

8.6 An operational amplifier uses an input resistor of 100 kΩ and a feedback capacitor of 0.1 μF. It is supplied with a square-wave input voltage which varies between + 1 V and – 1 V at a frequency of 1 kHz.

(a) If this voltage is applied at the commencement of the positive half-cycle describe the shape of the output waveform and calculate its maximum and minimum values.

(b) How would these limiting values be affected if the input wave were applied when it was halfway through its positive half-cycle?

8.7 Draw a suitable interconnection diagram for operational amplifiers to solve the following differential equation assuming that the y signal is available externally

$$y = A\frac{d^3x}{dt^3} - B\frac{d^2x}{dt^2} + C\frac{dx}{dt} + x$$

8.8 A batch of operational amplifiers of nominal gain 90 dB actually have production tolerances of ± 5 dB. Design a circuit to give a more stable nominal gain of 500 and calculate the gain spread of this circuit. Calculate also the input impedance and bandwidth of this circuit if the original amplifier had an input impedance of 200 kΩ and a gain-bandwidth product of 316 230 Hz.

8.9 A Wien bridge is powered from a source of angular frequency ω and consists of four arms AB, BC, CD and DA. AB consists of an unknown capacitor C_x of parallel loss resistance R_x. BC consists of a variable capacitor C_1 in series with a variable resistor R_1. Arms CD and DA each consist of fixed resistors R_2 and R_3. Show that, at balance, the loss resistance is given by the equation

$$R_x = \frac{R_3(1 + \omega C_1{}^2R_1{}^2)}{R_2\omega^2C_1{}^2R_1}$$

9 Data Display

9.1 Analogue or Digital?

Over the last decade digital display and recording equipment has become increasingly competitively priced. Previously, the choice of display equipment was largely limited to analogue or pointer instruments and the cathode-ray oscilloscope. Now however the engineer can make a conscious choice on nearly every occasion and determine whether digital or analogue displays are more appropriate.

Analogue displays take the form of a pointer moving across a calibrated scale and the observer is required to interpolate values between the scale graduations. This interpolation can lead to errors especially when the observer is tired or the environment is unpleasant (dust, fumes, etc.). A further source of error is the phenomenon of *parallax* which occurs if the line of sight is not normal to the plane of the scale. Analogue displays are unrivalled however in showing the trend of motion of a variable and in the comparison of a variable with a fixed value.

Digital displays are often more appropriate for untrained observers and for the presentation of values for recording in a written form.

9.2 Analogue Displays

9.2.1 Pointer Instruments

Over the many years during which they have been in use pointer-and-scale displays have been refined to minimise the major sources of error. Precision instruments are provided with a mirror-scale by means of which the observer may eliminate parallax errors. If the eye is aligned so that the pointer and its reflection coincide, the line of sight must be normal to the scale. Knife-edge pointers also minimise interpolation errors.

The scale of a precision pointer instrument can usually be read to an accuracy of at least one-half per cent over a scale length of some 10 cm. This usually corresponds to a pointer deflection across 75 - 80° of arc. Recently manufacturers of panel-mounted instruments have increased the effective scale length without increasing the over-all size by utilising scales having deflection arcs of 270°.

Among the more familiar pointer instruments are the electromechanical ones – the moving-iron and moving-coil meters, the electrodynamic or dynamometer meter, thermocouple and electrostatic types. The moving-coil instrument is by far the most common of these because its linear scale is much more easily read than the square-law scales of the others. Non-linear scales have their graduations cramped together at the lower end and this can cause reading errors.

Many electronic instruments present their output information by means of a moving-coil indicator. These instruments are capable of conditioning the variable before display and include a.c. and d.c. electronic voltmeters, Q-meters and true r.m.s. voltmeters. The outputs of null instruments, such as bridges and potentiometers, provide excellent opportunities for pointer displays to reveal trends towards or away from balance.

9.2.2 The Cathode-ray Oscilloscope (CRO)

The major disadvantage of electromechanical pointer displays such as moving-coil instruments arises out of their inertia. With a conventional pointer, the instrument is unlikely to be able to respond to frequencies higher than 25 Hz. Mirror galvanometers in which the pointer is replaced by a light-ray may extend this to 1 kHz maximum. The electron beam in a cathode-ray tube (section 3.2.2) has little inertia and is quite capable of being deflected at frequencies up to the gigahertz (10^9 Hz) region.

In order to display the waveshape of the measured voltage the cathode-ray tube must be supplemented by at least a timebase and an amplifier (figure 9.1). The waveform to be observed is applied to the y input terminals and, after amplification, is applied to the tube's vertical deflector plates. The timebase is essentially an oscillator producing a sawtooth waveform which sweeps the spot across the screen from left to right at uniform velocity. Its output is amplified and applied to the horizontal or x deflector plates. At the end of each sweep the spot 'flies back' almost instantaneously to the left-hand side. During this flyback period the timebase feeds a suppression pulse back to the grid of the electron gun. This darkens the spot during flyback to make it invisible.

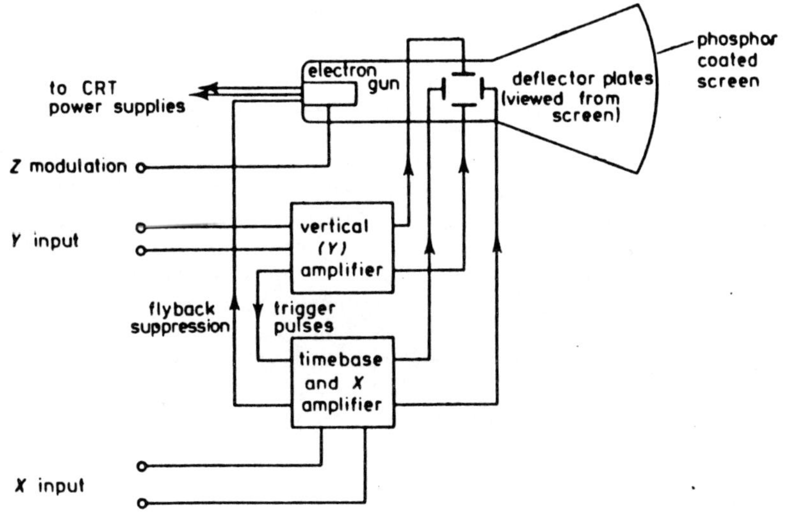

FIGURE 9.1

The cathode-ray oscilloscope

When viewing a repetitive waveform on an oscilloscope the observer is seeing many horizontal sweeps superimposed and it is the persistence of vision which creates the stationary image. For this to occur, all the sweeps must be superimposed exactly, that is, the sweep must always begin at the same point on the vertical waveform. A trigger pulse is fed from the y amplifier to the time-base to ensure correct initiation of the sweep. In some measurements it is convenient to be able to darken or brighten the trace during certain sections of the sweep. An external z modulation terminal is often provided by which external pulses may be applied to the tube grid to achieve this.

If the oscilloscope is merely to be used as an $x - y$ display, the timebase may be switched off. External waveforms now applied to the x input terminals will deflect the spot horizontally as required. Oscilloscopes are chiefly used in this mode to examine the frequency and phase relationships between two sinusoidal voltages by the Lissajous method.[5, 50]

If the waveshape to be examined is non-repetitive, the timebase may be set to give a single sweep in the 'one-shot' position. If such waveforms are slowly varying, their complete shape may be seen by using a tube with a long persistence (P2) phosphor which has an afterglow of several seconds. The observation of fast non-repetitive waveforms is more difficult. The classical method is to fit the oscilloscope with a camera and to leave the shutter open for the whole of the single sweep. A permanent record is thus made on film which can be viewed at leisure; 35 mm conventional film is most common but is gradually being replaced by Polaroid film. In spite of its expense the latter dispenses with the need for wet darkroom facilities.

The recent advent of the storage oscilloscope[51] has made the use of film techniques for nonrepetitive waveforms unnecessary. This oscilloscope uses a special storage cathode-ray tube which is capable of storing any image formed on its screen for periods of many hours during which it may be viewed at will before erasure by means of a switch. The cost of storage tubes is high however and their lifetime is limited.

Although the oscilloscope is indispensable for examining waveshapes, measurements made from the screen image are only accurate to ± 2.5 per cent on either the voltage or the time axis. It remains the most versatile of electrical instruments and at the time of writing is capable of responding to frequencies as high as 300 MHz.

As explained in section 8.2.3 any device employing electronic amplifiers is limited by the gain - bandwidth product criterion. This means that an oscilloscope having high gain (voltage sensitivity) and high bandwidth simultaneously will be very expensive. Accordingly many oscilloscopes are manufactured with modular plug-in amplifiers by which either high gain or high bandwidth are available separately for varying applications.

Double-beam oscilloscopes using multiple electron guns or electronic switching techniques are available for the simultaneous observation of two waveforms.

9.3 Analogue Recorders

In many instances the engineer requires a permanent record of some analogue data. The use of a camera attached to an oscilloscope has already been discussed. Alternative methods may be divided into magnetic and direct-writing techniques.

9.3.1 Magnetic Recording[52]

The data may be recorded directly on to the tape in the same manner as with domestic audio tape recorders. On replaying the record however, the amplitude of the various frequency components will be found to have changed. Although the amplitude stability is poor this technique is still employed where very wide bandwidth is required.

Most analogue instrumentation tape recorders now employ frequency-modulation techniques (see section 6.9.3). Bandwidths extending from zero to 80 kHz are available with amplitude accuracies of ± 5 per cent. Up to eight tracks may be simultaneously recorded across the 2.54 cm tape width. Tape speeds up to 152 cm/s are employed for the highest frequency signals.

Although the data is recorded permanently, magnetic recording does not produce a visible record of the signal. The user has to replay the tape into a direct-writing recorder to obtain a visual record. The possibility of recording and replaying at different speeds allows long events to be telescoped and fleeting events to be stretched for leisurely investigation.

Direct-writing Techniques

Direct-writing recorders employ a paper chart drawn at a constant speed by a d.c. servomotor beneath a pen or other marking device.

9.3.2 Galvanometric Recorders

The majority of direct-writing recorders fall into this category. Some form of pen is attached to a moving-coil (figure 9.2) galvanometer in place of the pointer.

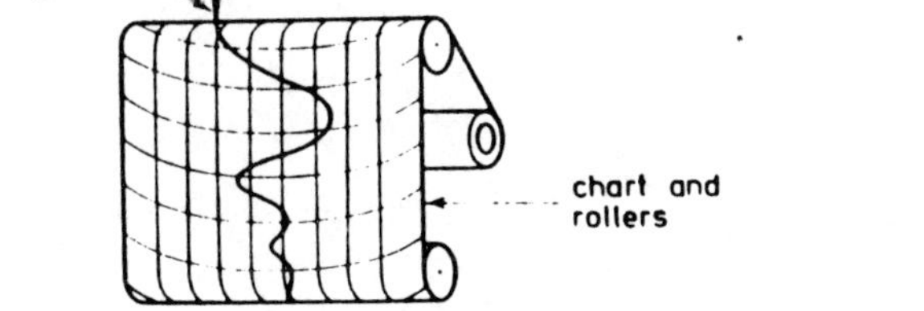

FIGURE 9.2

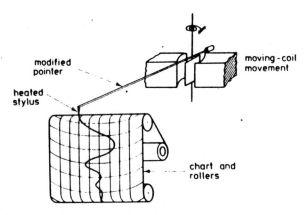

Pen-type galvanometric recorder

The simplest types employ a conventional ink pen fed from a stationary reservoir. The inertia of the moving parts limits the response to a few hertz. Inertia can be reduced by employing either a hot-wire stylus and heat-sensitive paper or an electric stylus which burns an image into the paper by sparking.

Frequencies up to 25 Hz may be recorded with these at an accuracy of 2 - 3 per cent of full-scale deflection. Multiple pens may be used for the simultaneous record of two or more signals but this sometimes reduces the usable chart width.

Where multiple channels are to be recorded simultaneously, galvanometric recorders using a light-beam as a pen are the most versatile. Several miniature mirror galvanometers are mounted (figure 9.3a) side by side in a common magnet block. The light source may be either a mercury-arc ultraviolet lamp or a quartz - iodine tungsten-filament lamp. In either case the light after reflection

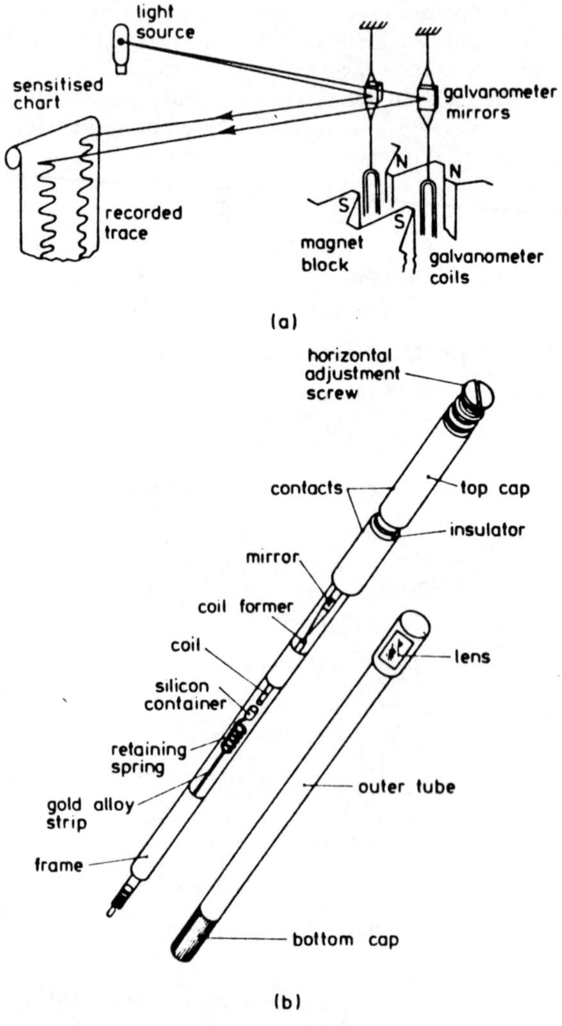

FIGURE 9.3

The principle of the light-beam recorder and its galvanometer construction is focused upon a sensitised paper. This paper produces a print after a few seconds' exposure to daylight. Up to 50 channels may be recorded simultaneously on the same 50 cm wide chart and all these channels may use the full chart width, crossing as desired. The galvanometers usually plug in to the magnet block which allows galvanometers of different bandwidths or sensitivity to be employed as required. Figure 9.3b shows the construction of these miniature galvanometers. The electrical connections to the coil are made via the tension spring and the upper bifilar suspension wires. The whole galvanometer is enclosed in a cylindrical case 3 mm diameter by 5 cm long.

The accuracy of such instruments is typically 2 per cent of full-scale deflection (f.s.d.) and with the lighter less-sensitive galvanometers frequency responses up to 5 kHz are possible. One disadvantage is the delay of several seconds until the paper develops the image. Such recorders are however much the cheapest method of recording many channels of data. Care must be taken when using these instruments at high frequencies near to their mechanical resonance frequency as significant errors can be introduced.[53]

The moving coils of galvanometric recorders have an impedance of only a few ohms which would heavily load many signal sources (see section 7.1.1). Such instruments are often provided with a set of galvanometer drive-amplifiers built into the unit to present a high impedance to the source. Such amplifiers are often emitter-follower stages (see section 8.1.4).

9.3.3 Potentiometric Recorders

These recorders offer the highest accuracy (0.2 per cent f.s.d) of any pen recorder and have been established for many years to measure variables such as thermocouple potentials. They rely for their action on the principle of the self-balancing potentiometer (see figure 9.4). The carriage is free to slide along the guide and makes contact with the resistive slidewire which is energised with d.c. The amplifier continually monitors any error between the input voltage to be measured and the slider potential. Such errors, after amplification, are used to

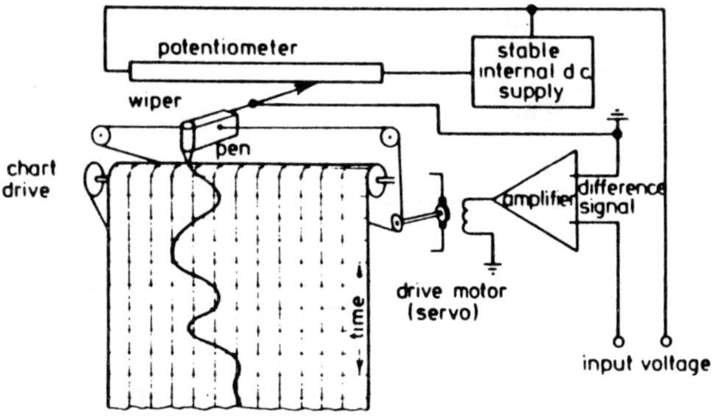

FIGURE 9.4

The self-balancing potentiometric recorder

drive the servomotor which moves the carriage via a system of wires and pulleys in the direction of balance. When the input voltage and the slider voltage are equal the motor stops. At fixed intervals other circuitry connects a standard cell to the input and automatically calibrates the slidewire.

Because of the high inertia of the motor, gearbox, carriage and linkages the maximum response speed is limited to 0.25 s for a full-scale deflection. The high input impedance of the amplifier means that these recorders present little loading to the measured variable. Such high-impedance sources as thermocouples may be directly connected to the input.

Multichannel potentiometric recorders usually work on a time-sharing system. They sample each variable in order, move to the balance position, and leave a distinguishing mark with a multi-colour pen.

Versions of potentiometric recorders having two slidewires, mutually at right-angles, are produced. These are known as x - y plotters and are used for recording two variables neither of which is time.

9.3.4 Fibre-optic Recorder

This newcomer to the recorder field[52] is unlikely to be met in general industrial use for some years. It is basically a cathode-ray oscilloscope using a cathode-ray tube with a fibre-optic screen. A chart of conventional ultraviolet recorder paper is held in contact with this screen and receives the imprint of any image. The negligible inertia of the tube's electron beam allows responses up to 5 kHz at present; future developments will undoubtedly extend this greatly. At present it is expensive and its use is confined to laboratory applications.

9.4 Digital Displays

The presentation of data directly as a decimal number has been common in scientific and industrial equipment for some years. The advent of semiconductor display devices has now made it economical for home consumer use. The pocket calculator is an outstanding example.

A. Gas-filled Devices

9.4.1 Alphanumeric Indicator Tube

This is basically a modified low-pressure neon tube. There is one anode and (usually) ten separate mutually-insulated cathodes. These are shaped into the numerals 0 to 9 and stacked one behind the other. The anode is held at a constant positive potential of about 150 V via a series safety resistor and whichever cathode is required is switched to earth potential (figure 9.5). These tubes have high reliability, high efficiency and low cost. They require however a high-voltage supply which is incompatible with semiconductor circuitry and are limited to the colour red. They are marketed under tradenames such as 'Numicator' and 'Nixie tubes'.

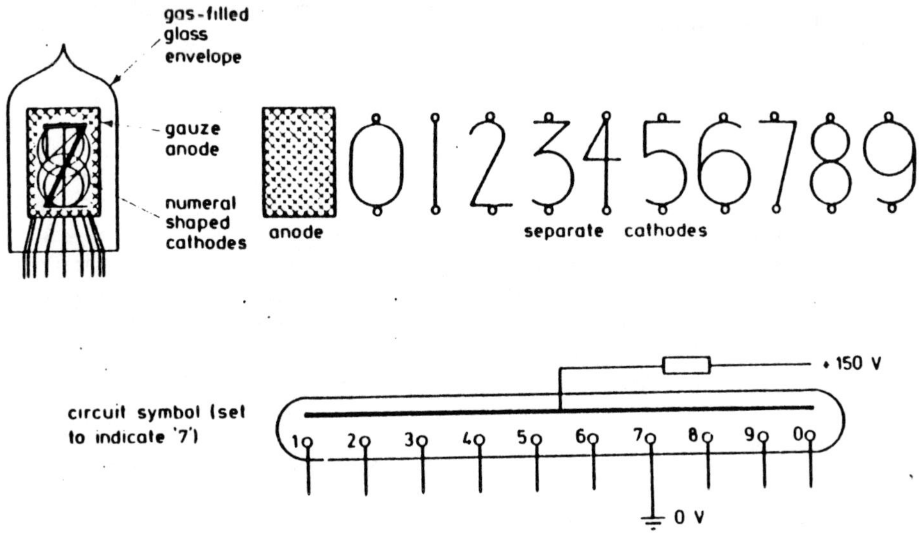

FIGURE 9.5

The alphanumeric tube, its components and circuit symbol

9.4.2 Plasma Panel[54]

This device is still in the development stage but will be widely adopted in some years' time. A matrix of holes is formed in a sheet of dielectric material and this is sandwiched between two other sheets to form isolated cells. These cells are filled with neon at low pressure and can be illuminated by applying an a.c. voltage between a vertical set of wires in front of the sandwich and a horizontal set at the rear. The cells to become ionised will be those indexing points where energised vertical and horizontal wires cross.

Such panels can display up to 400 characters, each requiring a 5 by 7 matrix of cells similar to figure 9.6c.

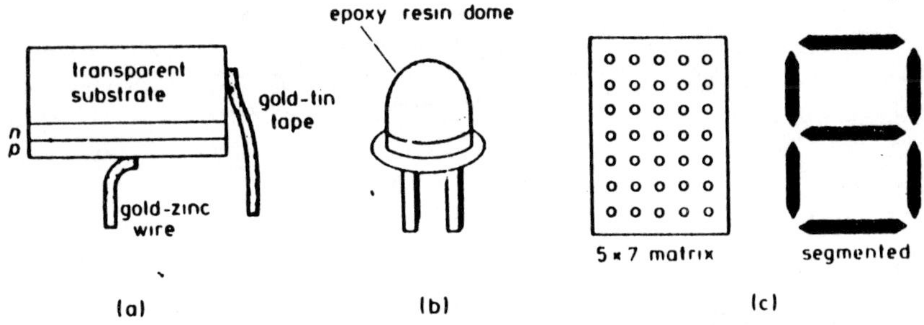

FIGURE 9.6

The light-emitting diode (LED)

B. Semiconductor Devices

9.4.3 The Light-emitting Diode (LED)

In some semiconducting materials the recombination of holes and electrons results in an emission of energy as light. The frequency and therefore the colour of the emitted light can be obtained from the familiar equation $hf = \delta E$ where δE is the emitted energy of recombination (see section 3.3.1). The introduction of impurities or dopes modifies these energies to yield a wide range of colours from green to red. The charge recombination occurs within a *pn* diode arranged as shown in figure 9.6a. They may be encapsulated individually as in figure 9.6b when used as indicator lamps or arranged as 5 by 7 matrix or a segmented matrix of chips to give a wide range of alphanumeric symbols (figure 9.6c).

As low-voltage indicator lamps LEDs are superior to filament lamps in that they do not run hot and are mechanically robust. They are very light in weight and almost instantaneous in response.

9.4.4 Electroluminescent Panels

Illuminated panels, at present of only low brightness, can be made by using a layer of [illegible]liating material between two electrodes as in figure 9.7. A high voltage ([illegible]0 to 600 V) has to be used but green, yellow and blue panels are easy to fabricate. Such panels can be used in displays on control consoles to great effect.

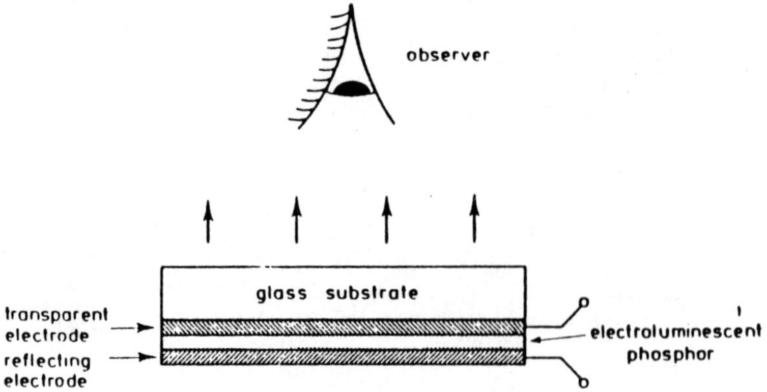

FIGURE 9.7

Cross-section of an electroluminescent panel

9.4.5 Liquid-crystal Displays (LCD)

The clock and watch industry are currently experimenting with liquid-crystal displays. A layer of organic material is sealed between two plates of electrically conducting glass. A voltage applied across the plates causes the material to become milky and to reflect ambient light whereas it is invisible under normal conditions.

Its great advantage is its versatility as devices can easily be fabricated in small quantities to any design. Characters up to 0.1 m high are currently available. The characters become brighter as the ambient-light level rises.

9.5 Digital Recorders

9.5.1 Paper Tape

This was perhaps the earliest and simplest form of recording digital information other than the automatic typewriter which is limited to speeds of about 10 characters per second. The width of the tape can usually accommodate five holes which allows for binary coding of the decimal numbers 0 to 31 (see table 7.1).

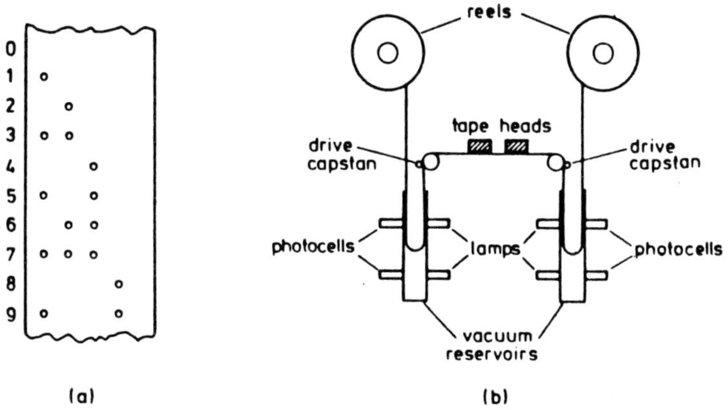

FIGURE 9.8

Digital data recording on tape

The form of the paper tape is shown in figure 9.8a. If the punch is capable of operating say twice per second, it will be seen that the rate of recording data is comparatively slow. When reading paper tape however five photocells, in line abreast, search the width of the tape for holes. Reading can thus be a speedy operation because no moving parts other than the tape are involved.

9.5.2 Magnetic Tape

Perhaps the most widely used method of recording digital information, especially in the computer field, is magnetic tape. Very sophisticated tape-transport systems have been evolved to cope with tape speeds of up to 9 m/s. Figure 9.8b shows how tape loops are provided on each side of the recording heads to minimise tape wear and inertia. The tape loops are held in vacuum reservoirs and the loop length is monitored by photocells whose output controls the speed of the feed and take-up reels. When making sudden changes in speed or direction only the negligible inertia of the tape loops is involved. The high inertia of the reels and their motors is overcome by their control systems.

Areas of zero, positive or negative residual magnetism are used to indicate digital 1s and 0s and vast amounts of data may be stored upon the reels which can carry hundreds of metres of tape. Up to eight tracks across the width of the tape allow eight sources of data to be monitored continuously and simultaneously. The chief disadvantage of magnetic methods is the complete lack of any visual indication on the tape.

9.5.3 The Data Logger

Where many sources of data have to be sampled and recorded at modest speeds some form of time-sharing is appropriate. Figure 9.9a illustrates the main components of a data logger. A multiplexer samples, in turn, the output from each transducer and feeds this information to the digitiser where it is binary-coded.

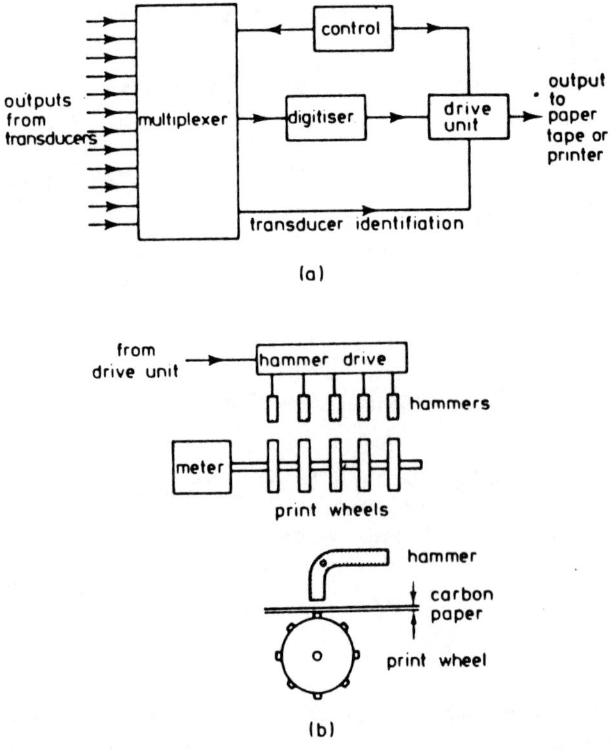

FIGURE 9.9

Data logger and line printer

The signals are now fed to some output recorder whose form depends upon the speed required. Paper tape may suffice for the lowest speeds but line printers, capable of recording up to 20 lines per second, may be used. Each line may contain 120 characters and a complete line is printed by the hammers in one rotation of the print wheels (figure 9.9b).

The use of a data logger is clearly appropriate where the outputs of many thermocouples or strain gauges on a structure are to be monitored and recorded.

References

1. H. Cotton, *Basic Electrotechnology* (Macmillan, London and Basingstoke, 1973).
2. J. A. Edminister, *Electric Circuits* (SI edition) (McGraw-Hill, New York, 1965).
3. W. D. Stanley, *Transform Circuit Analysis for Engineering and Technology* (Prentice Hall, Englewood Cliffs, N.J., 1968).
4. H. Cotton, *Advanced Electrical Technology* (Pitman, London, 1967).
5. H. Buckingham and E.M. Price, *Principles of Electrical Measurements* (English Universities Press, London, 1970).
6. F. Brailsford, *An Introduction to the Magnetic Properties of Materials* (Longmans, London, 1968).
7. P. Parker, *Electronics* (Arnold, London, 1950).
8. J. R. Ryder, *Electronic Fundamentals and Applications* (Pitman, London, 1970).
9. C. N. G. Matthews, *A Programmed Book on Semiconductor Devices* (Mullard, London, 1970).
10. A. T. Starr, *The Generation, Transmission and Utilisation of Electrical Power* (Pitman, London, 1961).
11. S. M. Shinners, *Control System Design* (Wiley, New York, 1964).
12. P. Atkinson, *Feedback Control Theory for Engineers* (Heinemann, London, 1968).
13. A. W. J. Griffin and R. S. Ramshaw, *The Thyristor and its Applications* (Chapman and Hall, London, 1965).
14. – – *A Review of the Principles of Electrical and Electronic Engineering*, vol. 1 Principles of Heavy Current Engineering, ed. L. Solymar (Chapman and Hall, London, 1974).
15. M. G. Say, *Magnetic Amplifiers and Saturable Reactors* (Newnes, London, 1954).
16. R. Kretzmann, *Industrial Electronics* (Philips, Eindhoven, 1953).
17. T. Roddam, *Transistor Inverters and Converters* (Iliffe, London, 1963).
18. – – *Power Engineering Using Thyristors*, vol. 1 Techniques of Thyristor Power Control, ed. M. J. Rose (Mullard, London, 1970).
19. – – *Electric Furnaces*, ed. C. A. Otto (Newnes, London, 1958).
20. A. G. E. Robbiette, *Electric Melting Practice* (Griffin, London, 1972).
21. E. O. Doebelin, *Measurement Systems: Application and Design* (McGraw-Hill, New York, 1966).
22. J. P. Froelich, *Information Transmittal and Communicating Systems* (Holt Reinhart and Winston, New York, 1969).
23. W. B. Davenport and W. L. Root, *Random Signals and Noise* (McGraw-Hill, New York, 1958).
24. J. F. Pierce, *Transistor Circuit Theory and Design* (Merrill, Columbus, Ohio, 1963).
25. F. R. Connor, *Signals* (Arnold, London, 1972).

26. R. E. Young, *Telemetry Engineering* (Iliffe, London, 1968).
27. W. Fraser, *Telecommunications* (Macdonald, London, 1968).
28. D. G. Tucker, *Modulators and Frequency Changers* (Macdonald, London, 1953).
29. K. W. Cattermole, *Principles of Pulse Code Modulation* (Iliffe, London, 1969).
30. G. C. Hartley et al., *Techniques of Pulse Code Modulation* (Cambridge University Press, 1967).
31. E. W. Golding and F. C. Widdis, *Electrical Measurement and Measuring Instruments* (Pitman, London, 1963).
32. R. C. Brewer, 'Transducers for Positional Measuring Systems', *Proc. Instn elect. Engrs*, 110 (1963) p. 1818.
33. A. R. Upson and J. H. Batchelor, *Synchro Engineering Handbook* (Hutchinson, London, 1966).
34. G. D. Radford, J. G. Rimmer and D. Titherington, *Mechanical Technology* (Macmillan, London and Basingstoke, 1969).
35. C. C. Perry and H. R. Lissner, *The Strain Gage Primer* (McGraw-Hill, New York, 1962).
36. H. Brandenburger, 'Oscillatom, Frequency Standards and Atomic Clocks', *Suisse horlog.*, 4 (1966) p. 36.
37. F. E. Smith, 'On Bridge Methods for Resistance Measurements of High Precision in Platinum Thermometry', *Phil. Mag.*, Series 6, 24 (1912) p. 541.
38. J. J. Hill and A. P. Miller, 'An Inductively Coupled Ratio Bridge for Precision Measurements of Platinum Resistance Thermometers', *Proc. Instn elect. Engrs*, 110 (1963) p. 453.
39. BS 1828 : 1961 Reference tables for copper *v*. constantan thermocouples; BS 4937 : Part 1 : 1973 Platinum - 10 per cent rhodium/platinum thermocouples; Part 2 : 1973 Platinum - 13 per cent rhodium/platinum thermocouples; Part 3 : 1973 Iron/copper-nickel thermocouples; Part 4 : 1973 Nickel - chromium/nickel - aluminium thermocouples.
40. H. C. Stansch, 'A Linear Temperature Transducer with Digital Output', 19th Annual Conference Instrument Society of America, New York (1966).
41. A. J. King, *The Measurement and Suppression of Noise* (Chapman and Hall, London, 1965).
42. D. F. A. Edwards, *Electronic Measurement Techniques* (Butterworths, London, 1971).
43. W. Kidwell, *Electrical Instruments and Measurements* (McGraw-Hill, New York, 1969).
44. J. D. Ryder, *Engineering Electronics* (McGraw-Hill, New York, 1957) pp. 133-4 and 160-76.
45. H. W. Bode, *Network Analysis and Feedback Amplifier Design* (Van Nostrand, Princeton, N. J., 1945).
46. J. V. Wait et al., *Introduction to Operational Amplifier Theory and Applications* (McGraw-Hill, New York, 1975).
47. M. Kahn, *The Versatile Op Amp* (Holt Reinhart and Winston, New York, 1970).
48. G. B. Clayton, *Operational Amplifiers* (Butterworths, London, 1971).
49. N. M. Morris, *Industrial Electronics* (McGraw-Hill, London, 1970).

50. – – *Frequency Comparisons Using Roulette Patterns* (Tektronix (U.K.) Ltd, 1960).
51. D. C. Calnon, 'Advances in Storage Oscilloscopes', *Electron. Inds, Philad.*, 25 (1966) pp. 109 - 11.
52. D. R. Baldwin, 'Analogue Recording Techniques', *Electron. Power*, 20 (1974) pp. 976 - 8.
53. B. A. Gregory, *An Introduction to Electrical Instrumentation* (Macmillan, London and Basingstoke, 1973) pp. 53 ff.
54. A. Burkitt, 'Display Innovations', *Electron. Engng*, 34 (1975) pp. 41 - 4.
55. R. Calvert, *The Transformer Ratio Arm Bridge* (Wayne Kerr Monograph no. 1, Bognor Regis, 1966).

Further Reading

British Standard Codes of Practice, Electrical Engineering : CP 1003 : Electrical apparatus for use in explosive atmospheres; CP 1004 : Street lighting; CP 1008 : Maintenance of switchgear; CP 1010 : Guide to loading oil-immersed transformers; CP 1011 : Maintenance of motor control gear; CP 1013 : Earthing; CP 1015 : Electrical equipment of industrial machines; CP 1017 : Distribution of electricity on construction and building sites.

British Standards Institution SL 26 : Electrical Engineering (a comprehensive list of Standards covering every aspect of applied electrical technology).

E. Hughes, *Electrical Technology* (Longmans, London, 1969).

N. M. Morris, *Advanced Industrial Electronics* (McGraw-Hill, London, 1974).

Answers to Problems

Chapter 1

1.2	1.25 A, 2.5 mC
1.3	+ 62.5 V.
1.4	1.0 V, 1.25 V, 1.556 V, 1.166 mC
1.5	15.8 A, 3.14 MΩ.

Chapter 2

2.1	50 V, 57.7 V, 1.154, 1.733
2.2	55.2 mH
2.3	22.35 V, 113.7 Hz, 0.14 H, 0.448 leading
2.4	15.36 A, 7.53 A, 10.71 A, 2360 W, 0.92 lagging
2.6	0.884 lagging, 364 μF
2.7	2360 W, 2830 VAr
2.8	50.6 pF, 209, 2309 Hz, 0.76 μA, 159 μA
2.9	$14.17\angle -34° 54'$ A
2.11	$0.476\angle 13° 8'$ A
2.12	0.85 lagging, 56.4 A, 25.8 kVAr
2.13	109 μF
2.14	4.00 W

Chapter 3

3.1	3.04×10^{-19} J, 6.02×10^{-19} J, no emission, 7.87×10^{5} m/s
3.2	30 mA/W
3.3	1.8 μA, 6.6 μA, 3.1

Chapter 4

4.2	$v_s = 500 \sin 100\pi t$
4.3	3600, 1800, 1200, 900, 720 and 600 rev/min
4.4	1.92 N m
4.5	2.93 N m
4.6	40 A, 695 rev/min
4.7	9.48 rev/s

Chapter 5

5.1	13 V
5.2	13.6 V

5.3 0.04

5.4 3.02 W

5.5 17.0 s

Chapter 6

6.1 33 min 20 s, 50.2 Hz

6.2 74 dB, 56 dB

6.3 (a) 1 kHz to 10 kHz (b) d.c. to 400 Hz (c) d.c. to 10 kHz

6.4 3.32 kHz

6.5 greater than 3.32 kHz

Chapter 7

7.1 200 Ω

7.2 1.5 V, 0.68 V

7.3 5.65 V (D positive), 3.96 V (E positive)

7.4 6 V, 1.65 mV, 272.7 kΩ

7.6 1.25×10^5 V/cm, 8.696×10^4 V/cm

7.7 2.04 mV

Chapter 8

8.1 (a) 120 mV, 16 V, 2.2 mA
(b) 19.8 mW, 35.2 mW
(c) 2.4 kΩ
(d) 17.7 mV
(e) 0.93 mA to 3.65 mA, 9 V to 21 V
(f) 54.4, 240
(g) 4.24 V

8.2 Typical but not unique values are R_L = 10 kΩ, R_e = 1333 Ω, R_1 = 184 kΩ and R_2 = 16.33 kΩ

8.3 0.835, 19.8 Ω, 12 kΩ, 83 kΩ

8.4 89.2 dB, 58.9 kHz, 0.13 Hz, 58.9 kHz, 170 MHz, 69.2 dB $\angle 84^\circ\ 18'$ lead
49.2 dB, $\angle 89^\circ\ 24'$ lag

8.5 − 0.7 V

8.6 (a) Triangular between 0 at t = 0 and − 50 mV at t = 500 μs
(b) − 25 mV at t = 250 μs, + 25 mV at 750 μs

8.8 R_F = 197 Ω, R_{in} = 100 kΩ; 494 to 503, 12.7 MΩ, 632 Hz

Index